Molecular Mechanisms of Xeroderma Pigmentosum

ADVANCES IN EXPERIMENTAL MEDICINE AND BIOLOGY

Molecular Mechanisms of Xeroderma Pigmentosum

Edited by

Shamim I. Ahmad, MSc, PhD

School of Science and Technology, Nottingham Trent University, Nottingham, England

Fumio Hanaoka, PhD

Graduate School of Frontier Biosciences, Osaka University, Osaka, Japan

Springer Science+Business Media, LLC
Landes Bioscience

Springer Science+Business Media, LLC
Landes Bioscience

Printed in the USA.

Springer Science+Business Media, LLC, 233 Spring Street, New York, New York 10013, USA
http://www.springer.com

Please address all inquiries to the publishers:
Landes Bioscience, 1002 West Avenue, 2nd Floor, Austin, Texas 78701, USA
Phone: 512/ 637 5060; FAX: 512/ 637 6079
http://www.landesbioscience.com

Molecular Mechanisms of Xeroderma Pigmentosum, edited by Shamim I. Ahmad and Fumio Hanaoka, Landes Bioscience / Springer Science+Business Media, LLC dual imprint / Springer series: Advances in Experimental Medicine and Biology

ISBN: 978-0-387-09598-1

Library of Congress Cataloging-in-Publication Data

Library of Congress Cataloging-in-Publication Data

Molecular mechanisms of xeroderma pigmentosum / edited by Shamim I. Ahmad, Fumio Hanaoka.
    p. ; cm. -- (Advances in experimental medicine and biology ; v. 637)
  Includes bibliographical references and index.
  ISBN 978-0-387-09598-1
  1. Xeroderma pigmentosum--Molecular aspects. 2. DNA repair. I. Ahmad, Shamim I. II. Hanaoka, Fumio, 1946- III. Series.
  [DNLM: 1. Xeroderma Pigmentosum--genetics. 2. DNA Repair Enzymes--genetics. 3. DNA Repair-Deficiency Disorders--genetics. 4. Xeroderma Pigmentosum--therapy. W1 AD559 v.637 2008 / WR 265 M718 2008]
  RL247.M67 2008
  616.5'15042--dc22
                                2008021297

DEDICATION

This book is dedicated to the sufferers of xeroderma pigmentosum and their parents and relations who painstakingly look after them throughout their suffering, and to the voluntary organizations and groups working tirelessly for xeroderma pigmentosum patients.

PREFACE

Xeroderma pigmentosum (XP), meaning parchment skin and pigmentary disturbance, is a rare and mostly autosomal recessive genetic disorder that was originally named by two dermatologists, the Austrian Ferdinand Ritter von Hebra and his Hungarian son-in-law Moritz Kaposi in 1874[1] and 1883.[2] The earliest published record (PubMed) available on the internet is a publication in 1949 by Ulicna-Zapletalova under the title, "Contribution to the pathogenesis of xeroderma pigmentosum".[3] It was in the late 1960s when James Cleaver (contributor of Chapter 1 of this book), at the University of California, San Francisco, while working on nucleotide excision repair (NER), read an article in a local newspaper about XP and soon after obtained a skin biopsy from a patient suffering from XP that showed that cells from it were deficient in NER. Thus, his studies led to the discovery that indeed this genetic defect was due to mutations in DNA repair genes that imbalance the NER pathway.[4,5] The discovery paved the way for further exploration of the link between DNA damage, mutagenesis, neoplastic transformation and DNA repair diseases. Since then, 4,088 papers, including excellent reviews, on XP are listed on the internet (PubMed data, February 2008), and an XP Society has been established in the USA (http://www.xps.org) and an XP Support Group in the United Kingdom (www.xpsupportgroup.org.uk).

Several clinical features of XP patients, associated with DNA repair deficiency, are highlighted in Chapter 2, most important being the severe photosensitivity (primarily to ultraviolet light) of XP patients and subsequent extremely high predisposition to development of malignant skin neoplasms, including basal cell carcinoma, squamous cell carcinoma and melanoma, especially on the sun-exposed areas. Additionally, cutaneous atrophy and actinic keratosis can occur. Other phenotypes includes neurological dysfunction (first identified by de Sanctis and Cacchione in 1932)[6] and ocular abnormalities. Studies show that 97.3% of XP patients suffer from ocular abnormalities, which include ocular neoplasm, photophobia, impaired vision, and corneal and conjunctival abnormalities.[7] It is likely that some patients suffer a variety of malignancies; thus, in a case study in India, it was observed that an XP patient suffered from multiple cutaneous malignancies including freckles, letigens, and keratosis, a non-tender ulcerated nodular lesion on the nose, a nodular ulcerated lesion on the right outer canthus of the conjunctiva, and a nodular growth on the cheek which turned out to be cancer of the skin of all types, squamous cell and basal cell carcinoma and malignant melanoma.[8]

It is likely that patients had been exposed to solar UV for long time to sustain these phenotypes. Neurological dysfunction is linked to abnormal motor activity, areflexia, impaired hearing, abnormal EEG and microcephaly.[9] Also slow growth, delayed secondary sexual development and abnormal speech prevail. Some patients are known to suffer from cancer at the tip of the tongue and the anterior part of the eye.[9]

In the recent past, much effort has gone into understanding the molecular pathogenesis of XP in terms of enhanced sensitivity and predisposition of sun-exposed areas of the body to erythema and various forms of skin cancers. Results of these studies have yielded exciting information on a multitude of protein interactions with various XP proteins involved in a number of activities associated with repair of UV photoproducts in DNA.

Written by the leading researchers and clinicians in the field, this book provides a comprehensive treatise on XP. It covers in detail what is known of the 8 XP complementation groups identified to date. These include XPA, B, C, D, E, F, G and XPV.

In Chapter 1, James E. Cleaver, one of the founding researchers of XP (and the winner of Career Award, 2006 from the American Skin Association) has highlighted the historical aspects of the development of research on XP and the discovery of mutation in humans affecting DNA repair. On the subject of clinical features of XP, the author in Chapter 2 has exhaustively covered all aspects of XP epidemiology and phenotypes including dermatological manifestations, other cancers in XP, and neurological and ophthalmological manifestations. Also management and prognosis of XP are highlighted. Although variation exists in sufferers of different complementation groups (Table 2, Chapter 2) such as skin cancer, opthalmological involvement commonly prevails in all the sufferers of XP, and the XPA and XPD groups are more prone to neurological disorders. Inherited polymorphisms of DNA repair genes contribute to variations in DNA repair ability and genetic susceptibility to different cancers. For example, in a recent study it has been shown that a polymorphism in XPD, codon 751, is associated with the development of maturity onset cataract and an increased risk of lung cancer.[10,11] Cancer of various kinds in those parts of the body exposed to UV light (primarily sun) is a major problem in XP and often leads to premature death; this issue has been described in Chapter 3. Discovering the link between various XP gene mutations and the phenotype may ultimately help define the complex cellular actions of the XP proteins. At the other end of the spectrum and the book, Chapter 16 details the possible preventive measures and treatments for XP. However, since the causative agent (solar exposure) for development of skin cancer in XP patients is well established, one of the most promising preventive measures is the complete isolation of patients from sunlight and artificially generated UV lights. Furthermore, in a recent paper, a new method, known as "gene through the skin", especially suitable for XP patients, has been proposed.[12] The authors carried out a successful in vitro study, correcting the genetic defect of cells from XP patients. Therefore the future holds out the possibility of gene therapy or protein replacement. On the other side of the spectrum is the warning from Reichrath[13] that XP patients completely protected from sunlight carry a risk of vitamin D deficiency and related problems. Hence additional care must be taken to monitor patients, providing supplements to prevent this deficiency and its consequences.

Sunlight emits three forms of ultraviolet light (UVA, UVB and UVC), UVC being the most potent of the three UV components for damaging DNA. Luckily this is completely absorbed by the stratospheric ozone layer and cannot reach us. Some UVB (about 5%) and most UVA (95%), however, penetrate this layer and therefore XP patients, if exposed to sunlight, receive some UVB and large amounts of UVA. Experimental data revealed that UVB can generate cyclobutane pyrimidine dimers (CPD) and pyrimidine 6-4 pyrimidone photoproducts (6-4PP), which are highly mutagenic and carcinogenic.[14] Little evidence is available to show that DNA absorbs UVA although recently it has been shown that CPD may also be induced by this waveband of light. Its mechanism of action is considered to be via free radical pathways rather than direct absorption of UVA by DNA.[15]

Over the last 60 years or so most researchers have focussed their studies on UVC, using a variety of living organisms including animal models, human cell lines and skin samples, and determining the damage to DNA and to cellular systems, and the repair mechanisms in the cells to combat the damage. No doubt, knowledge gained from the UVC studies has been extrapolated on UVB and UVA exposures of humans; nevertheless, it is also necessary that more attention be given to the importance of indirect actions of UVA exposure of humans and other living organisms. Furthermore, it is known that about 20-30% of XP patients also suffer from neurological abnormalities caused by neuronal death in the central and peripheral nervous systems. Since neural tissues are not exposed to sunlight, the reason for neurodegeneration in XP patients remains unclear and must be explored. A recent report, however, shows that 8,5'-cyclopurine-2'-deoxynucleosides (cPu), an oxidatively-induced DNA damage, is repaired by NER and that this lesion is responsible for neurodegeneration in XP patients lacking NER activity.[16]

Although the deficiency in NER in XP patients has been hypothesised as one reason why the oxidative DNA lesion, 8-oxoguanine and thymine glycol, cannot be removed by the same enzyme system responsible for removal of CPD and 6-4PP (see Chapter 12), a key question remains as to how UVA can generate these oxidatively induced DNA lesions when it cannot reach targets protected from sunlight. Some of the Editor's (SIA) own research shows that UVA photolysis of certain biological compounds generates free radicals and these may be responsible for damage to cellular systems. Free radicals such as superoxide anions (O_2^{-}), hydroxyl radicals ($^{\cdot}OH$) and singlet oxygen ($^1O_2$) may be produced.[17,18] Also a direct electron transfer (type 1) reaction has been proposed.[19] It is therefore likely that one or more of these reactions and/or resulting free radical formation (along with other lesions induced by UVB) may be responsible for UV-induced skin cancer in XP patients. Hence, more studies must be undertaken to explore the formation of free radicals by UVA photolysis of biological compounds and its effect on human health.

Of the eight XP genes so far discovered, seven of them, XP-A through XP-G, play roles in NER. This ubiquitously found repair process can recognize a variety of DNA damages including UV-induced CPD and 6-4PP and, using at least 28 NER enzymes including the seven XP gene products, the repair is carried out. Interestingly three NER genes are also part of the basal transcription factor TFIIH, and mutations in any one of 11 NER genes have been associated with clinical diseases with at least eight overlapping phenotypes.[20] The NER enzyme and the process of NER-induced DNA repair are covered in Chapter 12.

Chapter 4 introduces the *XPA* gene (located at 9q22), one of the key NER genes. Its biological role is to recognize and promote the binding of repair complex to DNA at the damaged site, including CPD and 6-4PP induced by UV light. The 31 kDa protein forms a core of the incision complex by interacting with a number of other XP proteins (see Fig. 2, Chapter 4). The complex is then responsible for recognizing kinks in DNA caused by a variety of other damaging agents (see refs. 16, 21, 75, 90 of Chapter 4) and makes an incision leading to removal of the damaged section and its replacement by polymerization of DNA with subsequent ligation of the newly synthesized strand. In a recent study, however, it has been shown that purified XPA and the minimal DNA-binding domain of XPA can, fully and preferentially, bind to mitomycin C-DNA crosslinks in the absence of other proteins from NER.[21] Another recent study shows that cells that overexpress the HMGA1 (high mobility group A1) proteins show deficiency in NER because these non-histone proteins are involved in inhibiting XPA expression, resulting in increased UV sensitivity.[22]

Chapter 5 describes *XPB* and *XPD* genes (located at 2q14 and 19q13, respectively), their products and biological roles; XPB is a helicase and together with XPD products constitute components of TFIIH. This complex is composed of 10 proteins, five of which (p8, p34, p44, p52, p62) make a tight complex along with XPB, and XPD is less tightly associated with the CAK subcomplex containing cyclin H, cdk7 and MAT1.[23] The DNA-dependent helicase activity of XPB and XPD is important in transcriptional initiation and the NER process although the molecular mechanism of these two gene products in NER is poorly understood. In a recent study it has been shown that the p52 subunit of TFIIH interacts with XPB and stimulates its ATPase activity.[24] XPB is a rare disease compared to other XP forms. Also mutation in XPD can lead to bladder cancer.[25]

XPC (gene location at 3p25) is the most commonly prevailing mutational site of all XP genetic defects. It is a key protein involved in DNA repair, damaged by a variety of agents including UV light. The primary role of this protein is to recognize the damage and allow NER to complete the repair process. In addition to participating in NER repair, XPC has also been shown to participate in BER (base excision repair), especially of those lesions induced oxidatively such as 7,8-dihydroxy-8-oxoguanine and other single base modifications that are repaired by BER.[26] In order to recognize damage, the protein forms a complex with Rad23p orthologs and centrin-2. Thus DNA damage in XPC patients is not recognized and remains unrepaired, and XPC phenotype prevails. In a recent study it has been shown that ING1b (a ubiquitous protein involved in a large number of biological activities including senescence, cell cycle arrest, apoptosis and DNA repair) enhances NER repair only in XPC proficient cells and via XPA, implying an essential role of ING1b in early and better access of NER machinery via association with chromatin to the sites of the lesion.[27] A recent study from China found a link between XPC deficiency and promotion and progression of bladder cancer, and also cellular development of resistance against anticancer drugs.[28] Further roles of XPC and its biological activities have been described in Chapter 6.

Patients suffering from a defect in XPE, on the other hand, are rare and the subgroup is considered to be one of the mildest of XP forms. XPE gene is located on chromosome 11, at 11q12-q13 for DDB1 and 11p11-p12 for DDB2; thus there are two components of the XPE protein, damaged DNA binding proteins 1 and 2. All XPE patients so far identified carry mutation in the DDB2 gene. It is suggested that this protein is involved in a variety of cellular functions including signal transduction, cell cycle regulation, RNA splicing, transcription and apoptosis, and as the name suggests, binds to a variety of DNA damaged sites. These interactions are eloquently described in Chapter 7.

A comprehensive review of the literature on XPF/ERCC4 is presented in Chapter 8. The protein (which authors prefer to call ERCC4) appears to be involved in NER and acts after forming a complex with other XP and non-XP proteins. It appears also to be involved in the repair of double strand DNA damage, in particular of inter-strand DNA cross-links, such as those caused by platinum compounds and PUVA (8-methoxypsoralen + UVA). This explains why certain XP group cell lines are sensitive to PUVA.[29] Although these researchers (one of them, W.C. Lambert, is the author of Chapter 14 of this book) had pointed to a role of XPA in the process, their current studies suggest that XPF also participates in the process, and in the absence of either functional protein the repair of double strand damage cannot take place.

Chapter 9 covers the *XPG* gene (chromosomal location, 13q33) and its biological roles. It is a rarer genetic defect relative to other XPs. In NER the XPG gene product has a role in assembly of the pre-incision complex and a catalytic role in making the incision at the site of DNA damage. Interestingly, Ito et al[30] have shown that XPG plays an important role in stabilizing TFIIH, which is active in both transcription and NER, and in allowing transactivation of nuclear receptors. A role in regulation of gene expression also has been proposed. Although some patients with XPG mutations display a milder phenotype of the disease, this genetic defect sometimes can overlap with the phenotypes of Cockayne syndrome (referred to as XP-G/CS complex, see Chapter 14), in which case the disease is more severe and complex.

XPV, the variant form of XP, is unusual in that the XPV gene polH (located at 6p21) and its product are not part of NER. XPV product is a polymerase, pol eta (η), whose role is to attend DNA damaged by UV and, interestingly, although in vitro studies have shown that the transdimer bypass process is error proof, this enzyme can introduce errors while replicating non-damaged sites of DNA. In cells lacking pol η, pol iota (ι) attends the UV-induced DNA lesions and introduces errors with high frequency. Abnormal spectra of UV-induced mutations are observed; and hence polymerase η deficient persons are more prone to UV-induced cancer. The mode of action of pol η and its association with other proteins are described in Chapter 10.

One of the most intriguing areas of study of XP is the interactions of a large number of proteins involved in NER. Chapter 11 tackles this issue and eloquently describes the mechanisms and interactions of XP/XP and XP/non-XP proteins. Cleverly organized in Table 2 of the chapter is a description of the XP proteins and their interactive non-XP proteins. For example, XPA protein interacts with another five non-XP proteins: ATR, RPA, TFIIH, XAB1, and XAB2 as well as with XPF. Likewise XPB proteins interact with six non-XP proteins and two XP proteins. A quick count of the non-XP proteins interacting with the XP proteins (in Table 1, Chapter 11) indicates 37 to date. Kraemer et al[20] have identified at least 28 proteins involved in NER. This shows that the repair pathways involved in DNA repair are not limited only to those proteins for which XP mutants are available, but a large number of additional proteins are involved whose presence and activities are essential for DNA repair to take place. It is likely that some or most of the non-XP proteins for which mutants are not found are essential for cell survival and mutation in any of these genes leads to fatality.

It has been indicated earlier that some XP patients suffer neurological disorders due to neurodegeneration and that endogenously produced free radicals are to be blamed for this. Free radicals such as $O_2^{-\cdot}$, $\cdot OH$, $^1O_2$, hydrogen peroxide (H_2O_2), nitrite ion (NO_2^-) and peroxynitrite anion ($ONOO^-$) are produced in biological systems. Although most biological systems have developed a number of defense mechanisms (known as antioxidants) such as catalase, superoxide dismutase and glutathione peroxidase, DNA is still subject to damage. Oxidatively damaged DNA is also repaired by NER, but when it cannot be repaired efficiently, such as in Cockayne syndrome and XP, the resulting accumulation of damage leads to skin cancer, progressive neurodegeneration and ocular abnormalities. In a recent study it has been proposed that 8,5'-cyclopurine-2'-deoxynucleoside is the most important oxidatively induced DNA lesion which, if not repaired due to NER deficiency (as in XP patients), leads to neurodegeneration.[31] Two criteria proposed for its a being premutational site are that: (i) it is chemically stable and (ii) it strongly blocks transcription by RNA polymerase II in XP cells. Chapter 13 investigates in depth the clinical features of neurodegenerative processes in specimens from XP patients and provides an insight into the molecular mechanisms.

Interestingly, XP shares a number of clinical and molecular features with a variety of other syndromes including trichothiodystrophy (XP/TDT), Cockayne syndrome (XP/CS) and progeroid syndromes. These syndromes or complexes can be divided into XP/TTD and XP/CS, which can further be subdivided into XPB/CS, XPD/CS, XPG/CS, XPH/CS and XP/CSB. XP/progeroid overlap is of one type, called XP-F/progeroid syndrome. In some overlap cases the same biochemical processes with the same gene(s) are involved and in some instances patients display more than one of these symptoms, which are clinically more heterogenous. A comprehensive investigation into these overlap syndromes and a new disease, UV sensitive syndrome (UVSS), are presented in Chapter 14. Further investigation of all these conditions will provide insights into the molecular mechanisms involved in each one and where they may interact and/or overlap.

To understand the molecular mechanisms of XP, XP mouse models have been used, and mice deficient in XPA, XPC, XPD, XPG, XPF and XPA/CSB have been produced and analysed. A recent elegant technique of targeting gene replacement in mouse embryonic stem cells has provided researchers with the ability to generate mutant mice defective in any specific gene(s).[32] Animals generated in this way display phenotypes and symptoms of XP patients and have provided valuable tools to understand how and where the deficiency in DNA repair may lead to tumor formation, and also in studies of developmental biology and the aging process. Mouse studies have recently contributed to our understanding of the role of ink4a-Arf in increasing the risk of melanoma photo-carcinogenesis in an XPC mutant background.[33]

As with many other genetic defects, the distribution of XP globally is not uniform. In most cases the frequency of mutation of a particular trait depends when and where a specific mutation arose, and the longer ago that is, the greater the frequency of mutation in the population unless some selective pressure prevailed. Another factor responsible for the high incidence of any mutation is consanguinity. Chapter 15 analyzes the world distribution of XP and shows that Japan has the highest incidence of XP and of varying complementation groups. After Japan perhaps Egypt suffers most from this inborn error. Here it is also shown that the most common complementation groups are XPA and XPC followed by XPV. XPB and XPE are least frequent. In a recent publication, however, 16 Japanese patients with XPV have been diagnosed and confirmed both clinically and at the cellular level.[34]

There is no evidence that interest in XP is waning, and this book should provide both the expert and novice researcher in the field with an excellent overview of the current status of research and pointers for future research goals.

Shamim I. Ahmad, MSc, PhD
Fumio Hanaoka, PhD

References

1. Hebra F, Kaposi M. On diseases of the skin, including the exanthemata. New Sydenham Society 1874; 61:252-258 (Translated by W. Tay, London).
2. Kaposi, M. Xeroderma pigmentosum. Ann Dermatol Venereol, 1883; 4:29-38.
3. Ulicna-Zapletalova M. Contribution to the pathogenesis of xeroderma pigmentosum. Cesk Dermatol 1949; 24:330-334.
4. Cleaver, JE. Defective repair replication of DNA in xeroderma pigmentosum. Nature 1968; 218:652-656.
5. Cleaver, JE. Xeroderma pigmentosum: a human disease in which an initial stage of DNA repair is defective. Proc Natl Acad Sci USA 1969; 63:428-435.
6. DeSanctis C, Cacchione A. L'idiozia xerodermica. Riv Sper Frentiatr Med Leg Alienazioni Ment 1932; 56:269-292.
7. Lichon V, Khachemoune A. Xeroderma pigmentosum: beyond skin cancer. J Drug Dermatol 2007; 6:281-288.
8. Mohanty P, Mohanty L, Devi BP. Multiple cutaneous malignancies in xeroderma pigmentosum. Indian J Dermatol Venereol Leprol 2001; 67:96-97.
9. Kraemer KH, Lee MM, Scotto J. Xeroderma pigmentosum. Cutaneous, ocular, and neurological abnormalities in 830 published case. Arch Dermatol 1987; 123:241-250.
10. Unal M, Guven M, Batar B et al. Polymorphism of DNA repair genes XPD and ERCC1 and risk of cataract development. Exp Eye Res 2007; 85(3):328-334.
11. Kiyohara C, Yoshimasu K. Genetic polymorphism in the nucleotide excision repair pathway and lung cancer risk: a meta analysis. Int J Med Sci 2007; 4:59-71.
12. Menck CF, Armelini MG, Lima-Bessa KM. On the search of skin gene therapy strategies of xeroderma pigmentosum disease. Curr Gene Ther 2007; 7:163-174.
13. Reichrath J. Sunlight, skin cancer and vitamin D: what are the conclusions of recent findings that protection against solar ultraviolet (UV) radiation causes 25-hydroxyvitamin D deficiency in solid organ transplant recipients, xeroderma pigmentosum, and other risk groups? J Steroid Biochem Mol Biol 2007; 103:664-667.
14. Cleaver JE, Crowley E. UV damage, DNA repair and skin carcinogenesis. Front Biosci 2002; 7:d1024-1043.
15. Mouret, S, Baudouin C, Charveron M et al. Cyclobutane pyrimidine dimers are predominant DNA lesions in whole human skin exposed to UVA radiation. Proc Natl Acad Sci USA 2006; 103:13765-13770.
16. Theruvatu JA, Jaruga P, Dizdaroglu M. The oxidatively induced DNA lesions 8,5'-cyclo-2'-deoxyadenosine and 8-hydroxy-2'-deoxyadenosine are strongly resistant to acid induced hydrolysis of glycosidic bond. Mech Ageing Dev 2007; 128(9):494-502.
17. Craggs J, Kirk SH, Ahmad SI. Synergistic action of near UV and phenylalanine, tyrosine or tryptophan on the inactivation of phage T7: role of superoxide radicals and hydrogen peroxide. J Photochem Photobiol 1994; 24:123-128.
18. Paretzoglou C, Stockenhuber S, Kirk SH et al. Generation of reactive oxygen species from the photolysis of histidine by near-ultraviolet light: effects on T7 as a model biological system. J Photochem Photobiol 1998; 43:101-105.

19. Ahmad SI, Hargreaves A, Taiwo FA et al. Near ultraviolet photolysis of L-mandelate, formation of reactive oxygen species, inactivation of phage T7 and implication on human health. J Photochem Photobiol B 2004; 77:55-62.
20. Kraemer KH, Patronas NJ, Schiffmann R. et al. Xeroderma pigmentosum, trichothiodystrophy and Cockayne syndrome: a complex genotype-phenotype relationship. Neurosci 2007; 145:1388-1396.
21. Mustra DJ, Warren AJ, Wilcox DE et al. Preferential binding of human XPA to the mitomycin C-DNA interstrand crosslink and modulation by arsenic and cadmium. Chem Biol Interact 2007; 168:159-168.
22. Adair JE, Maloney SC, Dement GA. High mobility group A1 proteins inhibit expression of nucleotide excision repair factor xeroderma pigmentosum group A. Cancer Res 2007; 67:6044-6052.
23. Giglia-Mari G, Coin F, Ranish JA et al. A new, tenth subunit of TFIIH is responsible for DNA repair syndrome Trichothiodystrophy group A. Nat Genet 2004; 36:714-719.
24. Coin F, Oksenych V, Egly JM. Distinct roles for the XPB/p52 and XPD/p44 subcomplexes of TFIIH in damaged DNA opening during nucleotide excision repair. Mol Cell 2007; 26:245-256.
25. Shao J, Gu M, Xu Z et al. Polymorphism of the DNA gene XPD and risk of bladder cancer in a Southeastern Chinese population. Cancer Genet Cytogenet 2007; 177:30-36.
26. Kassam SN, Rainbow AJ. Deficient base excision repair of oxidative DNA damage induced by methylene blue plus visible light in xeroderma pigmentosum group C fibroblasts. Biochem Biophys Res Commun 2007; 359:1004-1009.
27. Kuo WH, Wang Y, Wong RP et al. The ING1b tumor suppressor facilitates nucleotide excision repair by promoting chromatin accessibility to XPA. Exp Cell Res 2007; 313:1628-1638.
28. Chen Z, Yang J, Wang G et al. Attenuated expression of xeroderma pigmentosum group C is associated with critical events in human bladder cancer carcinogenesis and progression. Cancer Res 2007; 67:4578-4585.
29. Tsongalis GJ, Lambert WC, Lambert MW. Electroporation of normal human DNA endonucleases into xeroderma pigmentosum cells corrects their DNA repair defect. Carcinogenesis 1990; 11:499-503.
30. Ito S, Kuraoka I, Chymkowitch P et al. XPG stabilizes TFIIH, allowing transactivation of nuclear receptors: implications for Cockayne syndrome in XP-G/CS patients. Mol Cell 2007; 26:231-243.
31. Brooks PJ. The case for 8,5'-cyclopurine-2'-deoxynucleosides as endogenous DNA lesions that cause neurodegeneration in xeroderma pigmentosum. Neuroscience 2007; 145:1407-1417.
32. Yamazaki F, Okamoto H, Matsumora Y. Development of a new mouse model (xeroderma pigmentosum A-deficient, stem cell factor-transgenic) of ultraviolet B-induced melanoma. J Invest Dermatol 2005; 125:521-525.
33. Yang G, Curley D, Bosenberg MW et al. Loss of xeroderma pigmentosum C (Xpc) enhances melanoma photocarcinogenesis in ink4a-Arf-deficient mice. Cancer Res 2007; 67:5649-5657.
34. Tanioka M, Masaki T, Ono R et al. Molecular analysis of DNA polymerase eta gene in Japanese patients diagnosed as xeroderma pigmentosum variant type. J Invest Dermatol 2007; 127:1745-1751.

ABOUT THE EDITORS...

Shamim I. Ahmad and Fumio Hanaoka

SHAMIM I. AHMAD is a Senior Lecturer at Nottingham Trent University, Nottingham, England. After obtaining his MSc from Patna University, India, and his PhD from Leicester University, England, he joined Nottingham Polytechnic, which subsequently became Nottingham Trent University. For about three decades he has been working in the field of DNA damage and repair, particularly on near UV photolysis of biological compounds, production of free radicals and their implications on human health including skin cancer and xeroderma pigmentosum. He is also investigating compounds inducing double strand DNA damage, 8-methoxypsoralen +UVA, mitomycin C, and nitrogen mustard, including their importance in psoriasis treatment and Fanconi anemia. Additional research includes: thymineless death in bacteria, genetic control of nucleotides catabolism, development of an anti-AIDS drug, control of microbial infections of burns, phages of thermophiles and microbial flora of Chernobyl after the accident. In 2003 he received a prestigious "Asian Jewel Award" in Britain for "Excellence in Education". He is also the Editor of the book, *Molecular Mechanisms of Fanconi Anemia*, published by Landes Bioscience.

FUMIO HANAOKA is a Professor at the Graduate School of Frontier Biosciences, Osaka University and the Program Leader of the Solution Oriented Research for Science and Technology of the Japan Science and Technology Agency, Japan. He received his undergraduate and PhD degrees from the University of Tokyo and did his Postdoctoral at McArdle Laboratory for Cancer Research, University of Wisconsin, Madison, USA. He joined the University of Tokyo in 1980 and in 1989 moved to RIKEN Institute as the Head of the Radiation Research Laboratory. In 1995, he joined the Institute for Molecular and Cellular Biology (now known as Graduate School of Frontier Biosciences), Osaka University. His main research interests include the molecular mechanisms of DNA replication and repair in eukaryotes. He served as the President of Molecular Biology Society of Japan (2005-2007) and has been serving on several editorial boards, including *Journal of Biological Chemistry* and *Genes to Cells*.

PARTICIPANTS

Shamim I. Ahmad
School of Science and Technology
Nottingham Trent University
Nottingham
England

Brian D. Beck
Department of Biochemistry
    and Molecular Biology
Walther Cancer Institute
Indiana University School of Medicine
and
IU Cancer Research Institute
Indianapolis, Indiana
USA

Drew Bennett
Department of Pathology
The University of Iowa
    Roy J. and Lucille A. Carver
    College of Medicine
Iowa City, Iowa
USA

Abdul Manan Bhutto
Department of Dermatology
Chandka Medical College
Larkana
Pakistan

Ulrike Camenisch
Institute of Pharmacology
    and Toxicology
University of Zurich-Vetsuisse
Zurich
Switzerland

James E. Cleaver
Auerback Melanoma Laboratory
UCSF Cancer Center
University of California
San Francisco, California
USA

Chun Cui
Department of Anatomy
    and Developmental Neurobiology
The University of Tokushima
    School of Medicine
Tokushima
Japan

Leela Daya-Grosjean
Laboratory of Genetic Instability
    and Cancer
Institut Gustave Roussy
Villejuif
France

Steffen Emmert
Department of Dermatology
   and Venerology
Georg-August-University, Goettingen
Goettingen
Germany

Yoshihiro Fukui
Department of Anatomy
   and Developmental Neurobiology
The University of Tokushima
   School of Medicine
Tokushima
Japan

Claude E. Gagna
Department of Pathology
New York Institute of Pathology
New York, New York
USA

Alexei Gratchev
Department of Dermatology
University Medical Center Mannheim
Ruprecht-Karls University
   of Heidelberg
Mannheim
Germany

Yeunhwa Gu
Department of Radiological
   Technology
Suzuka University of Medical Science
Mie
Japan

Dae-Sik Hah
Department of Biochemistry
   and Molecular Biology
Walther Cancer Institute
Indiana University School
   of Medicine
Indianapolis, Indiana
USA

Fumio Hanaoka
Graduate School of Frontier
   Biosciences
University of Osaka
Osaka
Japan

Yoshi-Nobu Harada
Molecular Imaging Center
National Institute
   of Radiological Sciences
Chiba
Japan

Masaharu Hayashi
Department of Clinical
   Neuropathology
Tokyo Metropolitan Institute
   of Neuroscience
Tokyo
Japan

Ulrich R. Hengge
Department of Dermatology
Heinrich-Heine-University
Düsseldorf
Germany

Setsuji Hisano
Laboratory of Neuroendocrinology
Institute of Basic Medical Sciences
Graduate School of Comprehensive
   Human Sciences
University of Tsukuba
Tsukuba
Japan

Toshiki Itoh
Department of Pathology
The University of Iowa
   Roy J. and Lucille A. Carver
   College of Medicine
Iowa City, Iowa
USA

Sandra H. Kirk
School of Science and Technology
Nottingham Trent University
Nottingham
England

Muriel W. Lambert
Department of Pathology
UMDNJ-New Jersey Medical School
Newark, New Jersey
USA

W. Clark Lambert
Departments of Pathology
    and Dermatology
UMDNJ-New Jersey Medical School
Newark, New Jersey
USA

Suk-Hee Lee
Department of Biochemistry
    and Molecular Biology
Walther Cancer Institute
Indiana University School of Medicine
and
IU Cancer Research Institute
Indianapolis, Indiana
USA

Chikahide Masutani
Graduate School
    of Frontier Biosciences
University of Osaka
Osaka
Japan

Lisa D. McDaniel
Department of Pathology
University of Texas Southwestern
    Medical Center at Dallas
Dallas, Texas
USA

Hanspeter Nägeli
Institute of Pharmacology
    and Toxicology
University of Zürich-Vetsuisse
Zürich
Switzerland

Alain Sarasin
Laboratory of Genomes and Cancers
Institut Gustave Roussy
Villejuif
France

Roger A. Schultz
University of Texas Southwestern
    Medical Center at Dallas
Dallas, Texas
USA

Orlando D. Schärer
Department of Pharmacological
    Sciences and Chemistry
Stony Brook University
Stony Brook, New York
USA

Steven M. Shell
Department of Biochemistry
    and Molecular Biology
James H. Quillen College of Medicine
East Tennessee State University
Johnson City, Tennessee
USA

Kaoru Sugasawa
Biosignal Research Center
Organization of Advanced Science
    and Technology
Kobe University
and
SORST
Japan Science and Technology Agency
Hyogo
Japan

Xue-Zhi Sun
Regulatory Sciences Research Group
National Institute of Radiological
   Sciences
Chiba
Japan

Hidenori Yonehara
Department of Anatomy
   and Developmental Neurobiology
The University of Tokushima
   School of Medicine
Tokushima
Japan

Rui Zhang
Department of Anatomy
   and Developmental Neurobiology
The University of Tokushima
   School of Medicine
Tokushima
Japan

Yue Zou
Department of Biochemistry
   and Molecular Biology
James H. Quillen College of Medicine
East Tennessee State University
Johnson City, Tennessee
USA

CONTENTS

11. OTHER PROTEINS INTERACTING WITH XP PROTEINS 103

Steven M. Shell and Yue Zou

12. THE NUCLEOTIDE EXCISION REPAIR OF DNA IN HUMAN CELLS AND ITS ASSOCIATION WITH XERODERMA PIGMENTOSUM..113

Alexei Gratchev

13. ROLES OF OXIDATIVE STRESS IN XERODERMA PIGMENTOSUM.. 120

Masaharu Hayashi

14. XERODERMA PIGMENTOSUM: ITS OVERLAP WITH TRICHOTHIODYSTROPHY, COCKAYNE SYNDROME AND OTHER PROGEROID SYNDROMES 128

W. Clark Lambert, Claude E. Gagna and Muriel W. Lambert

15. POPULATION DISTRIBUTION OF XERODERMA PIGMENTOSUM.. 138

Abdul Manan Bhutto and Sandra H. Kirk

16. PROGRESS AND PROSPECTS OF XERODERMA PIGMENTOSUM THERAPY 144

Alain Sarasin

17. ANIMAL MODELS OF XERODERMA PIGMENTOSUM 152

Xue-Zhi Sun, Rui Zhang, Chun Cui, Yoshi-Nobu Harada, Setsuji Hisano,
  Yeunhwa Gu, Yoshihiro Fukui and Hidenori Yonehara

ACKNOWLEDGEMENTS

The Editors acknowledge the support of Nottingham Trent University, England, and Osaka University, Japan, for providing funding to carry out the work for this publication. We are also thankful to all our contributors for their highly skilled and knowledgeable production of the chapters without which it would not have been possible to produce this book.

CHAPTER 1

Historical Aspects of Xeroderma Pigmentosum and Nucleotide Excision Repair

James E. Cleaver*

"The Primal Light the whole irradiates
and is received therein as many ways
As there are splendours wherewithall it mates"

Dante A., Commedia III, El Paradiso, Canto XXIX
(14th Century) Translated by Dorothy L Sayers
(1962). Penguin Classics

Abstract

The discovery that xeroderma pigmentosum was a sun-sensitive hereditary human disease that was deficient in DNA repair was made when research into the fundamental mechanisms of nucleotide excision repair was in its infancy. The linkage between DNA damage, DNA repair and human cancer stimulated an enormous subsequent growth of the field of DNA repair and the identification of other repair deficient diseases and other repair pathways. This growth has established DNA repair as a central factor for maintaining genomic stability and preventing cancer, neurodegenerative disease and aging. The study of DNA repair impacts many other areas including human genetics, signal transduction, protein structure, DNA-protein interactions, DNA replication and recombination, transcription, telomere maintenance, development, differentiation, ecology and evolution.

Historical Aspects

Xeroderma pigmentosum (XP) was first named by Hebra and Kaposi in 1874,[1] and the association of XP with neurological dysfunction by de Sanctis and Cacchione in 1932.[2] There were many clinical descriptions of XP following these initial definitions, which amply showed it to be a rare sun-sensitive skin cancer syndrome, probably inherited, resulting in serious disfigurement unless stringent protection from sun exposure was practiced. The actual incidence of XP is difficult to establish, and has been estimated to be of the order of 1-5 cases per million of the population in the US or Europe.[3] The disease has been identified world-wide in all ethnic groups; some parts of the world show higher incidence, including Japan[4] and Egypt,[5] and several examples of very high incidence in isolated communities or common descent from a single ancestral case have been reported.[6,7] A whole village was recently identified in Guatemala founded by an XP carrier in which a large proportion of the children were severely affected.[8] The linkage of XP to a molecular genetic

*James E. Cleaver—Auerback Melanoma Laboratory, UCSF Cancer Center, University of California, San Francisco, CA, USA. Email: jcleaver@cc.ucsf.edu

Molecular Mechanisms of Xeroderma Pigmentosum, edited by Shamim I. Ahmad and Fumio Hanaoka. ©2008 Landes Bioscience and Springer Science+Business Media.

basis did not occur until the 1960s. Since then research into XP and into the molecular mechanisms of nucleotide excision repair (NER) have been inextricably entwined for the past 40 years.

This book summarizes the contributions of many major players in the DNA repair community that center around XP and related diseases of DNA repair. I have been privileged to be part of this community that grew from a very small group of interested parties to a wide-ranging, intensely active research field that overlaps innumerable other areas. The field of DNA repair was very small in the 1960s and the mechanisms seemed esoteric, remote from normal cell metabolism and with little impact. In retrospect, it now seems odd that the maintenance of the cell's most important molecule was not recognized as a critical function with its own dedicated metabolism; times have changed! DNA repair genes are now recognized as integral components of every genome and represent the core biochemical pathways that stabilize the genome against radiations and chemical onslaughts from internal and external sources. DNA repair now touches human genetics, cancer research, signal transduction, protein structure, DNA-protein interactions, DNA replication and recombination, telomere maintenance, neurodegeneration, development, ecology and evolution and many other fields.

It seems remarkable that 40 years have passed since XP was first identified as a repair deficiency disease. The experience for me has been like launching a raft that became a canoe, then a ship, a liner and finally a whole Navy! Hard to realize now, but the discovery of XP was one of the earliest forays into molecular genetics. The conjunction of clinical relevance with a basic discovery in DNA repair was an invaluable stimulation to the rapid development of the field of DNA repair. This association established the relevance of DNA repair to human health and attracted many investigators resulting in a rapid growth and excitement of the field of XP and DNA repair. Many justifications for a line of research or a grant proposal were based on XP.

The first flotation of a possible molecular basis for XP was Stanley Gartler's report from the International Congress of Human Genetics in 1964 showing XP cells were UV sensitive.[9] Charles J. Epstein told me personally that he investigated photoreactivation (PHR) in XP at about the same time. Ironically since then we have learnt that although PHR genes are absent from human cells,[10,11] homologs have evolved to regulate light-dependent diurnal functions.[12] In June 1967, I read an article by David Perlman, in the San Francisco Chronicle newspaper, that described a study by Henry Lynch of Creighton University of a familial skin cancer induced by sunlight, a disease XP. I mentioned this article to Bob Painter my senior colleague, saying, "Perhaps this is a mutation in the repair of UV damage like *Escherichia coli* B$_{s-1}$?" Bob's reply, "It's a crazy idea, but at your stage what have you got to lose!" Mentoring comes in various forms, sometimes it needs only to be a challenge and allowing a free hand!

With the help of Bill Epstein, John Epstein, Kimie Fukuyama and Bill Reed, of the UCSF Department of Dermatology, I was provided cell cultures from Californian XP patients that allowed us to show indeed that XP cells were repair deficient,[13,14] using techniques that Phil Hanawalt had developed in *E. coli*,[15,16] and Bob Painter had modified for mammalian cells.[17,18] Soon thereafter the repair deficiency was confirmed in vivo in human skin, an important experiment that extended the identification of repair deficiency direct to patients.[19] My work was first presented at a Radiation Research meeting in 1967, soon followed by an oral presentation at the National Institutes of Health, Bethesda, that was roundly criticized for its lack of statistical robustness. But, bolstered by Bob Painter's encouragement, I had nothing to lose by sticking to it and three papers were published in rapid succession: Nature 1968,[13] PNAS 1969[14] and JAMA in 1969.[20] Soon thereafter, XP was investigated by Setlow, Regan, German and Carrier,[21] and I awaited their results with some trepidation. I was much relieved to learn they had confirmed the NER deficiency in XP, from Jim Regan in the back of a cab in Rome traffic en route to an international meeting.

Harvey Blank, University of Miami sent me a telegram saying the identification of XP was "a landmark in Dermatology". Joshua Lederberg wrote an editorial in the Washington Post in 1968, using it as a plea for more funding for basic research, even in those halcyon days. Bob Haynes is known to have said, "Thank God we have a disease; now we can get money!" Mike Bishop, Nobelist for discovering oncogenes, wrote later "...While I was still in medical school, James Cleaver recognized xeroderma pigmentosum as a deficiency in the repair of DNA damage caused by ultraviolet light...I have been a believer in the somatic mutation hypothesis of cancer ever since."[22] As a result of this discovery I was thrust into the field of Dermatology, one from which I have drawn much personal pleasure from the opportunity to interact with physicians and patients.

While somatic mutation provided a framework for understanding cancer in XP patients, other clinical features of many XP patients and related diseases, especially neurodegeneration have been less easy to explain in molecular detail and may be associated with other functions of XP proteins than only DNA repair.[23] There is frequent use of a "default explanation" for neurodegeneration in repair deficient disease that says: the brain has high oxidative metabolism and high gene expression so oxidative damage to actively expressed genes may cause neuronal cell death. But, we have yet to identify the reason or source of damage, its molecular character, or the reason for specificity of symptoms to specific areas of the brain and central nervous system. Whereas both complementation groups of the human repair deficient neurodegenerative disease Cockayne syndrome (CS) are sensitive to UV light and oxidative damage,[24] in the mouse only CSB is sensitive to oxidative damage and CSA is resistant,[25] making simple correlations to deficient repair of oxidative damage difficult. We cannot even exclude that defective repair and neuronal degeneration may be but two of several different downstream manifestations of a common defect, or independent functions of a multi-functional biochemical system.[26]

The progress of research into XP and NER has had two major phases and we are now entering a third. The first phase started with the identification of the repair deficiency in XP in 1968[13] and ended in the period bracketed by the first cloning of a repair gene (*ERCC1*) by Rubin[27] and the demonstrations of repair in vitro in cell extracts by Wood and Lindahl in 1988[28] and Sancar in 1989.[29] During this phase complementation groups were defined and assumed to represent mutations in different genes;[30] the major photoproducts (cyclobutane pyrimidine dimers and [6-4] photoproducts) were clarified and rediscovered,[31] the kinetics of excision,[32,33] repair synthesis[34] and DNA chain elongation in UV damaged cells[35,36] were laid out. The number of complementation groups in XP eventually settled down to 8, A through G and V and hamster mutant cell lines corresponding to XP were isolated and characterized.[37,38]

Until XP genes were cloned, the assignment of cell lines to various groups depended on reference to a small number of cell lines that were arbitrarily designated sequentially as XPA, XPB, XPC, etc. Each new patient had to be assigned by autoradiography after fusion with representative cell lines from each of the predefined groups, The number of groups briefly rose to 10,[39] before later cell lines were reassigned to previously designated groups[40] and the number of complementation groups settled to an agreed set of 8. These groups represent a subset of NER genes that permit human viability when mutated or inactivated; many other NER genes may have more severe phenotypes. The assignment of cell lines from mildly affected patients to the XPE group, which had high residual levels of repair, resulted in some confusion because of the difficulty of resolving differences on a factor of 2 in autoradiographs of fused cells and made the molecular defects in this group appear heterogeneous.[41] This confusion was later resolved when the XPE gene and gene product were characterized and several cell lines were found to have been mis-assigned.[42]

The defects in XP were shown to represent multiple biochemical steps of incision and excision.[43-45] These defects resulted in increased UV sensitivity and mutagenesis,[46] which implied that the linkage between DNA repair and cancer was through increased somatic mutation rates. Toward the end of this period NER was found to occur at different rates in different parts of the genome.[47] These observations were to lead to a link between NER and blocked transcription and the concept of transcription coupled repair (TCR) as a distinct branch of NER.[48,49] TCR was later shown to be defective in the UV sensitive neurological disease CS.[50] The XP variant (XPV) was identified and named in 1972.[51] This group appeared to have normal NER but had a defect in DNA replication after UV damage that also resulted in an increased UV induced mutation rate.[52] Molecular resolution of the XPV group lay far in the future. One of the early reviews of XP, summarizing these early developments in the first phase of discovery was published in Advances in Genetics in 1975.[53]

The second phase ushered in the molecular dissection of repair with successive cloning of XP genes, the detailed assignment of their functions in the NER process, the demonstration of strong linkage between repair and transcription. This phase also saw the development of mouse models, which are now an extensive resource.[54] This phase can be thought of as ending in 1999 with the eventual cloning of the XP variant low fidelity polymerase.[55,56] This was cloned simultaneously by two independent groups using different techniques and finally wrapped up the near complete characterization of the molecular basis of NER deficient diseases. During both historic phases, additional mutagen-sensitive diseases were identified. These included the set of TCR deficient diseases: CS, trichothiodystrophy, cerebro-oculo-facio-skeletal syndrome (COFS) and the heterogenous UV sensitive syndrome.[3] The latter encompasses several different molecular defects associated with mild disease[57,58] and is less sensitive to oxidative damage than CS cells.[24] Other repair pathways were associated with ataxia telangiectasia (defective in double strand break signaling),[59] Fanconi anemia (defective in cross-link repair),[60,61] nonpolyposis colon cancer (defective in mismatch repair),[62] breast cancer (defective in homologous recombination)[63,64] and neurodegeneration (defective in single strand break repair and TCR)[50,65] to name a few. We are now in a phase where perhaps we can really make a difference in patients' lives by applying some of our knowledge of the exquisite molecular details of repair to improved diagnosis, care and treatment.

NER has become unexpectedly complicated. Our early ideas of "cut and patch", or "patch and cut", were epitomized by simple models elegantly set forth in an early 1967 Scientific American article by Hanawalt and Haynes.[66] NER was envisaged as a simple series of steps of damage recognition, excision, resynthesis and ligation. We thought that chromatin structure would complicate recognition, but could not imagine to what extent until recent technical developments have opened up new vistas in this area.

One surprise has been the incredible complexity of pyrimidine dimer damage recognition in DNA.[67] Recognition of pyrimidine dimers requires the functions of at least 8 separate protein subunits (DDB1 and DDB2[XPE], HR23B, XPC, XPA and RPA) subtly modified by p53 transactivation of DDB2 and XPC[68-70] and ubiquitination associated with the E3 ligase activity of DDB1.[71,72] Recognition of many other forms of chemical damage to DNA may involve only some or all of these gene products, many of which may have secondary functions. XPC expression is induced by ionizing radiation, but enigmatically appears to have limited functions in repair of X-ray damage.[73] The universal marker of chromatin modifications associated with repair, histone H2Ax phosphorylation (γH2Ax), plays a much more complex role during NER and DNA replication after UV damage to XP cells[74] than after DNA double strand breaks.[75] γH2Ax marks excision repair sites, arrested replication forks, apoptotic cells and associated conformational changes in chromatin structure.

The diversity of repair pathways was originally thought in terms of defined pathways: base excision repair, nucleotide excision repair, nonhomologous end joining, homologous recombination, mismatch repair and others. Recent research, exemplified by the chapters of this book, demonstrates that there is extensive cross-talk among these pathways. The components of repair mechanisms (DNA binding proteins, nucleases, helicases, etc) may consequently be viewed as members of a versatile tool bag. According to the cell type and particular injury, the cell selects and assembles these members into a machinery to achieve restoration of the original structures and functions. The traditional categories of NER, BER etc may simply be the more common assemblies, but whose components may perform other tasks in different combinations to rectify different injuries.

The XP variant complementation group was for a long time defined by default: a patient suffered from XP but NER was not defective.[51] A defect in DNA replication after UV irradiation was identified as early as 1975,[76] but it took 25 years before this, the last of the XP genes, was identified as a low fidelity polymerase, *Pol η*.[55,56] This discovery was especially remarkable, revealing the identification of new classes of DNA polymerases which fed back into the prokaryote field and aided identification of the mutasome system (UMUCD) with the bacterial class Y low fidelity polymerases.[77-79] Pol η is a remarkable polymerase: it is recruited to replication sites in response to many perturbations in DNA chain growth, one of the few exceptions possibly being ionizing radiation. The recruitment mechanism uses an exquisite balance of ubiquitination/deubiquitination by Rad6[80,81] of the anchoring protein, proliferating cell nuclear antigen (PCNA),[82,83] and activation of the S phase checkpoints, to say the least. Not only is it important for replication of UV damaged DNA, but it also participates in recruitment of other class Y polymerases,[84] replication during nutrient starvation,[85,86] mutagenesis,[52] cytoplasm-nuclear translocation,[87] recombination,[88-90] hypermutation,[91,92] and so on!

Yet many questions associated with XP remain unsolved, including this personal list of topics that need eventual answers, to which readers can add their own:

1. A complete analysis and description of the mechanism of photoproduct recognition and the mechanism of translocation of repair enzymes to sites of damage is still needed.
2. The structures of repair proteins and their protein-protein interactions promise still more surprises, especially understanding why so many transactions occur in exceptionally large nuclear macromolecular complexes.
3. Which selection pressures led to the evolutionary diversity of photoproduct recognition and repair in the three kingdoms: bacteria, archea and eukarya? Have possible mechanisms been exhausted, or will we discover more in parasites and in the enormous biological diversity of the prokaryotic worlds revealed by population sequencing?
4. What were the selection pressures that resulted in the evolutionary loss of photolyases from placental mammals but retention of homologs as diurnal regulators in eukaryotes?
5. How much tissue-specific variation occurs in NER, especially in differentiated tissues, what role does repair of oxidative damage play in neurodegeneration and is there any commonality between the NER-defective neurodegenerative diseases and more common disorders such as Parkinson's, ALS, Alzheimers, etc.
6. What are the roles of secondary modifications in the specificity and efficacy of repair including: phosphorylation and the relevant kinases and phosphatases; ubiquitination and deubiquitination; poly (ADP-ribosylation). Which enzymes promote and remove these modifications?
7. What underlies the species-specific differences among mammals, especially why do many strains of rodents have reduced pyrimidine dimer excision and why are XP and CS mice more cancer-prone than human XP patients but show less neurodegeneration?
8. How is NER integrated with the multiple components of signal transduction, transcription, replication and recombination, in regulating genomic stability?
9. What are the complete clinical spectra of human NER diseases and their correlation with gene-specific mutations?
10. What role does NER play in embryo development and maintenance of stem cell integrity and behavior?
11. How can we exploit NER and other repair systems in cancer therapy?
12. How can we improve cost-effective patient and prenatal diagnosis for XP and related diseases?
13. How can we provide improved protection and therapy for XP and related patients (e.g., estimation of "safe" solar exposures, predictability of course of disease, cost effective prophylactic care)?

This book contains outstanding contributions by many of the contemporary players in the field and provides a background of knowledge from which we can balance our present ignorance and the future challenges and marvel at the technologies available to those now joining the field, enabling them to do what we only hoped for decades ago.

Acknowledgements

I owe a debt of gratitude to colleagues, students and especially to XP and CS patients, families and their support groups for an enriching career and the opportunity to participate in a rewarding voyage of discovery. Financial support for many years (1966-1995) came from the US Department of Energy to the Laboratory of Radiobiology under the leaderships of

Harvey M. Patt and Sheldon Wolff, until the DOE abruptly abandoned support in 1996. I am very grateful for subsequent support from the March of Dimes, the American Cancer Society, the Academic Senate of the University of California San Francisco, the National Institutes of Environmental Health Sciences, the National Cancer Institute, the XP Society and the Luke O'Brien Fund. The references cited are, as far as possible, the earliest or key publications on individual topics; more comprehensive references will be found in the individual chapters. Other historical comments are drawn from my personal records and recollections and I take full responsibility for any errors, which are unintentional.

References

1. Hebra F, Kaposi M. In: Tay Wt, ed. London: The New Sydenham Society, 1874:252-258.
2. De Sanctis C, Cacchione A. Xerodermic idiocy. Riv Sper Freniat 1932; 56:269-292.
3. Cleaver JE, Kraemer KH. In: Scriver CR, Beaudet AL, Sly WS et al eds. The Metabolic and Molecular Bases of Inherited Disease New York: McGraw-Hill, 1995:4393-4419.
4. Hirai Y, Kodama Y, Moriwaki S et al. Heterozygous individuals bearing a founder mutation in the XPA DNA repair gene comprise nearly 1% of the Japanese population. Mutat Res 2006; 60:171-178.
5. Hashem N, Bootsma D, Keijzer W et al. Clinical characteristics, DNA repair and complementation groups in xeroderma pigmentosum patients from Egypt. Cancer Res 1980; 40:13-18.
6. Nishigori C, Zghal M, Yagi T et al. High prevalence of point mutations in exon 6 of xeroderma pigmentosum groupA-complementing (XPAC) gene in xeroderma pigmentosum group A patients in Tunisia. Am J Hum Genetics 1993; 53:1001-1006.
7. Nishigori C, Moriwaki S, Takebe H et al. Gene alterations and clinical characteristics of xeroderma pigmentosum group A patients in Japan. Archives of Dermatology 1994; 130:191-197.
8. Cleaver JE, Feeney L, Tang JY et al. Xeroderma pigmentosum group C in an isolated region of Guatemala. J Invest Dermatol 2007; 127:493-496
9. Gartler S. In: Fishbein M, ed. Second International Conference on Congenital Malformations, New York, 1964:94-97.
10. Cleaver JE. Photoreactivation:A radiation repair mechanism absent from mammalian cells. Biochemical Biophysical Research Communications 1966; 24:569-576.
11. Cook JS, McGrath JR. Photoreactivating-enzyme activity in metazoa. Proc Natl Acad Sci USA 1967; 58:1359-1365.
12. Hsu DS, Zhao X, Zhao S et al. Putative human blue-light photoreceptors hCRY1 and hCRY2 are flavoproteins. Biochemistry 1996; 35:13871-13877.
13. Cleaver JE. Defective repair replication in xeroderma pigmentosum. Nature 1968; 218:652-656.
14. Cleaver JE. Xeroderma pigmentosum:a human disease in which an initial stage of DNA repair is defective. Proc Natl Acad Sci USA 1969; 63:428-435.
15. Hanawalt PC. Normal replication of DNA after repair replication in bacteria. Nature 1967; 214:269-270.
16. Pettijohn D, Hanawalt P. Evidence for repair replication of ultraviolet damaged DNA in bacteria. J Mol Biol 1964; 9:395-410.
17. Rasmussen RE, Painter RB. Evidence for repair of ultraviolet damaged. deoxyribonucleic acid in cultured mammalian cells. Nature 1964; 203:1360-1362
18. Rasmussen RE, Painter RB. Radiation-stimulated DNA synthesis in cultured mammalian cells. J Cell Biol 1966; 29:11-19.
19. Epstein JH, Fukuyama K, Reed WB et al. Defect in DNA synthesis in skin of patients with xeroderma pigmentosum demonstrated in vivo. Science 1970; 168:1477-1478.
20. Reed WB, Landing B, Sugarman G et al. Xeroderma pigmentosum. Clinical and laboratory investigation of its basic defect. JAMA 1969; 207:2073-2079.
21. Setlow RB, Regan JD, German J et al. Evidence that xeroderma pigmentosum cells do not perform the first step in the repair of UV damage to their DNA. Proc Natl Acad Sci USA 1969; 64:1035-1041.
22. Bishop M. How to Win the Nobel Prize. Cambridge, MA: Harvard University press, 2003.
23. Cleaver JE. Cancer in xeroderma pigmentosum and related disorders of DNA repair. Nat Rev Cancer 2005; 5:564-573.
24. Spivak G, Hanawalt PC. Host cell reactivation of plasmids containing oxidative DNA lesions is defective in Cockayne syndrome but normal in UV-sensitive syndrome fibroblasts. DNA Repair (Amst) 2006; 5:13-22.
25. de Waard H, de Wit J, Andressoo JO et al. Different effects of CSA and CSB deficiency on sensitivity to oxidative damage. Mol Cell Biol 2004; 24:7941-7948.
26. Friedberg EC. Cockayne syndrome-a primary defect in DNA repair, transcription, both or neither? Bioessays 1996; 18:731-738.

27. Rubin JS, Joyner AL, Bernstein A et al. Molecular identification of a human DNA repair gene following DNA-mediated gene transfer. Nature 1983; 306:206-208.
28. Wood RD, Robins P, Lindahl T. Complementation of the xeroderma pigmentosum DNA repair defect in cell-free extracts. Cell 1988; 53:97-106.
29. Sibghatullah HI, Carlton W, Sancar A. Human nucleotide excision repair in vitro:repair of pyrimidine dimers, psoralen and cisplatin adducts by HeLa cell-free extract. Nucleic Acids Res 1989; 17:4471-4484.
30. De Weerd-Kastelein EA, Keijzer W, Bootsma D. Genetic heterogeneity of xeroderma pigmentosum demonstrated by somatic cell hybridization. Nat New Biol 1972; 238:80-83.
31. Franklin WA, Lo KM, Haseltine WA. Alkaline lability of fluorescent photoproducts produced in ultraviolet light-irradiated DNA. J Biol Chem 1982; 257:13535-13543.
32. Regan JD, Trosko JE, Carrier WL. Evidence for excision of ultraviolet-induced pyrimidine dimers from the DNA of human cells in vitro. Biophys J 1968; 8:319-325.
33. Zelle B, Lohman PH. Repair of UV-endonuclease-susceptible sites in the 7 complementation groups of xeroderma pigmentosum A through G. Mutat Res 1979; 62:363-368.
34. Painter RB, Cleaver JE. Repair replication, unscheduled DNA synthesis and the repair of mammalian DNA. Radiat Res 1969; 37:451-466.
35. Cleaver JE, Thomas GH. Single strand interruptions in DNA and the effects of caffeine in Chinese hamster cells irradiated with ultraviolet light. Biochem Biophys Res Commun 1969; 36:203-208.
36. Lehmann AR. Postreplication repair of DNA in ultraviolet-irradiated mammalian cells. J Mol Biol 1972; 66:319-337.
37. Thompson LH, Rubin JS, Cleaver JE et al. A screening method for isolating DNA repair-deficient mutants of CHO cells. Somatic Cell Genet 1980; 6:391-405.
38. Busch DB, Cleaver JE, Glaser DA. Large-scale isolation of UV-sensitive clones of CHO cells. Somatic Cell Genet 1980; 6:407-418.
39. Fischer E, Keijzer W, Thielmann HW et al. A ninth complementation group in xeroderma pigmentosum, XP I. Mutat Res 1985; 145:217-225.
40. Robbins JH. Xeroderma pigmentosum complementation group H is withdrawn and reassigned to group D. Hum Genet 1991; 88:242.
41. Chu G, Chang E. Xeroderma pigmentosum group E cells lack a nuclear factor that binds to damaged DNA. Science 1988; 242:564-567.
42. Itoh T, Linn S, Ono T et al. Reinvestigation of the classification of five cell strains of xeroderma pigmentosum group E with reclassification of three of them. J Invest Dermatol 2000; 114:1022-1029.
43. Gianelli F, Croll PM, Lewin SA. DNA repair synthesis in human heterokaryons formed by normal and UV-sensitive fibroblasts. Exp Cell Res 1973; 78:175-185.
44. Gianelli F, Pawsey SA. DNA repair synthesis in human heterokaryons. II. a test for heterozygosity in xeroderma pigmentosum and some insights into the structure of the defective enzyme. J Cell Sci 1974; 15:163-176.
45. Gianelli F, Pawsey SA. DNA repair synthesis in human heterokaryons. The rapid and slow complementing varieties of xeroderma pigmentosum. J Cell Scie 1976; 20:207-213.
46. Maher VM, McCormick JJ. Cytotoxic and mutagenic effects of carcinogenic aromatic amides and polycyclic hydrocarbons and ultraviolet irradiation in normally repairing and repair-deficient (xeroderma pigmentosum) diploid human skin fibroblasts. Basic Life Sci 1975; 58:785-787.
47. Mansbridge JN, Hanawalt PC. In: Friedberg EC, Bridges BR, eds. Cellular responses to DNA damage. UCLA Symposium on molecular and cellular biology, New series. New York: Alan R. Liss, 1983:195-207.
48. Bohr VA. Gene specific DNA repair. Carcinogen. 1991; 12:1983-1992.
49. Hoeijmakers JHJ, Egly JM, Vermeulen W. TFIIH: a key component in multiple DNA transactions. Current Opin Genet Dev 1996; 6:26-33.
50. Venema J, Mullenders LH, Natarajan AT et al. The genetic defect in Cockyne syndrome is associated with a defect in repair of UV-induced DNA damage in transcriptionally active DNA. Proc Natl Acad Sci USA 1990; 87:4707-4711.
51. Cleaver JE. Xeroderma pigmentosum:variants with normal DNA repair and normal sensitivity to ultraviolet light. J Invest Dermatol 1972; 58:124-128.
52. Maher VM, Oulette LM, Curren RD et al. Frequency of ultraviolet light-induced mutations is higher in xeroderma pigmentosum variant cells than in normal human cells. Nature 1976; 261:593-595.
53. Cleaver JE, Bootsma D. Xeroderma pigmentosum: biochemical and genetic characteristics. Annu Rev Genet 1975; 9:19-38.
54. Friedberg EC, Meira LB, Cheo DL. Database of mouse strains carrying targeted mutations in genes affecting cellular responses ot DNA damage. Version 2. Mutat Res 1998; 407:217-226.
55. Johnson RE, Kondratick CM, Prakash S et al. hRAD30 mutations in the variant form of xeroderma pigmentosum. Science 1999; 264:263-265.

56. Masutani C, Kusumoto R, Yamada A et al. The XPV (xeroderma pigmentosum variant) gene encodes human DNA polymerase η. Nature 1999; 399:700-704.
57. Itoh T, Fujiwara Y, Ono T et al. UVs syndrome, a new general category of photosensitive disorder with defective DNA repair, is distinct from xeroderma pigmentosum variant and rodent complementation group I. Am J Hum Gene 1995; 56:1267-1276.
58. Horibata K, Iwamoto Y, Kuraoka I et al. Complete absence of Cockayne syndrome group B gene product gives rise to UV-sensitive syndrome but not Cockayne syndrome. Proc Natl Acad Sci USA 2004; 101:15410-15415.
59. Paterson MC, Smith PJ. Ataxia-telangiectasia:an inherited genetic disease demonstrating sensitivity to ionizing radiation and related DNA damaging chemicals. Annu Rev Genet 1979; 13:291-318.
60. Ahmad SI, Hanaoka F, Kirk SH. Molecular biology of Fanconi anemia-an old problem, a new insight. BioEssays 2002; 24:439-448
61. Ahmad SI, Kirk SH, eds. Molecular Mechanisms of Fanconi Anemia. Austin: Landes Bioscience, 2006; ISBN 0-387-31972-7.
62. Fishel R, Lescoe MK, Rao MRS et al. The human mutator gene homolog MSH2 and its association with hereditary nonpolyposis colon cancer [published erratum appears in Cell 77, 167]. Cell 1993; 75:1027-1038.
63. Gudmundsdottir K, Ashworth A. The roles of BRCA1 and BRCA2 and associated proteins in the maintenance of genomic stability. Oncogene 2006; 25:5864-5874.
64. Katagiri T, HS, Shinohara A et al. Multiple possible sites of BRCA2 interacting with DNA repair protein RAD51. Genes Chromosomes Cancer 1998; 21:217-222.
65. Caldecott KW. DNA single-strand breaks and neurodegeneration. DNA Repair (Amst) 2004; 3(8-9):875-882.
66. Hanawalt PC, Haynes RH. The repair of DNA. Sci Am 1967; 16:36-43
67. Hoeijmakers JH. Genome maintenance mechanisms for preventing cancer. Nature 2001; 411:366-374.
68. Tan T, Chu G. p53 Binds and activates the xeroderma pigmentosum DDB2 gene in humans but not mice. Mol Cell Biol 2002; 22:3247-3254.
69. Fitch ME, Nakajima S, Yasui A et al. In vivo recruitment of XPC to UV-induced cyclobutane pyrimidine dimers by the DDB2 gene product. J Biol Chem 2003; 276:46909-46910.
70. Ford JM. Role of p53 in the mammalian cellular response to UV damage. Photochem Photobiol 1998; 67:73S-74S.
71. Sancar A, Lindsey-Boltz LA, Unsal-Kacmaz K et al. Molecular mechanisms of mammalian DNA repair and the DNA damage checkpoints. Annu Rev Biochem 2004; 73:39-85.
72. Wood RD, Mitchell M, Sgouros J et al. Human DNA repair genes. Science 2001; 291:1284-1289.
73. Tusher VG, Tibshirani R, Chu G. Significance analysis of microarrays applied to the ionizing radiation response. Proc Natl Acad Sci USA 2001; 98:5116-5121.
74. Marti TM, Hefner E, Feeney L et al. H2AX phosphorylation within the G1 phase after UV irradiation depends on nucleotide excision repair and not DNA double strand breaks. Proc Natl Acad Sci USA 2006; 103:9891-9896.
75. Lowndes NF, Toh GW-L. DNA repair:the importance of phosphorylating histone H2AX. Curr Biol 2005; 15:R99-R102.
76. Lehmann AR, Kirk-Bell S, Arlett CA et al. Xeroderma pigmentosum cells with normal levels of excision repair have a defect on DNA synthesis after UV-irradiation. Proc Natl Acad Sci USA 1975; 72:219-235.
77. Hubscher U, Maga G, Spadari S. Eukaryotic DNA polymerases. Annu Rev Biochem 2002; 71:133-163.
78. Ohmori H, Friedberg EC. Fuchs RPP et al. The Y-Family of DNA Polymerases. Molecular Cell 2001; 8:7-8.
79. Woodgate R, Levine AS. Damage inducible mutagenesis:recent insights into the activities of the Umu family of mutagenesis proteins. Cancer Surv 1996; 28:117-140.
80. Hoege C, Pfander B, Moldovan GL et al. RAD6-dependent DNA repair is linked to modification of PCNA by ubiquitin and SUMO. Nature 2002; 419:135-141.
81. Friedberg EC. Reversible monoubiquitination of PCNA:A novel slant on regulating translesion DNA synthesis. Mol Cell 2006; 22:150-152.
82. Kannouche PL, Wing J, Lehmann AR. Interaction of human DNA polymerase eta with monoubiquitinated PCNA:a possible mechanism for the polymerase switch in response to DNA damage. Mol Cell 2004; 14:491-500.
83. Huang TT, Nijman SM, Mirchandani KD et al. Regulation of monoubiquitinated PCNA by DUB autocleavage. Nat Cell Biol 2006; 8:339-347.
84. Kannouche P, Fernandez de Henestrosa AR, Coull B et al. Localization of DNA polymerases eta and iota to the replication machinery is tightly co-ordinated in human cells. EMBO J 2003; 22:1223-1233.

85. Godov VG, Jarosz DF, Walker FL et al. Y-family DNA polymerases respond to DNA damage-independent inhibition of replication fork progression. EMBO J 2006; 25:868-879.
86. de Feraudy S, Limoli CL, Giedzinski E et al. Pol η is required for DNA replication during nucleotide deprivation by hydroxyurea. Oncogene 2007 (in press).
87. Kannouche P, Broughton BC, Volker M et al. Domain structure, localization and function of DNA polymerase eta, defective in xeroderma pigmentosum variant cells. Genes and Development 2001; 15:158-172.
88. Thakur M, Wernick M, Collins C et al. DNA polymerase η undergoes alternative splicing, protects against UV sensitivity and apoptosis and suppresses Mre11-dependent recombination. Genes Chromosomes Cancer 2001; 32:222-235.
89. Limoli CL, Giedzinski E, Cleaver JE. Alternative recombination pathways in UV-irradiated XP Variant cells. Oncogene 2005; 24:3708-3714.
90. McIlwraith MJ, Vaisman A, Liu Y et al. Human DNA polymerase eta promotes DNA synthesis from strand invasion intermediates of homologous recombination. Mol Cell 2005; 20:783-792.
91. Delbos F, De Smet A, Ahmad Faili A et al. Contribution of DNA polymerase η to immunoglobulin gene hypermutation in the mouse. J Exp Med 2005; 201:1191-1196.
92. Yavuz S, Yavuz AS, Kraemer KH et al. The role of polymerase eta in somatic hypermutation determined by analysis of mutations in a patient with xeroderma pigmentosum variant. J Immunol 2002; 169:3825-3830.

CHAPTER 2

Clinical Features of Xeroderma Pigmentosum

Ulrich R. Hengge* and Steffen Emmert

Background

Xeroderma pigmentosum (XP) was first described in 1874 by Hebra and Kaposi. Albert Neisser was the first to report neurological abnormalities associated with XP in 1883. XP is an autosomal recessive disease with defective nucleotide excision repair (NER). It is characterized by easily recognizable clinical hallmarks (Table 1). These manifestations are due to cellular hypersensitivity to ultraviolet (UV) radiation resulting from a defect in DNA repair. Two types of NER exist: global genome (GG-NER) and transcription coupled (TC-NER). Eight complementation groups, XPA-XPG, corresponding to defects in the corresponding gene products of XPA-XPG genes and XP-variant, have been described. These entities occur with different frequencies (e.g., XPA is relatively common, whereas XPE is fairly rare) and they differ with respect to disease severity (e.g., XPG is severe, whereas XPF is mild) and involvement of skin, central nervous system and opthalmological manifestations (Table 2). Cockayne syndrome rarely occurs together with XPB, XPD and XPG.

In addition to the DNA repair defects, UV radiation also exerts pronounced immunosuppressive effects that are likely to be involved in the pathogenesis of XP. Although typical symptoms of immune deficiency, such as multiple infections, are not usually observed in patients with XP, prominent depletion of Langerhans cells, induced by UV radiation, has been described in XP patients.[1] Various other defects in cell-mediated immunity such as impaired cutaneous responses to recall antigens, impaired lymphocyte proliferative responses to mitogens and decreased production of interferon as well as reduced natural killer cell activity have been detected in XP patients.

Epidemiology

The frequency of XP in the United States is about 1 case/250.000 inhabitants. Not uncommonly, parental consanguinity and familiarity are present in patients with XP.[2,3]

XPC is the most common group in the United States, constituting almost 1/3 of XP patients. The unscheduled DNA synthesis is usually between 15-30% of normal. Symptoms for neurological disorders are rare in XP-C. XPD is the second most common type of XP in the United States and accounts for the majority of US patients with symptoms for neurological disorder being present in about half of all those patients, while the cutaneous and immunologic presentations are quite heterogeneous.

Internationally, the incidence of XP is about the same in Europe, whereas it is higher (1:40.000) in Japan, where XPA is the most common group. In Europe XPA and XPC are the two most prevalent forms of XP. There is no gender preference. As an autosomal recessive disorder, there is usually no positive family history as the heterozygous parents are clinically healthy.

* Corresponding Author: Ulrich R. Hengge —Department of Dermatology, Heinrich-Heine-University, Düsseldorf, Germany. Email: ulrich.hengge@uni-duesseldorf.de

Molecular Mechanisms of Xeroderma Pigmentosum, edited by Shamim I. Ahmad and Fumio Hanaoka. ©2008 Landes Bioscience and Springer Science+Business Media.

Table 1. Clinical hallmarks of xeroderma pigmentosum

- Severe photosensitivity (painful sunburns in early childhood)
- Poikiloderma
- Dryness (xerosis)
- Premature skin aging
- Malignant tumors (squamous cell cancers, basal cell cancers and melanoma), most often on face, head and neck
- Various neurological and ophthalmological symptoms and manifestations

Table 2. Characteristics of xeroderma pigmentosum complementation groups

Comple-mentation Group	Frequency (%)	Skin Cancer	Neurological Involvement	Ophthalmological Involvement	Gene Defect
A	30	+	+++	+	XPA
B	0.5	+	+	+	XPB/ERCC3
C	27	+	−	+	XPC
D	15	+	+++	+	XPD/ERCC2
E	1	+	−	+	DDB2/XPE/p48
F	2	+	−	+	XPF/ERCCC4
G	1	+	+	+	XPG/ERCC5
Variant	23.5	+	−	+	XPV/hRAD30

Dermatological Manifestations

In general, skin problems proceed neurological and ophthalmological symptoms. Several key cutaneous features are usually found. While babies are normal at birth, in the first years of life, diffuse erythema, scaling and pronounced freckle-like pigmentation develop (Fig. 1). In accordance with the increased light sensitivity, changes are seen over light-exposed areas, in particular face, head and neck and in severe cases they subsequently appear in the lower legs and even the trunk. One needs to become alert, when babies present with severe solar dermatitis/sunburn, often associated with constant crying, for which no other explanation can be found. The sunburn will usually persist for extended periods of time, not uncommonly, for several weeks and may include blister formation upon minimal sun exposure.

Once the erythema has resolved, multiple freckles on sun-exposed skin areas cause mottled pigmentation, telangiectasias and actinic damage. XP is one of the few diseases that can cause poikiloderma at an early age (Fig. 2). Poikiloderma is characterized by erythema, hyper- and hypopigmentation as well as scarring and telangiectasias. As the skin suffers actinic damage, the surface become atrophic and dry, which has led to the term "xeroderma" (dry skin) for this condition (Fig. 3).

The incidence of tumors is about 1000-fold increased as compared with the normal population.[4] The process of malignant transformation in XP has been estimated to be around 8 years as compared with 60 years for a representative control cohort. The mean patient age of developing skin cancer is 8 years in XP patients and for the onset of actinic damage around 1-2 years of age.

Characteristically, children or adolescents develop large areas with field cancerization including multiple actinic keratoses, in situ squamous cell cancer and malignant skin tumors (Fig. 3). Upon the malignant tumors, the UV light-induced cancers like squamous cell cancer, basal cell cancer and lentigo maligna melanomas are predominating[4,5] (Fig. 4). Especially for melanoma, the role of UVB is widely accepted.[6] Interestingly, the total UV-dose plays an important role in lentigo malignant

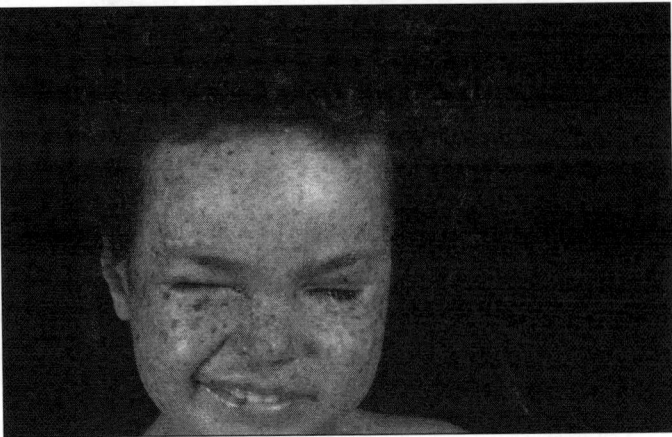

Figure 1. Pronounced mottled pigmentation of the face and lips with mild erythema and scarring on the forhead, cheeks and nose of a 4-year-old girl.

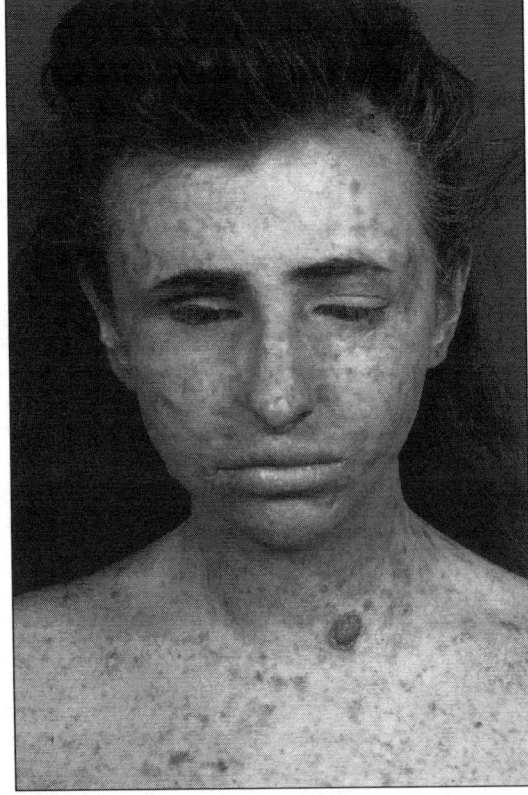

Figure 2. Severe poikiloderma with disfiguring scars of the face and neck. Also note the perioral xerosis and the patient's photophobia. Exophytic, flesh-coloured squamous cell cancer on the neck.

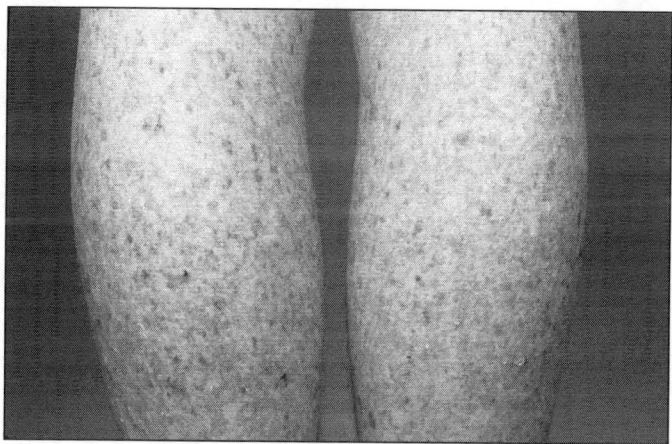

Figure 3. Severe actinic damage (field cancerization), pigment changes and teleangiectasias on the lower legs of a 14-year old girl.

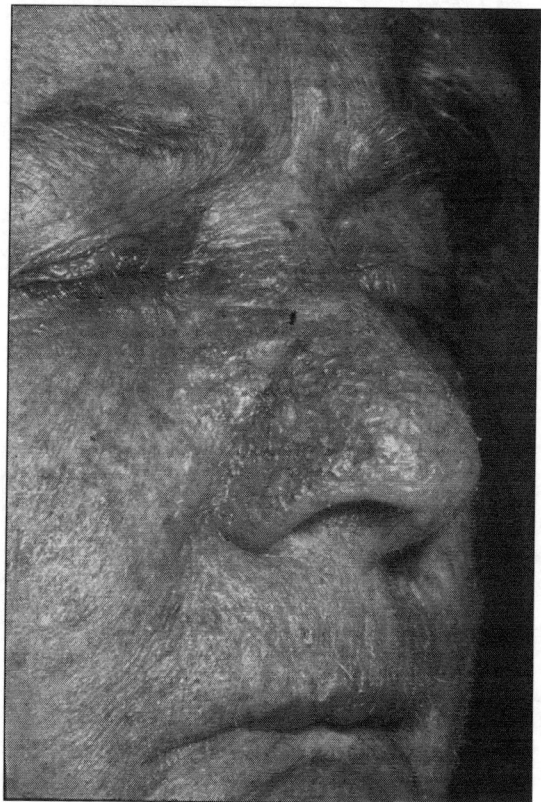

Figure 4. Poikiloderma and several basal cell cancers on the nose of a 45 year-old patient with mild XPC.

melanoma in contrast to other malignant melanomas. The important role of UV light has also been confirmed in XP-A transgenic mice, where an about 1000-fold increased risk of developing UV-induced cancers has been documented.[7] In addition, UV radiation has immunosuppressive effects, which are important for the control of premalignant and malignant cells.[8] Despite the fact that tumor cells express tumor-associated antigens, that are recognized by specific T-cells, tumor progression occurs. The danger associated with the development of malignant melanoma and squamous cell carcinoma is their early metastases.

Other Cancers in Xeroderma Pigmentosum

Besides the above mentioned cancers, keratoacanthomas and sarcomas including fibrosarcomas and angiosarcomas have been described.[9] The incidence of tumors of the oral mucosa (inner lips, tongue, gingival; Fig. 5) and other organs (brain and lung cancer and leukemia) is also increased. Particularly, the occurrence of leukemias has been reported in XP.[10] Other oral manifestations include caries of primary dentation.

Neurological Manifestations

Neurological abnormalities have been described in 18% of 830 patients in the largest study available to date[2] with mental retardation being present in 80% of these individuals. The median intelligence quotient score was 45 with a range of 15-81. The second most frequent neurological abnormality was spasticity or ataxia (30% of subjects with neurological involvement), followed by microcephaly (24%).[2] Patients with neurological symptoms can be classified on the basis of age at the onset of neurological symptoms: juvenile (before age 20 years) and adult (after 20 years). As a general rule, the presence of neurological abnormalities correlates with the degree of NER defects; patients with the greatest impairment of DNA repair (i.e., complementation groups XPA and XPD; Table 2) are more prone to develop neural degeneration. As many as 50% of patients with XPD manifest neural deterioration, while on the contrary, neurological involvement is rare in patients with XPC, which is the most common complemention group in the United States. In general, Central Nervous System (CNS) involvement is due to premature neuronal death due to damage resulting from DNA damage from reactive oxygen species or other free radicals. Neurodegeneration probably results from accumulating mutations due to cells'

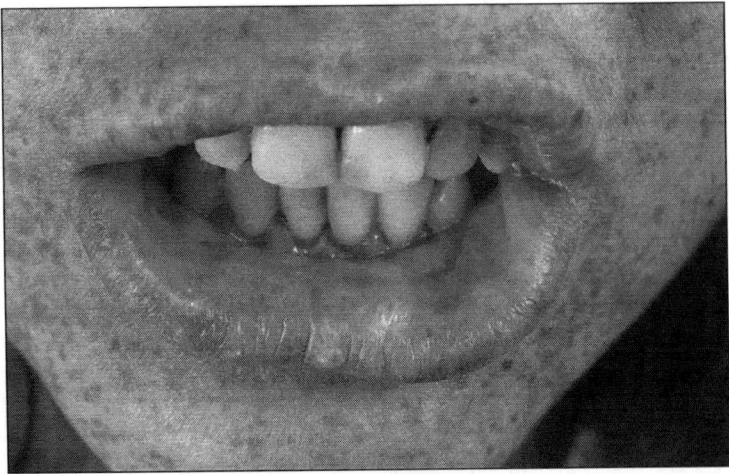

Figure 5. One basal cell cancer on the inner side and one squamous cell cancer on the outer side of the lower lip in a 20-year old woman.

inability to repair DNA damage in response to oxidative stress (see Chapter 13). Neurological symptoms are progressive and irreversible. Therefore, many patients develop severe disability in the sense that they become bed-ridden and incontinent with infections, sepsis and aspiration pneumonia as potential complications.

Neurological problems include reduced intelligence (80%) with a median IQ test result of 45, abnormal motor activity (30%) with hyporeflexia, areflexia, ataxia, chorea, spasticity as well as peripheral neuropathy and impaired hearing.[11-14] For example, in 32 Japanese patients with XPA, the most profound DNA repair defect, mental retardation, microcephaly, nystagmus, dysarthria, ataxia and short stature were described as the most prominent neurological manifestations.[2,15] Half of the patients with neurological abnormalities have skin cancer. Sometimes, these neurological manifestations are more severe than the cutaneous symptoms. Not uncommonly, sexual development is delayed or absent in 12 % in XP patients with neurological disorders.

The incidence of central nervous system tumors is about 10-fold higher than in normal individuals. The neurological tumors include astrocytoma, meduloblastoma, glioblastoma and malignant schwannoma.[16]

The de Sanctis-Cacchione-syndrome refers to cutaneous XP, neurological abnormalities, hypogonadism and dwarfism.[17] These children are characterized by small stature, spasticity and debility. Sometimes, additional symptoms such as segmental demyelinisation, microcephaly, sensory deafness and epilepsy may also occur.

Ophthalmological Manifestations

About 40% of XP patients suffer from ophthalmological problems. The photosensitivity in XP is variable, but usually more pronounced in the range of 290-320 nm (UVB range). However, the minimal erythema dose is generally lower than normal at other wavelengths. Among the ocular tissues, the lids, conjunctiva and cornea receive substantial amounts of UV light and are therefore important target sites in XP patients. Important ocular problems include photophobia, conjunctivitis (with conjunctival infection?) and corneal abnormalities. Blepharitis, ectropion, symblepharon, loss of eyelashes, atrophy and scarring represent some less common ophthalmological features in XP patients.[2,3] Corneal abnormalities include corneal clouding, vascularization and corneal ulcers causing impaired vision in about 15% are some less common ophthalmological features in XP patients.[2,3]

The risk of ocular neoplasias such as basal cell cancer and squamous cell cancer is increased about 2000-fold. These cancers occur in up to 15% of patients and are most often localized on the cornea and conjunctiva, while malignant melanoma occurred in about 5%.[2,18] In addition, fibrovascular pannus of the cornea and epitheliomas of the lids and conjunctivae may occur.

Differential Diagnoses

Differential diagnoses of XP include several other rare syndromes, which can be ruled out by serologic, genetic and metabolic tests (Table 3). Some overlaps may be seen with Cockayne syndrome (CS) and trichothiodystrophy (TTS). Both diseases are of autosomal recessive inheritance and are characterized by increased X-rays (or light) sensitivity, while CS patients present with characteristic neurological symptoms, ataxia, mental retardation and sensory deafness, distinct facial features with large, halonated eyes, prominent nose, progressive cachexia, myopathy, short stature, microcephaly and pigmented degeneration of the retina as well as optic atrophy and cataract.[19,20]. Recently, some overlaps have been detected between CS and TTD with XP.[21,22] In CS, there is a defect in TCR, but the global genome repair is functional. CS shows no increased risk of skin cancer.[23]

The characteristic hallmarks of TTD are short and fragile hair, ichthyosis and short stature. Interestingly, the hair shows a deficit of sulfur-rich proteins.[24] The hair has a characteristic tiger pattern under the polarizing microscope. Additional symptoms include reduced intelligence and compromised fertility and certain defects in XP-B and XP-D genes.

Table 3. Differential diagnoses

Ephelids are frequent cutaneous findings; concern arises only when they are very numerous; subacute cutaneous lupus erythematosus is infrequent and basal cell nevus syndrome is rare. The other listed differentials are extremely rare.

- Ephelids	- LEOPARD syndrome
- Subacute cutaneous lupus erythematosus	- Hartnup disease
- Basal cell nevus syndrome	- Progeria (Werner syndrome)
- Erythropoietic porphyria	- Rothmund-Thomson syndrome
- Cockayne syndrome	- Bloom syndrome (congenital teleangiectatic
- Trichothiodystrophy	erythema)
- Hydroa vacciniforme	

Diagnosis of XP

Dermatological manifestations serve to make the diagnosis. The clinical features including the pathological photosensitivity at an early age are the hallmarks of the disease. No consistent routine laboratory abnormalities are present in XP. Specialized laboratories may analyze the sensitivity of cultured fibroblasts to UV radiation and chromosomal breakage, complementation studies and gene sequencing to identify the specific molecular defect. Ideally, the chromosomal breakage is compared with the patients' parents cells as they are obligate heterozygotes for XP. Prenatal diagnosis is possible; however, due to the recessive nature of the disease, there may be no suggestion of such disease in the family history. For prenatal testing, unscheduled DNA synthesis serves as the classical method for the diagnosis of XP.

There is no specific histology of XP; usually melanocytic nevi are numerous and increased melanin pigments are present in the basal cell layer together with a chronic inflammatory infiltrate in the upper dermis. At later stages, the histological picture of poikiloderma may be seen including hyperkeratosis, atrophy, hyperpigmentation and telangiectasias. The dermis usually shows elastotic changes. Likewise, there is no specific test for CNS and eye involvement. Even in patients with neurological symptoms, the electroencephalographic findings may be normal. Electromyography (EMG) and nerve conduction studies (NCS) are helpful because axonal polyneuropathy is common in XP.[11] In addition, several metabolic studies are recommended to exclude neurological manifestations mimicking XP and CNS involvement. Therefore, serum and urine amino acids, serum copper, lysosomal enzymes, cholesterol esterification, mucopolysaccharides as well as long-chain fatty acid analyses should be performed besides lactate and pyruvate concentration to rule out other CNS conditions.

Management

As a curative therapy such as gene therapy is currently not available, the overall preventive measure is to keep the patient away from sunlight. This includes shifting the daily activities into the night as much as possible; this has led to the term "moon babies". In addition, protective clothing, hats and appropriate eye care can serve to minimize UV-induced damage. Sunscreens should be regularly applied to all exposed surfaces, including the hands and the lower limbs. Preferably, physical and chemical sunscreens are to be used simultaneously and around the entire year and during evening as well as early morning hours.

Patient education about effective sun protection and early recognition of skin cancers is also helpful. Counseling can also be offered by the XP society (www.xps.org). Prevention of skin cancer in XP patients has also been achieved to some degree with the use of oral isotretinoin.[25]

Recent attempts to repair DNA damage after UV exposure have been made by topical delivery of DNA repair enzymes to the skin by means of specially engineered liposomes (T4 endonuclease-V).[26] Careful dermatologic examinations are mandatory at regular intervals of 3-6 months.

Prognosis

Until today, early diagnosis and consequent avoidance of sunlight as well as regular dermatological screening have helped to increase the life expectancy. The prognosis is significantly impaired as fewer than 40% of patients survive beyond 20 years of age.[27] However, some individuals with milder disease may survive to about 40 years.

Ken Kraemer and colleagues have constructed Kaplan-Meier survival curves from 830 published XP patients, describing a 90% probability of surviving to age 13 and 80% of surviving to age 28 and a 70% probability surviving to 40 years. Overall, the life expectancy for XP patients was reduced by 30 years.[2] Patients are likely to die from cancer (33%), infections (11%) and various other diseases.[2] Heterozygote carriers have no increased risk to develop skin tumors. Subsequently to the chronic occurrence of the aforementioned skin cancers, some degree of mutilation may develop, which may be severe in certain patients (Fig. 1).

References

1. Jimbo T, Ichihashi M, Mishima Y et al. Role of excision repair in UVB-induced depletion and recovery of human epidermal Langerhans cells. Arch Dermatol 1992; 128:61-67.
2. Kraemer KH, Lee MM, Scotto J. Xeroderma pigmentosum. Cutaneous, ocular and neurologic abnormalities in 830 published cases. Arch Dermatol 1987; 123:241-250.
3. Goyal JL, Rao VA, Srinivasan R et al. Oculocutaneous manifestations in xeroderma pigmentosa. Br J Ophthalmol 1994; 78:295-297.
4. English JS, Swerdlow AJ. The risk of malignant melanoma, internal malignancy and mortality in xeroderma pigmentosum patients. Br J Dermatol 1987; 117:457-461.
5. Kobayashi M, Satoh Y, Irimajiri T et al. Skin tumors of xeroderma pigmentosum (I). J Dermatol 1982; 9:319-322.
6. Kraemer KH, Lee MM, Scotto J. Early onset of skin and oral cavity neoplasms in xeroderma pigmentosum. Lancet 1982; 1:56-57.
7. van Steeg H, Mullenders LH, Vijg J. Mutagenesis and carcinogenesis in nucleotide excision repair-deficient XPA knock out mice. Mutat Res 2000; 450:167-180.
8. Morison WL, Bucana C, Hashem N et al. Impaired immune function in patients with xeroderma pigmentosum. Cancer Res 1985; 45:3929-3931.
9. De Silva BD, Nawroz I, Doherty VR. Angiosarcoma of the head and neck associated with xeroderma pigmentosum variant. Br J Dermatol 1999; 141:166-167.
10. Schroeder TM. Relationship between chromosomal instability and leukemia. Hamatol Bluttransfus 1974; 14:94-96.
11. Hakamada S, Watanabe K, Sobue G et al. Xeroderma pigmentosum: neurological, neurophysiological and morphological studies. Eur Neurol 1982; 21:69-76.
12. Robbins JH. Xeroderma pigmentosum. Defective DNA repair causes skin cancer and neurodegeneration. JAMA 1988; 260:384-388.
13. Robbins JH, Brumback RA, Moshell AN. Clinically asymptomatic xeroderma pigmentosum neurological disease in an adult: evidence for a neurodegeneration in later life caused by defective DNA repair. Eur Neurol 1993; 33:188-190.
14. Rolig RL, McKinnon PJ. Linking DNA damage and neurodegeneration. Trends Neurosci 2000; 23:417-424.
15. Mimaki T, Itoh N, Abe J et al. Neurological manifestations in xeroderma pigmentosum. Ann Neurol 1986; 20:70-75.
16. Nakamura T, Ono T, Yoshimura K et al. Malignant schwannoma associated with xeroderma pigmentosum in a patient belonging to complementation group D. J Am Acad Dermatol 1991; 25:349-353.
17. De Sanctis C, Cacchione A. L'idiozia xerodermica. Riv Sper Freniat 1932; 56:269-292.
18. Vivian AJ, Ellison DW, McGill JI. Ocular melanomas in xeroderma pigmentosum. Br J Ophthalmol 1993; 77:597-598.
19. Nance MA, Berry SA. Cockayne syndrome: review of 140 cases. Am J Med Genet 1992; 42:68-84.
20. Lehmann AR. DNA repair-deficient diseases, xeroderma pigmentosum, Cockayne syndrome and trichothiodystrophy. Biochimie 2003; 85:1101-1111.
21. Berneburg M, Lowe JE, Nardo T et al. UV damage causes uncontrolled DNA breakage in cells from patients with combined features of XP-D and Cockayne syndrome. EMBO J 2000; 19:1157-1166.
22. Broughton BC, Berneburg M, Fawcett H et al. Two individuals with features of both xeroderma pigmentosum and trichothiodystrophy highlight the complexity of the clinical outcomes of mutations in the XPD gene. Hum Mol Genet 2001; 10:2539-2347.

23. Scott RJ, Itin P, Kleijer WJ et al. Xeroderma pigmentosum-Cockayne syndrome complex in two patients: absence of skin tumors despite severe deficiency of DNA excision repair. J Am Acad Dermatol 1993; 29:883-889.
24. Itin PH, Sarasin A, Pittelkow MR. Trichothiodystrophy: update on the sulfur-deficient brittle hair syndromes. J Am Acad Dermatol 2001; 44:891-920.
25. Kraemer KH, DiGiovanna JJ, Moshell AN et al. Prevention of skin cancer in xeroderma pigmentosum with the use of oral isotretinoin. N Engl J Med 1988; 318:1633-1637.
26. Yarosh D, Klein J, O'Connor A et al. Effect of topically applied T4 endonuclease V in liposomes on skin cancer in xeroderma pigmentosum: a randomised study. Xeroderma Pigmentosum Study Group. Lancet 2001; 357:926-929.
27. Kraemer KH, Lee MM, Scotto J. Xeroderma pigmentosum. Cutaneous, ocular and neurologic abnormalities in 830 published cases. Arch Dermatol 1987; 123:241-50.

CHAPTER 3

Xeroderma Pigmentosum and Skin Cancer

Leela Daya-Grosjean*

Abstract

The hypersensitivity of DNA repair deficient xeroderma pigmentosum (XP) patients to solar irradiation results in the development of high levels of squamous and basal cell carcinomas as well as malignant melanomas in early childhood. Indeed, XP presents a unique model for analysing the effects of unrepaired DNA lesions in skin carcinogenesis. The skin cancer predisposition, observed in XP patients, is due to the mutator gene activity of XP cells which lead to high levels of UV specific modifications of crucial regulatory genes in skin cells leading to cancer. Thus, the high levels of UV specific mutations, seen in oncogenes and tumor suppressor genes, which have been characterized in XP tumors, clearly demonstrate the major role of the UV component of sunlight in skin cancer development. The UV specific C to T and the tandem CC to TT UV signature transition mutations found in XP tumors are located at bipyrimidine sequences, the preferred UV targets in DNA. The same UV specific alterations are seen in key regulatory genes in sporadic skin cancers but at lower frequencies than those found in XP tumors.

Introduction

A predisposition to skin cancers is the cardinal feature of xeroderma pigmentosum (XP) and is the grave phenotypic consequence of the failure of crucial cellular pathways required to maintain genome integrity.[1,2] Among these cellular systems, the importance of DNA repair in protection against environmental genotoxic agents is exemplified by classical XP patients deficient in nucleotide excision repair (NER) (80%) and XP-variants, defective in postreplication repair (20%).[3,4] A wide range of helix distorting lesions including ultraviolet (UV) photoproducts, caused by sunlight, are dealt with by NER; sunlight being a major environmental carcinogen implicated in skin cancer (the most common cancer in fair skinned populations). Indeed, the clinical and cellular outcome of repair deficiency in XP patients is manifested as a hypersensitivity to UV light and more than a 1000 fold increase in cutaneous tumors on sun-exposed areas of the body compared to normal individuals.[4,5] It is important to remember that, following DNA damage, many cellular responses are triggered in cells which can modify the normal cell phenotype through genetic and epigenetic changes. Thus, modulation of cytoskeleton structure, transcription factors and signal transduction can lead to cell cycle arrest, DNA repair or apoptosis via p53 regulated genes and transient immune suppression.[6,7] The UV mutator phenotype of XP cells is a consequence of replication through unrepaired DNA lesions resulting in the accumulation of numerous genetic alterations. When these modifications arise in important regulatory genes, they can lead to activation of proto-oncogenes and inactivation of tumor suppressor genes, which are known to play an important role in the multi-step skin cancer progression.[8] Indeed, XP has provided a unique

*Leela Daya-Grosjean—Genomes and Cancer, FRE2939/CNRS, IFR54, Institute Gustave Roussy, 39, rue Camille Desmoulins, 94805 Villejuif Cedex, France. Email: daya@igr.fr

Molecular Mechanisms of Xeroderma Pigmentosum, edited by Shamim I. Ahmad and Fumio Hanaoka. ©2008 Landes Bioscience and Springer Science+Business Media.

human model for our understanding of the genetic basis of skin carcinogenesis mainly through molecular analysis of the germinal variants of the XP NER genes and XP skin tumors in which genetic modifications, essential for tumor development, are accentuated.

XP Genetics and Skin Cancer

The heterogeneous phenotype of classical XP patients is a result of defects in any one of the seven NER involved XP genes A to G, which occur with different frequency, geographic distribution and show varying clinical features.[9,10,3] The phenotype of XPA and XPC patients are entirely due to defects in NER, as these genes are not involved in other processes. Severe NER deficiency is manifested by the XPA patients because the XPA protein is indispensable for both global and transcription coupled repair (TCR). The majority of XPA individuals characterized are Japanese with half of the patients being homozygous for the same carrier mutation.[11] XPA patients present pronounced central nervous system disorders as well as acute clinical skin defects and are extremely prone to UV induced skin tumors.[9] The XPC complementation group represents one of the largest groups worldwide and XPC patients who are deficient only in global genome repair, are free from neurological defects and present classical XP symptoms with severe skin abnormalities and are highly predisposed to UV induced skin cancers.[12,9]

The XPB and XPD proteins are DNA helicases which are components of the transcription factor TFIIH and are implicated in both TCR and global NER. Defects in the XPB and XPD genes can result in the XP as well other UV hypersensitive clinical phenotypes, trichothiodystrophy and Cockayne's syndrome. XPB patients are extremely rare and all present phenotypic heterogeneity associated with the specific mutation type. XPD patients represent 20% of XP cases with heterogeneous clinical phenotypes. The XPD patients with classical XP features have moderate skin cancer predisposition with mild to severe neurological disorders.[10,13] Patients belonging to the rarer complementation groups XPE, XPF and XPG, generally present mild but variable clinical features. Thus, the XPE patients, mutated in the DNA damage binding gene, DDB2, which is implicated in global repair of cyclobutane pyrimidine dimers (CPD), do not show neurological abnormalities and have mild skin disorder with late onset of cutaneous cancers. The XPF protein is implicated in the incision step of NER as a hetero-dimer with ERCC1 and most XPF patients also present a mild XP phenotype with cancers usually developing late in life.[3,12] Among the few XPG patients described worldwide, about half carry missense XPG mutations and present a mild classic XP clinical phenotype. The XPG mutations which give rise to truncated proteins are associated with severe neurological symptoms associated with a mixed XP/Cockayne's syndrome phenotype.

The variant XP group, XPV, has mutations in the low fidelity DNA polymerase η required to bypass replication of lesions and are normal for NER. XPV patients are found worldwide and represent the second largest subgroup of XP individuals. They present mild to severe skin disease and show a predisposition to skin tumors which, however, appear later in life (usually after 15 years of age) than seen with classic XP patients.[3]

Cutaneous Malignancies in XP Patients

The nonmelanoma skin cancers (NMSC), which account for mainly basal cell carcinomas (BCC) and squamous cell carcinomas (SCC) develop as multiple primary tumors in XP patients. NMSC are derived from the basal layer of the epidermis with BCC occurring in the hair-growing epithelium whereas SCC derive from inter-follicular cells. BCC arise de novo as slow growing, locally invasive tumors which metastasize rarely. SCC which develop with a multistep progression, are often derived from precancerous skin lesions, actinic keratoses (AK) and tend to be more aggressive and frequently metastasize. In the general population BCC are predominant whereas SCC seem to prevail in XP patients, but the data are often contradictory between studies analyzing different cohorts of XP patients.[5,14-16,17-19] NMSC occur mainly on sun-exposed body sites of XP patients as early as 3-5 years of age in striking contrast with the general population where the mean age for

NMSC development is around 50-60 years. The frequency of NMSC is 50% in young XP patients aged under 10 years and the SCC metastasize with a higher probability (4%) compared to under 2% observed in the general population.[4,20] Malignant melanoma (MM) derive from melanocytes and its development can be associated with severe painful sunburn which is characteristic in young XP patients.[21] In general, about 15-20% of MM occurs at atypical melanocytic nevi associated with UV exposure but the role of UV in the induction of malignant melanoma remains unclear. Nevertheless, XP patients develop melanomas mainly on sun exposed areas of body, 65% being located on the head and neck, 28% on the arms and legs and 7% on the rest of the body. The incidence of melanomas in XP patients aged under 20 years is up to 2000 fold higher than in an age matched population in which only 4% of sporadic cancers are melanomas compared to up to 22% in XP patients.[22,23]

The Etiology of XP Skin Cancers and UV Hallmark Mutations

The UV component of sunlight is the major etiological factor implicated in the development of skin cancer in normally healthy populations and its effect is exacerbated in NER deficient XP patients. Short wavelength UVC (200-280nm) is filtered out by ozone in the stratosphere and only 1-10% of UVB (280-320nm) and up to 90-99% of UVA (320-400nm) reaches the earth's surface. UVB has immunomodulatory effects in the skin and induces gene modifications at dipyrimidine sites which are targets for UV induced DNA damage which result mainly in C to T transition mutations.[24] The most significant alterations due to UVB are the tandem CC to TT transitions considered to be the veritable UV signature mutation. Indeed, the analysis of XP tumors shows that un-repaired DNA lesions result in higher levels of C to T and the UV signature tandem mutation CC to TT, in the key genes responsible for skin cancer development, described later. UVA on its own is a weak carcinogen but contributes to inflammation, immunosuppression and gene mutation.[25] UVA induces more oxidative damage than UVB mainly through production of reactive oxygen species (ROS) which can produce single strand breaks, DNA-protein crosslinks as well as base alterations such as 8-hydroxydeoxyguanine (8-OHdG) which are mutagenic.[26] It is important to note that XP fibroblasts are markedly deficient in the antioxidant enzyme catalase activity which may be an important factor contributing to the high levels of UV induced skin carcinogenesis seen in XP patients.[27,28]

It is clear that UV plays an important role in initiation, promotion and progression during skin carcinogenesis. Childhood exposure to UV and intense intermittent UV exposure are major risk factors for BCC formation whereas SCC development is particularly associated with chronic cumulative UV exposure. Melanoma can be associated with intense intermittent UV exposure during childhood as well as acute sunburn in adults.[21] Thus, UVB induced initiation is due to the formation of DNA adducts, which are mainly CPD and 6-4 photoproducts. These adducts give rise to UV signature mutations in skin cells with significantly higher levels in repair defective XP cells.[29] Repeated exposure of initiated, mutation bearing epidermal cells to UVB and UVA components of sunlight could stimulate uncontrolled proliferation of cells carrying mutated proto-oncogenes or tumor suppressor genes and result in clonal expansion. Continued UV exposure allows selective development of clones insensitive to UV induced apoptosis. Further genetic alterations then allow progression towards the formation of genetically unstable malignant skin tumors which form more quickly and in greater numbers in XP patients. It is important to note that uncontrolled activation of the mitogenic sonic hedgehog (SHH) signaling pathway plays a central role in the genesis of BCC, whereas specific pathways involved in the development of SCC and melanoma have not been defined as yet. UV induced mutations have, nevertheless, been found in genes associated with different cellular pathways leading to the activation of proto-oncogenes and modification of tumor suppressor genes in SCC, BCC and melanoma. The following sections describes the UV modified gatekeeper genes controlling cell growth and death and the caretaker genes essential for maintaining the integrity of the genome, which have been found to contribute to skin cancer development in XP patients.[30]

Tumor Suppressor Genes

The major functions of tumor suppressor genes are to inhibit cancer development by triggering of cellular responses to DNA damage. These are the gatekeeper genes which control cell cycle arrest to allow time for repair of damaged DNA or induce apoptosis which eliminates cells carrying too much damage. The p53 gene has been recognized as one of the most important tumor suppressor genes as it is found mutated in over half of all human cancers.[31] Significantly, up to 90% p53 mutations are observed in XP skin tumors compared to over 50% p53 mutations seen in sporadic skin cancers (Fig. 1). In fact, p53 mutation is an early event in the development of skin cancers as attested by p53 mutated clones of cells found in normal sun-exposed skin and the high levels of p53 alterations observed in benign actinic keratosis, precursors of SCC.

Analysis of XP NMSC has shown that 75% of p53 mutations are missense UV specific C to T or CC to TT transitions located at bipyrimidine sites. In sporadic NMSC, C to T transitions predominate with only 10% of tandem CC to TT substitutions, compared to more than 60% seen in XP NMSC. The distribution of p53 mutation spectra hot spots (located mainly in the highly conserved exons 5-9), is different in NMSC and in internal malignancies but the codon 248 hotspot, which disrupts p53 DNA binding and trans-activation functions, is common to all tumor types.[32] Interestingly, p53 mutation hotspots are different between BCC, SCC and melanoma. In XP more SCC have been characterized and these present five distinct p53 mutation hot spots at codons 179, 196, 248 and 282, different from those found in sporadic skin cancers.[32] In melanomas from the general population and XPV patients, mutations in the p53 gene are rare (10%) compared to the high frequency (>50%) found in melanomas of classic XP patients. Alterations in XP melanomas are mainly C to T transitions with a preponderance of the UV signature tandem CC to TT mutations.[33,34]

The two tumor suppressor genes p16INK4a, an inhibitory protein of CDK4 and p14ARF which stabilizes p53, map at the INK4a-ARF locus on chromosome 9p21 and play essential roles in cell cycle regulation. The INK4a-ARF locus is the second most commonly altered locus found in human cancers after p53, occurring through homozygous deletion, point mutation or methylation.[35] INK4a-ARF modifications are seen in a high proportion (<76%) of sporadic SCC compared to BCC (3.5%) and promoter methylation is the predominant mechanism for inactivation of this locus.[36-38] Compared to 25% UV specific INK4a-ARF alterations seen in sporadic NMSC, 43% are found in XP NMSC (33% in SCC and 20% in BCC).[39] Interestingly, multiple mutations of the locus are found in XP skin tumors that are not observed in sporadic tumors. Moreover, 60% of XP tumors show a positive association of INK4a-ARF and p53 mutations rarely found in the sporadic tumors.[39] The de-stabilisation of the INK4a-ARF/p53 pathway must enhance tumor progression as is observed in XP patients. Melanomas from XP patients have not been analysed for alterations of the INK4a-ARF locus but it should be noted that 20% of familial melanoma in the general population is associated with germline mutations at this locus.[40]

Oncogene Activation in XP Skin Tumors

Many of the caretaker genes responsible for controlling normal cell growth, differentiation and apoptosis are classed as proto-oncogenes which can be activated by point mutations, amplification and rearrangement. These genes include growth factors and receptors, proteins involved in many different signaling pathways and nuclear transcription factors. In skin tumors, activation of the ras family of proto-oncogenes, N-ras, Ki-ras and Ha-ras, encoding the GTP binding proteins, results in constitutive activation of ras signal transduction. Different studies analyzing XP skin tumors, show varying mutation frequencies of ras probably due to differences in methodology. A comparative study of skin cancers from our laboratory has allowed us to establish a more than two-fold higher ras mutation frequency in XP tumors compared to sporadic skin cancers.[41] Modification at codon 12 of all three ras genes was found but with a preponderance for N-ras alterations. These mutations were again associated with DNA lesions at bipyrimidine sequences. Moreover, in XP skin tumors we find high levels of amplification and rearrangement of Ha-ras and the mitogenic c-myc oncogene (Fig. 1). This is probably the consequence of several rounds of abortive re-initiation of replication at

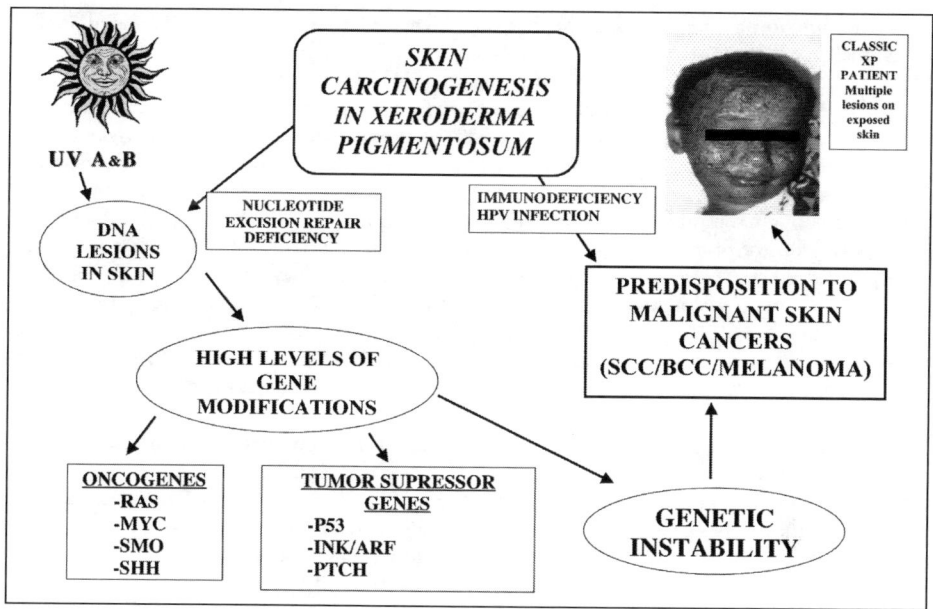

Figure 1. Unrepaired UV induced DNA lesions in xeroderma pigmentosum patients, deficient in NER, result in genetic instability leading to a predisposition to skin cancers from early childhood.

unrepaired UV lesions. Two key genes of the SHH pathway, implicated in BCC development are proto-oncogenes, SHH and smoothened, which are modified in XP BCC.[42,43] In sporadic melanoma, N-ras and ras regulated BRAF mutations have been frequently observed but little data are available for XP melanoma. Interestingly, analysis of a cell line, established from testicular metastasis of a malignant melanoma of an XP patient, showed activation of the N-ras gene at codon 61.[44]

Modifications of the SHH Signaling Pathway Genes in XP BCC

The analysis of the autosomal dominant disorder called Gorlin or nevoid basal cell carcinoma (NBCC) syndrome, characterized by the development of multiple BCC has lead to the identification of the SHH pathway. This pathway plays a key role in skin and hair follicle growth as well as morphogenesis and is found to be abrogated in the majority of BCC.[45] The SHH signaling pathway is activated by binding of the secreted signal pepetide of the SHH glycoprotein to its cell surface receptor patched (PTCH). In the absence of SHH, PTCH inhibits smoothened (SMO) a G-protein-coupled-like receptor. Inactivation of PTCH by binding of the SHH ligand relieves repression of SMO and initiates the signal transduction cascade allowing activation of downstream transcription factors such as the Gli family of proteins. Deregulation of this pathway by mutation of one or other of the SHH pathway genes, alters cell cycle arrest and differentiation and induces hair follicle tumors, the BCC. Analysis of XP BCC has revealed high levels of patched, smoothened and sonic hedgehog gene mutations (Fig. 1) compared to those found in sporadic BCC.[42,43,46] No modifications of SHH pathway have been found in SSC, indicating the specificity of this pathway in BCC development. The major role of solar UV in XP BCC development is again evident in the very high levels of UV specific mutations found in the altered SHH pathway genes compared to those in sporadic BCC.[45]

It is highly significant that the analysis of XP BCC in our laboratory has been able to firmly establish a role for SHH, a postulated proto-oncogene in human skin carcinogenesis. To date the study of 110 sporadic BCC has revealed only one SHH gene mutation. However, six mutations in the SHH gene were characterized in 33 XP BCC, indicating that the SHH protein plays an oncogenic function in the development of BCC.[43] The SHH mutations are all located at the N-terminal signaling domain and are all situated on one face of the compact globular protein, suggesting they may effect interactions with specific SHH partners. The majority of germline patched gene mutations are rearrangements producing truncated proteins. The PTCH mutations characterized in sporadic and XP BCC show fewer (30% and 20%, respectively) rearrangements and the remaining are mainly UV specific point mutations. The point mutations are distributed at random in the intra- and extra-cellular domains and patched gene mutations found in the large extracellular loops may modify interaction with the SHH ligand and prevent signal transduction.[46] Mutations of the proto-oncogene smoothened in XP BCC, as in sporadic BCC, are mainly located in the genetic region determining extra-membrane domains and may be important for binding small SMO modulating proteins.[42] Significantly, a G to T transversion at codon 535, located in the trans membrane domain, found in XP is a mutation hot spot observed in sporadic BCC that can arise from replication of 8-OHdG, a base modification created by UVA induced ROS. Interestingly more than one member of the SHH pathway could be found modified in the same BCC.[45]

Impaired Immune Response in XP Patients and Human Papilloma Virus

Solar UV radiation, as well as being mutagenic, is also important in modulating the immune responses of the skin.[47] Various immunological defects have been reported in XP patients, including reduced natural killer cell activity, diminished lymphocyte production of α and γ interferons and impaired UV-induced cytokine production.[48,49] These are all important modifications which, by lowering immuno-surveillance, may be implicated together with the DNA repair defect in the UV-induced skin cancer proneness of XP patients. Human papillomaviruses (HPVs) are common infections of the skin which are often found associated to benign lesions and nonmelanoma skin cancers (NMSC), mainly squamous cell carcinomas (SCC) and basal cell carcinomas (BCC). Interestingly, immunodeficient Epidermodysplasia Verruciformi (EV) patients develop numerous SCCs on sun-exposed body sites associated with EV HPV.[50] Moreover, immunosuppressed solid organ transplant patients also show an increased risk of HPV infections and present a high incidence of skin cancers particularly SCC on sun-exposed sites.[51] Thus, XP, EV and organ transplant recipient patients, all have a predisposition for the development of NMSC on sun-exposed body sites with an inversed SCC/BCC ratio compared to that found in the normal population. The precise role of HPVs in skin carcinogenesis is speculative and the unique study analysing HPV in XP skin cancers recently carried out in our laboratory shows an age-related significant association between EV HPV and SCCs.[52] Thus, HPV DNAs are found more frequently in SCCs (20/40) than in BCCs (4/27) or normal skin (2/9) of XP patients. The HPV spectra, identified in the XP tumours, include 22 different epidermodysplasia verruciformis (EV) HPV types, which predominate in SCCs (48%) compared to BCCs (15%) and normal skin (22%). The data, showing an association between EV HPV and SCCs from young XP patients, are comparable to that found for NMSC from adult immunosuppressed organ transplant patients and raise the question of the importance of HPV infection in skin carcinogenesis. Thus, one wonders whether the immunological defects reported in XP patients play an important role in HPV infection of XP patients and what is the impact of HPV on skin tumour development.

Conclusions

The analysis of skin cancers from the repair deficient XP patients has greatly enhanced our understanding of the genetic basis of skin carcinogenesis, an important issue today as the incidence of skin cancers in fair skinned populations is increasing at an alarming rate. In fact skin cancers are the most prevalent cancers in humans today due to changes in lifestyle, aging populations and increased UV fluence caused by ozone depletion. The majority of populations present few skin cancers up to 50 years of age after which time individual risk factors such as skin type, smoking, diet, occupational exposures, medication and more importantly, lifetime sun exposure patterns can all contribute to an increase in incidence of skin cancers. The data obtained by molecular analyses of XP skin cancers has allowed the characterisation of crucial regulatory genes which lead to

Table 1. Gene modifications in XP skin tumors

Genes	Gene Alterations Non-Melanoma Skin Cancers	Melanoma
Oncogenes:		
RAS	53%	#
SMO*	30%	nd
SHH*	18%	nd
c-MYC§	<70%	nd
Ha-RAS§	<50%	nd
Tumor suppressor genes:		
p53	<90%	>50%
INK4a-ARF	43%	-
PTCH*	<90%	-

Table 1. The majority of the gene alterations seen in XP skin tumors are UV specific.
#N-RAS has been found mutated in one XP melanoma metastase.
§c-MYC and Ha-ras alterations are mainly amplifications and rearrangements.
Specific differences are seen between basal cell carcinomas (BCC) and squamous cell carcinomas (SCC) in XP patients. Higher levels of p53 modifications are found in BCC as well as the sonic hedgehog pathway genes (*). RAS and INK4a-ARF modifications are prevalent in SCC.

malignant cell transformation and skin cancer development in general (Table 1). Indeed, enhanced mutation levels in XP cells can help identify and/or confirm the importance of different genes in skin tumor development. Analyses of BCC and SCC in XP have contributed to our knowledge of the essential role of aberrant SHH pathway signaling in BCC formation. Many of the genes involved in skin carcinogenesis in general, including the ras oncogenes, the p53 and INK/ARF tumor suppressor genes, are found modified in SCC but the causal association of a specific gene or cellular pathway in SCC development has yet to be defined. The increased risk of SCC in XP patients requires further analysis to obtain insight on the molecular mechanisms implicated in the pathogenesis of SCC in general. There is also growing evidence indicating the importance of oxidative stress in photo-carcinogenesis. Further analysis of different signaling pathways activated by UVA in XP cells should help elucidate their role in skin tumor development, considering the deficiency in the ROS scavenger enzyme, catalase, exhibited by all XP cells. It is important to remember that XP mouse models have also been extremely useful models in providing significant information on our understanding of skin carcinogenesis. In conclusion, ongoing research on the XP syndrome continues to improve our knowledge on the impact of UV on DNA damage and mutation in skin cells. Understanding the many different factors implicated in skin tumor induction and progression is essential for prevention and development of new targeted therapies.

References

1. Friedberg EC, Aguilera A, Gellert M et al. DNA repair: from molecular mechanism to human disease. DNA Repair (Amst) 2006; 5(8):986-996.
2. Cleaver JE. Cancer in xeroderma pigmentosum and related disorders of DNA repair. Nat Rev Cancer 2005; 5(7):564-573.
3. Lehmann AR. DNA repair-deficient diseases, xeroderma pigmentosum, Cockayne syndrome and trichothiodystrophy. Biochimie 2003; 85(11):1101-1111.
4. van Steeg H, Kraemer KH. Xeroderma pigmentosum and the role of UV-induced DNA damage in skin cancer. Mol Med Today 1999; 5(2):86-94.
5. Kraemer KH, Lee MM, Scotto J. Xeroderma pigmentosum. Cutaneous, ocular and neurologic abnormalities in 830 published cases. Arch Dermatol 1987; 123(2):241-250.
6. Boukamp P. Nonmelanoma skin cancer: what drives tumor development and progression? Carcinogenesis 2005; 26(10):1657-1667.

7. Melnikova VO, Ananthaswamy HN. Cellular and molecular events leading to the development of skin cancer. Mutat Res 2005; 571(1-2):91-106.
8. Daya-Grosjean L, Sarasin A. The role of UV induced lesions in skin carcinogenesis: an overview of oncogene and tumor suppressor gene modifications in xeroderma pigmentosum skin tumors. Mutat Res 2005; 571(1-2):43-56.
9. Kraemer KH. Nucleotide excision repair genes involved in xeroderma pigmentosum. Jpn J Cancer Res 1994; 85(2):inside front cover.
10. Copeland NE, Hanke CW, Michalak JA. The molecular basis of xeroderma pigmentosum. Dermatol Surg 1997; 23(6):447-455.
11. Hirai Y, Kodama Y, Moriwaki S et al. Heterozygous individuals bearing a founder mutation in the XPA DNA repair gene comprise nearly 1% of the Japanese population. Mutat Res 2006; 601(1-2):171-178.
12. Kraemer KH, Levy DD, Parris CN et al. Xeroderma pigmentosum and related disorders: examining the linkage between defective DNA repair and cancer. J Invest Dermatol 1994; 103(5 Suppl):96S-101S.
13. Oh KS, Khan SG, Jaspers NG et al. Phenotypic heterogeneity in the XPB DNA helicase gene (ERCC3): xeroderma pigmentosum without and with Cockayne syndrome. Hum Mutat 2006; 27(11):1092-1103.
14. Moussaid L, Benchikhi H, Boukind EH et al. Cutaneous tumors during xeroderma pigmentosum in Morocco: study of 120 patients. Ann Dermatol Venereol. 2004; 131(1 Pt 1):29-33.
15. Khatri ML, Bemghazil M, Shafi M et al. Xeroderma pigmentosum in Libya. Int J Dermatol 1999; 38(7):520-524.
16. Khatri ML, Shafi M, Mashina A. Xeroderma pigmentosum. A clinical study of 24 Libyan cases. J Am Acad Dermatol 1992; 26(1):75-78.
17. Bouadjar B, Ait-Belkacem F, Daya-Grosjean L et al. Xeroderma pigmentosum. A study in 40 Algerian patients. Ann Dermatol Venereol 1996; 123(5):303-306.
18. Thielmann HW, Popanda O, Edler L et al. Clinical symptoms and DNA repair characteristics of xeroderma pigmentosum patients from Germany. Cancer Res 1991; 51(13):3456-3470.
19. Thielmann HW. Xeroderma pigmentosum patients from Germany (the Mannheim XP collection): clinical and biochemical characteristics. Recent Results Cancer Res 1993; 128:275-297.
20. Moriwaki S, Kraemer KH. Xeroderma pigmentosum—bridging a gap between clinic and laboratory. Photodermatol Photoimmunol Photomed 2001; 17(2):47-54.
21. Kennedy C, Bajdik CD, Willemze R et al. The influence of painful sunburns and lifetime sun exposure on the risk of actinic keratoses, seborrheic warts, melanocytic nevi, atypical nevi and skin cancer. J Invest Dermatol 2003; 120(6):1087-1093.
22. Takebe H, Nishigori C, Tatsumi K. Melanoma and other skin cancers in xeroderma pigmentosum patients and mutation in their cells. J Invest Dermatol 1989; 92(5 Suppl):236S-238S.
23. Kraemer KH, Lee MM, Andrews AD et al. The role of sunlight and DNA repair in melanoma and nonmelanoma skin cancer. The xeroderma pigmentosum paradigm. Arch Dermatol 1994; 130(8):1018-1021.
24. Pfeifer GP, You YH, Besaratinia A. Mutations induced by ultraviolet light. Mutat Res 2005; 571(1-2):19-31.
25. Halliday GM. Inflammation, gene mutation and photoimmunosuppression in response to UVR-induced oxidative damage contributes to photocarcinogenesis. Mutat Res 2005; 571(1-2):107-120.
26. Cadet J, Sage E, Douki T. Ultraviolet radiation-mediated damage to cellular DNA. Mutat Res 2005; 571(1-2):3-17.
27. Vuillaume M, Daya-Grosjean L, Vincens P et al. Striking differences in cellular catalase activity between two DNA repair-deficient diseases: xeroderma pigmentosum and trichothiodystrophy. Carcinogenesis 1992; 13(3):321-328.
28. Arbault S, Sojic N, Bruce D et al. Oxidative stress in cancer prone xeroderma pigmentosum fibroblasts. Real-time and single cell monitoring of superoxide and nitric oxide production with microelectrodes. Carcinogenesis 2004; 25(4):509-515.
29. Daya-Grosjean L, Dumaz N, Sarasin A. The specificity of p53 mutation spectra in sunlight induced human cancers. J Photochem Photobiol B 1995; 28(2):115-124.
30. Ehrhart JC, Gosselet FP, Culerrier RM et al. UVB-induced mutations in human key gatekeeper genes governing signalling pathways and consequences for skin tumourigenesis. Photochem Photobiol Sci 2003; 2(8):825-834.
31. Olivier M, Hussain SP, Caron de Fromentel C et al. TP53 mutation spectra and load: a tool for generating hypotheses on the etiology of cancer. IARC Sci Publ 2004; (157):247-270.
32. Giglia-Mari G, Sarasin A. TP53 mutations in human skin cancers. Hum Mutat 2003; 21(3):217-228.
33. Spatz A, Giglia-Mari G, Benhamou S et al. Association between DNA repair-deficiency and high level of p53 mutations in melanoma of Xeroderma pigmentosum. Cancer Res 2001; 61(6):2480-2486.

34. D'Errico M, Calcagnile A, Canzona F et al. UV mutation signature in tumor suppressor genes involved in skin carcinogenesis in xeroderma pigmentosum patients. Oncogene 2000; 19(3):463-467.
35. Liggett WH, Jr., Sidransky D. Role of the p16 tumor suppressor gene in cancer. J Clin Oncol 1998; 16(3):1197-1206.
36. Soufir N, Moles JP, Vilmer C et al. P16 UV mutations in human skin epithelial tumors. Oncogene 1999; 18(39):5477-5481.
37. Soufir N, Ribojad M, Magnaldo T et al. Germline and somatic mutations of the INK4a-ARF gene in a xeroderma pigmentosum group C patient. J Invest Dermatol 2002; 119(6):1355-1360.
38. Brown VL, Harwood CA, Crook T et al. p16INK4a and p14ARF tumor suppressor genes are commonly inactivated in cutaneous squamous cell carcinoma. J Invest Dermatol 2004; 122(5):1284-1292.
39. Soufir N, Daya-Grosjean L, de La Salmoniere P et al. Association between INK4a-ARF and p53 mutations in skin carcinomas of xeroderma pigmentosum patients. J Natl Cancer Inst 2000; 92(22):1841-1847.
40. Lynch HT, Shaw TG, Lynch JF. Inherited predisposition to cancer: a historical overview. Am J Med Genet C Semin Med Genet 2004; 129(1):5-22.
41. Daya-Grosjean L, Robert C, Drougard C et al. High mutation frequency in ras genes of skin tumors isolated from DNA repair deficient xeroderma pigmentosum patients. Cancer Res 1993; 53(7):1625-1629.
42. Couve-Privat S, Bouadjar B, Avril MF et al. Significantly high levels of ultraviolet-specific mutations in the smoothened gene in basal cell carcinomas from DNA repair-deficient xeroderma pigmentosum patients. Cancer Res 2002; 62(24):7186-7189.
43. Couve-Privat S, Le Bret M, Traiffort E et al. Functional analysis of novel sonic hedgehog gene mutations identified in basal cell carcinomas from xeroderma pigmentosum patients. Cancer Res 2004; 64(10):3559-3565.
44. Keijzer W, Mulder MP, Langeveld JC et al. Establishment and characterization of a melanoma cell line from a xeroderma pigmentosum patient: activation of N-ras at a potential pyrimidine dimer site. Cancer Res 1989; 49(5):1229-1235.
45. Daya-Grosjean L, Couve-Privat S. Sonic hedgehog signaling in basal cell carcinomas. Cancer Lett 2005; 225(2):181-192.
46. Bodak N, Queille S, Avril MF et al. High levels of patched gene mutations in basal-cell carcinomas from patients with xeroderma pigmentosum. Proc Natl Acad Sci USA 1999; 96(9):5117-5122.
47. Nishigori C, Yarosh DB, Donawho C et al. The immune system in ultraviolet carcinogenesis. J Investig Dermatol Symp Proc 1996; 1(2):143-146.
48. Gaspari AA, Fleisher TA, Kraemer KH. Impaired interferon production and natural killer cell activation in patients with the skin cancer-prone disorder, xeroderma pigmentosum. J Clin Invest 1993; 92(3):1135-1142.
49. Suzuki H, Kalair W, Shivji GM et al. Impaired ultraviolet-B-induced cytokine induction in xeroderma pigmentosum fibroblasts. J Invest Dermatol 2001; 117(5):1151-1155.
50. Majewski S, Jablonska S. Epidermodysplasia verruciformis as a model of human papillomavirus-induced genetic cancer of the skin. Arch Dermatol 1995; 131(11):1312-1318.
51. Bordea C, Wojnarowska F, Millard PR et al. Skin cancers in renal-transplant recipients occur more frequently than previously recognized in a temperate climate. Transplantation 2004; 77(4):574-579.
52. Luron L, Avril MF, Sarasin A et al. Prevalence of human papillomavirus in skin tumors from repair deficient xeroderma pigmentosum patients. Cancer Lett 2006.

CHAPTER 4

XPA Gene, Its Product and Biological Roles

Ulrike Camenisch and Hanspeter Nägeli*

Abstract

The 31 kDa XPA protein is part of the core incision complex of the mammalian nucleotide excision repair (NER) system and interacts with DNA as well as with many other NER subunits. In the absence of XPA, no incision complex can form and no excision of damaged DNA damage occurs. A comparative analysis of the DNA-binding properties in the presence of different substrate conformations indicated that XPA protein interacts preferentially with kinked DNA backbones. The DNA-binding domain of XPA protein displays a positively charged cleft that is involved in an indirect readout mechanism, presumably by detecting the increased negative potential encountered at sharp DNA bends. We propose that this indirect recognition function contributes to damage verification by probing the susceptibility of the DNA substrate to be kinked during the assembly of NER complexes.

Introduction

Mammalian cells employ the nucleotide excision repair (NER) system to eliminate DNA photoproducts, generally known as cyclobutane pyrimidine dimers and (6-4) photoproducts, induced by exposure to short-wavelength UV light, as well as bulky DNA adducts generated by many electrophilic chemicals (reviewed in refs. 16,21,75,90). The same NER pathway also processes a subset of oxidative lesions and protein-DNA crosslinks.[37,68,71] This extremely versatile DNA repair complex operates by a simple cut and patch mechanism that is initiated by bulky lesion recognition, followed by dual incision of damaged strands, excision of the injured residue as the component of a single-stranded DNA segment of 24-32 nucleotides in length,[28] restoration of the duplex structure through DNA repair synthesis and, finally, ligation of the newly synthesized repair patch to the preexisting strand.[1,3,55]

The cellular NER reaction can be broken down into two branches that differ primarily in the initial recognition of bulky base lesions. Although the order of arrival and departure of each subunit is still debated,[70,88] a favored model for the "global genome repair" (GGR) pathway involves the initial recognition of bulky lesions by a heterotrimeric complex composed of xeroderma pigmentosum group C (XPC) protein, a human homolog of RAD23 and centrin 2, a centrosomal protein.[2,60,80,87] UV-damaged DNA binding protein (UV-DDB) is additionally required to detect cyclobutane pyrimidine dimers[18] and, as an accessory factor, to facilitate the recognition of (6-4) photoproducts.[54] Although UV-DDB and XPC are essential for DNA damage recognition in the GGR response, these two subunits are dispensable for the "transcription-coupled repair" (TCR) pathway, which is specialized on the removal of DNA lesions from the transcribed strand of active genes.[22,50] In fact, the

*Corresponding Author: Hanspeter Nägeli—Institute of Pharmacology and Toxicology, University of Zürich-Vetsuisse, Winterthurerstrasse 260, CH-8057 Zürich, Switzerland. Email: u.camenisch@access.unizh.ch

Molecular Mechanisms of Xeroderma Pigmentosum, edited by Shamim I. Ahmad and Fumio Hanaoka. ©2008 Landes Bioscience and Springer Science+Business Media.

progression of RNA polymerase II along transcribed strands is inherently sensitive to the presence of DNA damage, leading to the hypothesis that, in the TCR process, the initial damage recognition occurs through elongating RNA polymerases that are blocked at DNA lesions.[39,76]

The subsequent NER steps are the same for GGR and TCR and involve the sequential recruitment of transcription factor IIH (TFIIH), replication protein A (RPA), XPA, XPG and XPF-ERCC1 (a dimer composed of XPF and excision repair cross complementing 1 protein). TFIIH, RPA and XPA participate in the formation of a key NER intermediate characterized by transient unwinding of the duplex substrate, thus generating a partially open DNA structure with "Y-shaped" double-stranded to single-stranded transitions flanking the damaged base.[17,75] These double-stranded to single-stranded junctions define the sites of cleavage, on either side of the adduct, by structure-dependent endonucleases. XPG is responsible for the 3' incision, which occurs 15-25 nucleotides away from the damaged residue, whereas XPF-ERCC1 makes the second incision 3-9 nucleotides away from the damaged base on the 5' side.[72,87] Finally, the NER pathway relies on the redundancy of the genetic code in the two complementary strands of the double helix to direct the synthesis of repair patches. Because only the damaged strand is subjected to double DNA incision, the intact opposing strand can be used as a template for DNA repair synthesis mediated by replication factor C (RFC), proliferating cell nuclear antigen (PCNA), a DNA polymerase (δ or ϵ) and DNA ligase I.[1,3,55]

The critical requirement for XPA not only in GGR but also in the TCR, pathway placed this protein at the forefront of DNA damage recognition.[35] The goal of this chapter is to summarize the knowledge gained in recent years as to how the relatively small XPA subunit contributes to the specificity and selectivity of the mammalian NER complex. It is generally thought that, unlike many other core NER factors,[26] XPA is not involved in separate DNA transactions such as recombination, transcription or replication. However, recent findings indicate that activation of the S-phase checkpoint in response to UV irradiation is dependent on XPA without any requirement for other NER players.[8] Because further NER processing is apparently not necessary for the S-phase checkpoint signal, a proposed model for this additional function is that the binding of XPA protein to UV-damaged sites may promote the stalling of replicative DNA polymerases at lesion sites.

XPA Gene

XPA was identified as one of eight genes underlying the human XP disorder.[20] The mammalian *XPA* gene was first cloned by phenotypic complementation of the UV radiation sensitivity of XP-A cells transfected with mouse genomic DNA.[83] The human complementary DNA encodes a polypeptide of 273 amino acids with about 95% identity to the mouse homolog, which consists of 303 amino acids.[82] Homologous factors with closely similar sequences have subsequently been identified across all eukaryotic species examined.

The human *XPA* gene is located on chromosome 9 (9q34.1) and is organized in six exons distributed over 25 kb of genomic DNA.[77] The genomic structure of the *XPA* gene correlates well with the tertiary structure elements of the protein product (Fig. 1). The first exon is essential for the nuclear localization of XPA protein but not for DNA repair activity when the factor is expressed at high levels.[53] The other five exons are all necessary for the DNA repair function. Exon 3 codes for a Cys2-Cys2 zinc finger that determines the overall protein fold. Mutation of any of the four cysteine residues of this motif to a serine results in the loss of secondary structure, generating a vastly different protein conformation that fails to support DNA repair.[4,53] The promoter activity of the mammalian *XPA* gene appears to be extraordinarily low in fibroblasts, generating a steady-state mRNA level of only 5 to 8 transcripts per cell.[41] Nevertheless, typical human cell lines contain up to 150,000-200,000 XPA protein molecules/cell.[36]

XPA Protein and Interactions with Other NER Factors

XPA is a zinc-binding protein with an isoelectric point of 7.5, a Stokes radius of 33 Å and an S value of 2.2.[33] A single zinc atom is associated with the central motif $Cys105-X_2-Cys108-X_{17}-Cys$

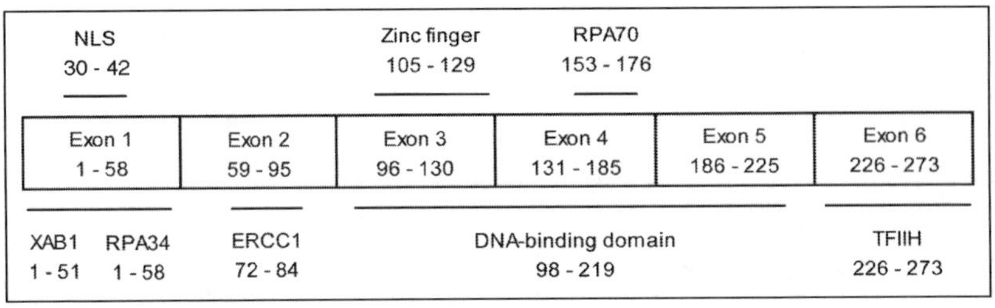

NLS 30 - 42		Zinc finger 105 - 129	RPA70 153 - 176		
Exon 1 1 - 58	Exon 2 59 - 95	Exon 3 96 - 130	Exon 4 131 - 185	Exon 5 186 - 225	Exon 6 226 - 273
XAB1 RPA34 1 - 51 1 - 58	ERCC1 72 - 84	DNA-binding domain 98 - 219			TFIIH 226 - 273

Figure 1. Domain structure of human XPA protein.

126-X$_2$-Cys129.[25] When migrating in sodium dodecyl sulfate-polyacrylamide gels, XPA protein forms several bands with an apparent mass of 38-42 kDa. The discrepancy between the calculated molecular mass for XPA (31 kDa) and its apparent mass on denaturing gels has been ascribed to the presence of disordered regions in the polypeptide fold. Accordingly, the presence of multiple XPA bands in denaturing polyacrylamide gels is thought to reflect distinct conformations of the protein.[29] The disordered regions are evolutionary conserved, suggesting that the exceptional molecular flexibility of XPA protein may be related to its versatile function in accommodating a disparate variety of bulky DNA lesions within the growing NER complex.[30] Conceivably, internal motions of the protein may alter the nucleic acid interaction surface to fit the structure of various kinds of damaged DNA substrates.[13] In vivo studies monitoring the mobility of individual repair factors in mammalian cells indicate that the majority of XPA molecules diffuse rapidly in monomeric form within the nuclear compartment.[67] However, the retention time of recombinant human XPA protein in gel filtration experiments suggested that it may be able to form homodimers in solution.[93] In vitro, the interaction of XPA with DNA is stabilized by the presence of a second XPA molecule in a cooperative manner.[47]

A potential nuclear localization signal has been mapped to the basic amino acid region corresponding to codons 30-42 of exon 1 (Fig. 1). This sequence is required for the trafficking of XPA protein from the cytoplasm to the nucleus.[53] Interestingly, a cytoplasmic GTPase, XAB1, has been shown to interact with the N-terminal region of XPA encoded by exon 1, suggesting that accessory factors such as XAB1 may modulate the nuclear import of XPA.[61] More recent studies indicated that the translocation of XPA from the cytoplasm to the nucleus is enhanced following UV irradiation and it appears that this nuclear targeting in response to DNA damage is dependent on ATR (Ataxia Telangiectasia mutated and Rad3-related). In fact, treatment of human lung carcinoma cells with ATR kinase inhibitors, or knock-down of the ATR expression by transfection with specific small interfering RNA, reduced the UV-induced nuclear import of XPA protein.[91] However, direct phosphorylation by the ATR kinase is unlikely to mediate this translocation process as the formation of a small fraction of phosphorylated XPA does not correlate with the trafficking of XPA protein in time course experiments.[91,92]

Considering the relatively small size of the protein, XPA interacts with an astounding number of other NER subunits (Fig. 2). Several interaction domains have been identified by deletions studies (Fig. 1). The N-terminal portion encoded by exon 1 (residues 1-58) as well as a central region encoded by exon 4 (residues 153-176) contain sequences for binding to RPA. The association of XPA with the 70-kDa subunit of RPA (RPA70) is essential for the NER function. In contrast, binding to the 34-kDa subunit of RPA (RPA34) stimulates this activity without being absolutely required for the NER response.[44] The C-terminal region encoded by exon 6 (residues 226-273) has been shown to bind to TFIIH.[63] The central domain encoded by exons 3-5 (residues 98-219) contains the Cys2-Cys2 zinc finger, is required for the interaction with RPA70 and coincides with a minimal polypeptide fragment necessary for binding to DNA.[38] Finally, XPA also interacts with the ERCC1-XPF heterodimer and sequences essential for the association with ERCC1 are the polyglutamic acid cluster (residues 78-84) as well as a nearby tetrapeptide (residues 72-75) in exon 2.[43,64] A deletion mutant lacking the tetrapeptide sequence is unable to complement the

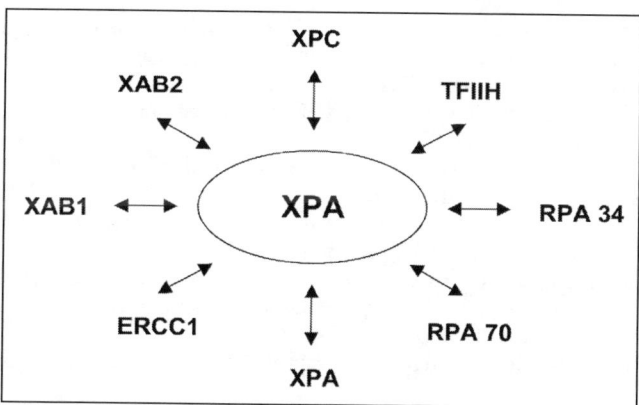

Figure 2. Interactosome map of human XPA.

repair defect of XP-A cells.[45] Also, this deletion mutant inhibits DNA repair synthesis in wild-type cell extracts, indicating that it displays a dominant negative effect. This dominant phenotype can be explained by a mechanism in which the deletion protein retains DNA-binding activity and interactions with other NER factors, but fails to attract the ERCC1-XPF endonuclease complex to the position of 5' incision.[45] The second endonuclease, XPG, is recruited to the location of 3' incision via an association with RPA.[15] Thus, both endonuclease subunits are dependent on XPA or its interaction partner RPA for the correct recruitment to their sites of action in the vicinity of DNA insults.

Additional interaction partners, referred to as XPA-binding protein 1 and 2 (XAB1, XAB2) have been pulled out in a yeast two-hybrid screen.[61] XAB1 consists of 374 amino acids, is localized mainly in the cytoplasm and, as mentioned before, displays an intrinsic GTPase activity that is possibly involved in the nuclear localization of XPA. The XAB2 gene product is composed of 855 amino acids and contains a series of 15 tetratricopeptide repeats.[59] The presence of these multiple motifs suggests that XAB2 may function as a scaffold protein in nucleic acid transactions such as transcription, cell cycle control or RNA processing. Indeed, subsequent immunoprecipitation experiments demonstrated that a fraction of XAB2 can be isolated as a subunit of larger complexes comprising RNA polymerase II and the Cockayne syndrome complementation group A and B proteins.[19,59] Furthermore, the microinjection of antibodies raised against XAB2 inhibited transcripton in human fibroblasts, suggesting that XAB2 is an interaction partner of XPA that has a structural role in the assembly of the transcription machinery.[59]

DNA-Binding Activity of XPA Protein

XPA protein has no inherent catalytic properties, but numerous studies have provided evidence that it displays a binding preference for UV- or chemical carcinogen-damaged DNA[40,73] (reviewed in ref. 84). Compared to UV-DDB and XPC, however, the affinity of XPA for damaged duplexes is orders of magnitude lower.[84] For example, Jones and Wood[33] estimated the binding constant of XPA for duplexes containing (6-4) photoproducts to be $\sim 3 \times 10^{-6}$ M, whereas the reaction constant for binding of the UV-DDB complex to the same substrate is > 5,000-fold greater.[69] Also, it appears that XPA alone recognizes only the (6-4) photoproduct and has little affinity for cyclobutane pyrimidine dimers, which are the predominant UV-induced lesions in DNA.[33] Moreover, attempts to produce specific footprints of XPA protein on damaged DNA by Dnase I protection or other techniques have failed. Nevertheless, a DNA damage "verification" function for XPA has been proposed by several authors, who observed that the DNA-binding affinity of XPA is increased by the formation of complexes with other NER partners. In particular, it has been observed that the interaction with damaged DNA is enhanced in the presence of RPA[23,44,52,65] or ERCC1.[58,74] These larger complexes generated by XPA with either RPA or ERCC1 have a much higher affinity for damaged DNA than each protein alone, indicating that the different subunits may cooperate to form a tight complex with target sites. XPA also interacts with XPC protein, but this association fails to stabilize XPA-DNA complexes. Instead, XPC is displaced from damaged DNA fragments by the combined action of XPA and RPA.[85,94]

In vitro binding assays demonstrated that XPA protein displays an affinity for distorted DNA structures carrying mismatches, loops or bubbles, even if no actual DNA lesion has been introduced into the substrate.[52] XPA turned out to have a particularly strong preference for binding to artificially distorted DNA molecules, like three- or four-way DNA junctions, that share the architectural feature of presenting two double strands emerging from a central bend.[12,52] On the basis of these findings, it has been proposed that the function of XPA protein is to recognize a DNA kink that may be introduced as an obligatory intermediate in both GGR and TCR. This indirect damage recognition function of XPA will be discussed in more detail below.

Recruitment of XPA to Active NER Complexes

In mammalian cells, NER is executed by the sequential recruitment of individual factors to DNA lesion sites, rather than by the action of a preassembled "repairosome". A functional NER repairosome would achieve a mass of nearly 3 MDa, but the nuclear mobility of repair factors, measured by fluorescence recovery after photobleaching, are not compatible with such large machines and instead support a scenario in which NER factors are incorporated one-by-one into growing repair complexes.[27] The mathematical modeling of different NER mechanisms suggests that a strictly sequential assembly of freely diffusing factors is the more efficient strategy compared to alternative scenarios such as random assembly of individual factors at lesion sites, or the complete preassembly of a very large repairosome.[66]

At what stage is XPA incorporated into the nascent excision complex? Two opposing models have been proposed for the initial lesion recognition step: "XPC first" or "XPA first".[70] In the "XPC first" model, the XPC complex constitutes the primary sensor that binds to damaged sites and initiates the NER pathway by recruiting TFIIH and the other successive factors.[5,57,81] This scenario is supported by competition experiments aimed at determining the order in which NER proteins are recruited to lesion sites. Such competition experiments showed that damaged plasmids preincubated with the XPC complex are more rapidly repaired in cell extracts than those preincubated with XPA and RPA,[81] indicating that the NER system works more efficiently in vitro when XPC is allowed time to bind to DNA before addition of the remaining repair subunits.

Another experimental strategy to study the order of assembly of the human NER complex is based on the nuclear trafficking of each core subunit in intact living cells. To analyze the movement of NER factors to sites of bulky lesion formation, cell monolayers were exposed to UV light through filters with small pores, thereby generating localized foci of DNA damage and repair. The translocation of XPC and XPA from unirradiated regions of the nuclei to these damaged foci was monitored by staining the factors with fluorescently tagged antibodies.[87] Interestingly, XPC turned out to accumulate in DNA repair foci in both wild-type and XP-A cells, whereas XPA protein did not move to the damaged foci in XP-C cells. These results are consistent with XPC being the first factor that recognizes damaged DNA sites. In contrast, XPA is not recruited to DNA lesions without prior recognition of the damaged DNA by the XPC subunit. Similarly, using XP cells of other complementation groups, these focal UV irradiation studies were extended to demonstrate that the recruitment of TFIIH and the 3' endonuclease XPG depends on XPC but not on XPA protein. This subcellular imaging approach also showed that XPA is not required for the damage-specific recruitment of RPA, implying that the majority of XPA and RPA molecules are not interacting in the absence of DNA damage and that an association between these two subunits only occurs at their site of action in the ultimate excision complex.[67]

Role of XPA-RPA Interactions

The independent recruitment of XPA and RPA to lesion sites raised the question of why the interaction between these two partners is essential for the NER reaction. RPA represents the most abundant single-stranded DNA binding factor in human cells.[89] It's binding to single DNA filaments is mainly mediated by two distinct oligonucleotide/oligosaccharide-binding folds (OB-folds) located in the 70-kDa subunit.[7] Each RPA monomer occupies approximately 30 nucleotides,[34] which corresponds roughly to the length of the gapped DNA intermediate generated in the NER process. This coincidence between the size of excision products and the occupancy length on single-stranded DNA suggests that RPA might be necessary to protect the undamaged intact strand from inadvertent nuclease attack.[14,24] In DNA binding experiments using purified factors and short DNA fragments, RPA interacts preferentially with the undamaged strand, while avoiding the damaged strand and this bias is further increased by the addition of its interaction partner XPA.[42,48] Taken together these results converge on the hypothesis that XPA and RPA may cooperate during assembly of the NER complex to verify the need for processing of the damaged strand and, concomitantly, protect the undamaged complementary strand from inappropriate nuclease activity.[52] Although XPA and RPA have a cooperative "double-check" role that is necessary to stabilize the preincision complex, these two factors adopt distinct molecular functions on the DNA substrate. The lesion verification function is attributed to XPA, whereas RPA is recruited to the undamaged strand to protect the native template. Once RPA is bound to single-stranded DNA opposite the lesion, it provides a docking platform in conjuction with XPA to position the structure-specific endonucleases XPG (which makes the 3' incision) and XPF-ERCC1 (which makes the 5' incision) in a proper orientation with respect to their sites of cleavage.[6,16,49,62,79]

How Does XPA Protein Contribute to DNA Damage Recognition?

A putative nucleic acid interaction domain has been identified by nuclear magnetic resonance (NMR) spectroscopy of the minimal DNA-binding fragment of XPA, which consists of 122 amino acids and ranges from residue Met98 to Phe219.[10] This solution structure analysis revealed that the acidic zinc-containing subdomain (residues 105-129), which has a compact globular appearance, is accompanied by a C-terminal subdomain (residues 138-209) that forms a positively charged cleft on the protein surface (Fig. 3). In addition, the superficial cleft has the appropriate curvature and size to accommodate single-stranded or double-stranded DNA.[9,11,31] Further chemical shift perturbation experiments conducted in the presence of either a DNA fragment or a short RPA peptide sequence led to the surprising finding that the zinc finger domain of XPA (residues 105-129) is not involved in DNA binding instead, is required for the association with RPA.[9,31]

Positively charged surfaces are especially prominent in proteins that bind to DNA with no sequence specificity.[56] There is a striking absence of point mutations among the pathologic variants in XP-A patients that would result in amino acid changes within the basic DNA-binding cleft. Thus, the structural information disclosed by NMR analyses of XPA protein prompted a systematic mutational screen to determine the functional role of each basic residue in the presumed DNA-binding cleft (Fig. 3). The mapping of this region by electrophoretic mobility shift, photocrosslinking and host-cell reactivation assays demonstrated that a cluster of positively charged side chains on the surface of this area is indeed required for the efficient interaction with target DNA. In particular, two neighboring basic residues (Lys179 and Lys141), on the N-terminal side of the DNA-binding cleft, form a critical hotspot for recognition of the nucleic acid substrate (Camenisch et al manuscript submitted). Other positively charged side chains in the DNA-binding cleft are less important for substrate binding. When tested in host-cell reactivation assays, single Lys->Glu mutations at positions 179 and 141 impaired the ability of mutant XPA proteins to complement the DNA repair defect of XP-A fibroblasts, indicating that each single glutamic acid substitution generates a repulsive effect in the vicinity of the negatively charged DNA backbone.

A Lys179Glu/Lys141Glu tandem mutant has been purified and subjected to biochemical characterization because this double mutation confers a considerably stronger DNA repair defect than single amino acid mutations in the DNA-binding cleft.[12] Other combinations of double amino acid

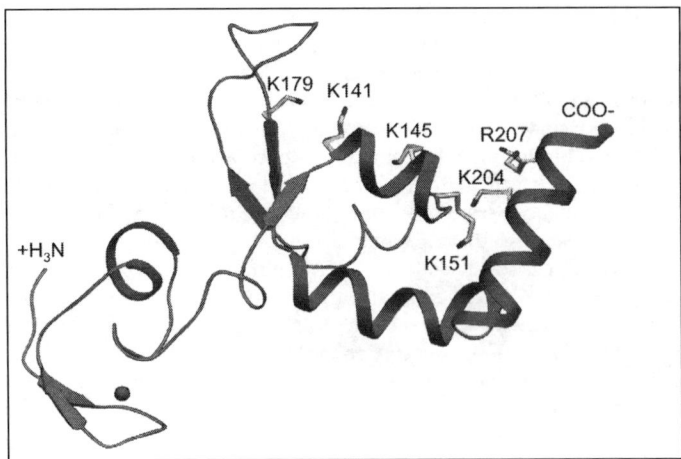

Figure 3. NMR model of the minimal DNA-binding fragment of human XPA protein.

substitutions throughout the positively charged DNA-binding surface resulted in more moderate excision repair defects. Like the respective single mutants, the double Lys179Glu/Lys141Glu derivative failed to interact with linear DNA fragments but was still able to bind to 4-way DNA junction molecules that were used as a surrogate for distorted reaction intermediates. Such junction molecules are thought to mimic the DNA deformations, involving sharp backbone kinks, that are often encountered when large multiprotein complexes assemble on DNA. The transient appearance of a site-specific kink during the processing of damaged DNA substrates is also suggested by the paradigm of NER in prokaryotic model organisms.[78] Surprisingly, the Lys179Glu/Lys141Glu tandem mutant binds to four-way DNA junctions with exactly the same affinity as wild-type XPA, although it generates different nucleoprotein products that migrate faster in native gels than the control complexes elicited by wild-type protein. Subsequent photocrosslinking experiments revealed that the subtle molecular defect underlying the formation of such abnormal complexes resides in the inability of the Lys179Glu/Lys141Glu double mutant to undergo close contacts with the kinked junction region of the tested 4-way DNA structures. Finally, the aberrant complexes formed by the interaction of this tandem mutant with 4-way junctions were unable to recruit the XPF-ERCC1 endonuclease,[12] indicating that the assembly of productive excision intermediates, which include XPF-ERCC1, is dependent on the proper interaction of XPA protein with a narrow bending angle in the DNA substrate.

A Hypothesis for the Mechanism of Damage Verification by XPA

XPA is one of several NER proteins that exhibit a moderate preference for binding to damaged DNA over the native duplex. However, none of these proteins, individually, bind modified DNA with sufficient selectivity to account for the high specificity of lesion removal by the NER system. Instead, damage recognition in the NER process is likely to be achieved by a discrimination cascade, in which different successive steps of mediocre specificity lead to a target selectivity that is comparable to that of sequence-specific transcription factors.[46] A major decision point during the sequential assembly of NER complexes is related to the question of how the system "knows" whether cleavage should occur, which is appropriate only if a lesion is actually present. Also, it is not clear how the NER machinery "knows" which of the two DNA strands should be subjected to dual cleavage.

What is the role of XPA protein in this damage recognition cascade? The strict requirement for positively charged side chains on its DNA-binding surface indicates that XPA may represent a molecular sensor of abnormal electrostatic potential along the DNA substrate. The case of UV endonuclease V, for which detailed crystallographic information is available, illustrates that a cluster of basic amino acids may participate in DNA damage recognition through electrostatic interactions with the unique backbone deformation induced by DNA kinks. This mechanism exploits the shortened spacing between adjacent phosphate residues, occurring at sites of sharp DNA bends, which increase the local negative electrostatic potential.[86] An electrostatic sensor mechanism for the detection of DNA kinks is further supported by the finding that the distance between adjacent basic headgroups in the DNA-binding cleft of XPA protein is actually shorter than the respective distance between phosphate groups in the backbone of linear DNA.[9] XPA protein is known to have a strong preference for artificially kinked substrates or other distorted DNA molecules that share the architectural feature of presenting two double strands emerging from a central bend[12,52] thus providing additional evidence for a function of XPA in detecting the increased electrostatic potential generated by the closer spacing of phosphate moieties at narrow DNA backbone angles.

How does the affinity for kinked backbones lead to a damage verification function by XPA protein? It has been pointed out that the general structural feature of DNA containing UV photoproducts or other bulky adducts is an increased local flexibility. Base stacking is the predominant energetic force leading to the intrinsic rigidity of DNA,[51] but the loss of base stacking is a common consequence of bulky lesion formation and weakened base stacking results in a flexible hinge.[32] This generic property of damaged DNA containing bulky lesions implies that XPA protein may exploit its affinity for sharply bent DNA backbones to carry out a universal damage verification function through an indirect mechanism, i.e., by probing the susceptibility of the DNA substrate to be kinked during the assembly of NER complexes. In this model, the critical role of XPA is associated with its ability to act as a bridging factor that "hands over" the recognition intermediate to the ultimate excision complex. It is likely, that XPA plays a regulatory role not only by verifying the presence of DNA damage but, together with its interaction partner RPA, also in monitoring the correct three-dimensional arrangement of the NER complex before activating the endonuclease subunits. This double-check mechanism by XPA and RPA avoids futile repair events on undamaged DNA and protects the intact complementary strand from inappropriate cleavage.

Acknowledgements

Work in the authors laboratory is supported by the Swiss National Science Foundation Grant 3100A–113694 and by the Krebsliga Schweiz.

References

1. Aboussekhra A, Biggerstaff M, Shivji MK et al. Mammalian DNA nucleotide excision repair reconstituted with purified protein components. Cell 1995; 80:859-868.
2. Araki M, Masutani C, Takemura M et al. Centrosome protein centrin 2/caltractin 1 is part of the xeroderma pigmentosum group C complex that initiates global nucleotide excision repair. J Biol Chem 2001; 276:18665-18672.
3. Araujo SJ, Tirode F, Coin F et al. Nucleotide excision repair of DNA with recombinant human proteins: definition of the minimal set of factors, active forms of TFIIH and modulation by CAK. Genes Dev 2000; 14:349-359.
4. Asashina H, Kuraoka I, Shirakawa M et al. The XPA protein is a zinc metalloprotein with an ability to recognize various kinds of DNA damage. Mutat Res 1994; 315:229-237.
5. Batty D, Rapic'-Otrin V, Levine AS et al. Stable binding of human XPC complex to irradiated DNA confers strong discrimination for damaged sites. J Mol Biol 2000; 300:275-290.
6. Bessho T, Sancar A, Thompson LH et al. Reconstitution of human excision nuclease with recombinant XPF-ERCC1 complex. J Biol Chem 1997; 272:3833-3837.
7. Bochkarev A, Pfuetzner RA, Edwards AM et al. Structure of the single-stranded-DNA-binding domain of replication protein A bound to DNA. Nature 1997; 385:176-181.
8. Bomgarden RD, Lupardus PJ, Soni DV et al. Opposing effects of the UV lesion repair protein XPA and UV bypass polymerase η on ATR checkpoint signaling. EMBO J 2006; 25:2605-2614.
9. Buchko GW, Daughdrill GW, de Lorimier R et al. Interactions of human nucleotide excision repair protein XPA with DNA and RPA70 Delta C327: chemical shift mapping and 15N NMR relaxation studies. Biochemistry 1999; 38:15116-15128.

10. Buchko GW, Ni S, Thrall BD et al. Human nucleotide excision repair protein XPA: expression and NMR backbone assignements of the 14.7 kDa minimal damaged DNA binding domain (Met98-Phe219). J Biomol NMR 1997; 10:313-314.
11. Buchko GW, Ni S, Thrall BD et al. Structural features of the minimal DNA binding domain (M98-F219) of human nucleotide excision repair protein XPA. Nucleic Acids Res 1998; 26:2779-2788.
12. Camenisch U, Dip R, Schumacher SB et al. Recognition of helical kinks by xeroderma pigmentosum group A protein triggers DNA excision repair. Nat Struct Mol Biol. 2006; 13:278-284.
13. Cleaver JE, States JC. The DNA damage-recognition problem in human and other eukaryotic cells: the XPA damage binding protein. Biochem J 1997; 328:1-12.
14. De Laat WL, Appeldoorn E, Jaspers NG et al. DNA structural elements required for ERCC1-XPF endonuclease activity. J Biol Chem 1998; 273:7835-7842.
15. De Laat WL, Appeldoorn E, Sugasawa E et al. DNA binding polarity of human replication protein A positions nucleases in nucleotide excision repair. Genes Dev 1998; 12:2598-2609.
16. De Laat WL, Jaspers NG, Hoeijmakers JHJ. Molecular mechanism of nucleotide excision repair. Genes Dev 1999; 13:768-785.
17. Evans E, Fellows J, Coffer A et al. Open complex formation around a lesion during nucleotide excision repair provides a structure for cleavage by human XPG protein. EMBO J 1997; 16:625-638.
18. Fitch ME, Nakajima S, Yasui A et al. In vivo recruitment of XPC to UV-induced cyclobutane pyrimidine dimers by the DDB2 gene product. J Biol Chem 2003; 278:46906-46910.
19. Fousteri M, Vermeulen W, van Zeeland AA et al. Cockayne syndrome A and B proteins differentially regulate recruitment of chromatin remodeling and repair factors to stalled RNA polymerase II in vivo. Mol Cell 2006; 23:471-482.
20. Friedberg EC, Aquilera A, Gellert M et al. DNA repair: from molecular mechanism to human disease. DNA Rep 2006; 5:986-996.
21. Gillet LC, Scharer OD. Molecular mechanisms of mammalian global genome nucleotide excision repair. Chem Rev 2006; 106:253-276.
22. Hanawalt PC. Subpathways of nucleotide excision repair and their regulation. Oncogene 2002; 21:8949-8956.
23. He Z, Henricksen LA, Wold MS et al. RPA involvement in the damage-recognition and incision steps of nucleotide excision repair. Nature 1995; 374:566-569.
24. Hermanson-Miller IL, Turchi JJ. Strand-specific binding of RPA and XPA to damaged duplex DNA. Biochemistry 2002; 41:2402-2408.
25. Hess NJ, Buchko GW, Conradson SD et al. Human nucleotide excision repair protein XPA: extended X-ray absorption fine-structure evidence for a metal-binding domain. Prot Sci 1998; 7:1970-1975.
26. Hoogstraten D, Nigg AL, Heath H et al. Rapid switching of TFIIH between RNA polymerase I and II transcription and DNA repair in vivo. Mol Cell 2002; 10:1163-1174.
27. Houtsmuller AB, Rademakers S, Nigg AL et al. Action of DNA repair endonuclease ERCC1/XPF in living cells. Science 1999; 284:958-961.
28. Huang JC, Svoboda D, Reardon JT et al. Human nucleotide excision nuclease removes thymine dimers from DNA by incising the 22nd phosphodiester bond 5' and the 6th phosphodiester bond 3' to the photodimer. Proc Natl Acad Sci USA 1992; 89:3664-3668.
29. Iakoucheva LM, Kimzey AL, Masselon CD et al. Aberrant mobility phenomena of the DNA repair protein XPA. Prot Sci 2001; 10:1353-1362.
30. Ikegami T, Kuraoka I, Saijo M et al. Resonance assignments, solution structure and backbone dynamics of the DNA- and RPA-binding domain of human repair factor XPA. J Biochem 1999; 125:495-506.
31. Ikegami T, Kuraoka I, Saijo M et al. Solution structure of the DNA- and RPA-binding domain of the human repair factor XPA. Nat Struct Biol 1998; 5:701-706.
32. Isaacs RJ, Spielmann HP. A model for initial DNA lesion recognition by NER and MMR based on local conformational flexibility. DNA Rep 2004; 3:455-464.
33. Jones CJ, Wood RD. Preferential binding of the xeroderma pigmentosum group A complementing protein to damaged DNA. Biochemistry 1993; 32:12096-12104.
34. Kim C, Paulus BF, Wold MS. Interactions of human replication protein A with oligonucleotides. Biochemistry 1994; 33:14197-14206.
35. Kobayashi T, Takeuchi S, Saijo M et al. Mutational analysis of a function of xeroderma pigmentosum group A (XPA) protein in strand-specific repair. Nucleic Acids Res 1998; 26:4662-4668.
36. Köberle B, Roginskaya V, Wood RD. XPA protein as a limiting factor for nucleotide excision repair and UV sensitivity in human cells. DNA Rep 2006; 5:641-648.
37. Kuraoka I, Bender C, Romieu A et al. Removal of oxygen free-radical-induced 5', 8-purine cyclodeoxy-nucleosides from DNA by the nucleotide excision-repair pathway in human cells. Proc Natl Acad Sci USA 2000; 97:3832-3837.

38. Kuraoka I, Morita EH, Saijo M et al. Identification of a damaged-DNA binding domain of the XPA protein. Mutat Res 1996; 362:87-95.
39. Lainé JP, Egly JM. Initiation of DNA repair mediated by a stalled RNA polymerase IIO. EMBO J 2006; 25:387-397.
40. Lao Y, Gomes XV, Ren Y et al. Replication protein A interactions with DNA. Part III. Molecular basis of recognition of damaged DNA. Biochemistry 2000; 39:850-859.
41. Layher SK, Cleaver JE. Quantification of XPA gene expression levels in human and mouse cell lines by competitive RT-PCR. Mutat Res 1997; 383:9-19.
42. Lee JH, Park CJ, Arunkumar AI. NMR study on the interaction between RPA and DNA decamer containing cis-syn cyclobutane pyrimidine dimer in the presence of XPA: implication for damage verification and strand-specific dual incision in nucleotide excision repair. Nucl Acids Res 2003; 31:4747-4754.
43. Li L, Elledge SJ, Peterson CA et al. Specific association between the human DNA repair proteins XPA and ERCC1. Proc Natl Acad Sci USA 1994; 91:5012-5016.
44. Li L, Lu X, Peterson CA et al. An interaction between the DNA repair factor XPA and replication protein A appears essential for nucleotide excision repair. Mol Cell Biol 1995; 15:5396-5402.
45. Li L, Peterson CA, Lu X et al. Mutations in XPA that prevent association with ERCC1 are defective in nucleotide excision repair. Mol Cell Biol 1995; 15:1993-1998.
46. Lin JJ, Sancar A. (A)BC excinuclease: the Escherichia coli nucleotide excision repair enzyme. Mol Microbiol 1992; 6:2219-2224.
47. Liu Y, Liu Y, Yang Z et al. Cooperative interaction of human XPA stabilizes and enhances specific binding of XPA to DNA damage. Biochemistry 2005; 17:7361-7368.
48. Liu Y, Yang Z, Utzat CD et al. Interactions of human replication protein A with single-stranded DNA adducts. Biochem J 2006; 385:519-526.
49. Matsunaga T, Park CH, Bessho T et al. Replication protein A confers structure-specific endonuclease activities to the XPF-ERCC1 and XPG subunits of human DNA repair excision nuclease. J Biol Chem 1996; 271:11047-11050.
50. Mellon I. Transcription-coupled repair: a complex affair. Mutat Res 2005; 577:155-161.
51. Mills JB, Hagerman PJ. Origin of the intrinsic rigidity of DNA. Nucleic Acids Res 2004; 32:4055-4059.
52. Missura M, Buterin T, Hindges R et al. Double-check probing of DNA bending and unwinding by XPA-RPA: an architectural function in DNA repair. EMBO J 2001; 20:3554-3564.
53. Miyamoto I, Miura N, Niwa H et al. Mutational analysis of the structure and function of the xeroderma pigmentosum group A complementing protein. Identification of essential domains for nuclear localization and DNA excision repair. J Biol Chem 1992; 267:19786-19789.
54. Moser J, Volker M, Kool H et al. The UV-damaged DNA binding protein mediates efficient targeting of the nucleotide excision repair complex to UV-induced photo lesions. DNA Rep 2005; 4:571-582.
55. Mu D, Park CH, Matsunaga T et al. Reconstitution of human DNA repair excision nuclease in a highly defined system. J Biol Chem 1995; 270:2415-2418.
56. Nadassy K, Wodak SJ, Janin J. Structural features of protein-nucleic acid recognition sites. Biochemistry 1999; 38:7199-7126.
57. Naegeli H. Mechanisms of DNA damage recognition in mammalian nucleotide excision repair. FASEB J 1995; 9:1043-1050.
58. Nagai A, Saijo M, Kuraoka I et al. Enhancement of damage-specific DNA binding of XPA by interaction with the ERCC1 DNA repair protein. Biochem Biophys Res Commun 1995; 211:960-966.
59. Nakatsu Y, Asahina H, Citterio E et al. XAB2, a novel tetratricopeptide repeat protein involved in transcription-coupled DNA repair and transcription. J Biol Chem 2000; 275:34931-34937.
60. Nishi R, Okuda Y, Watanabe E et al. Centrin 2 stimulates nucleotide excision repair by interacting with xeroderma pigmentosum group C protein. Mol Cell Biol 2005; 25:5664-6574.
61. Nitta M, Saijo M, Kodo N et al. A novel cytoplasmic GTPase XAB1 interacts with DNA repair protein XPA. Nucl Acids Res 2000; 28:4212-4218.
62. O'Donovan A, Davies AA, Moggs JG et al. XPG endonuclease makes the 3' incision in human DNA nucleotide excision repair. Nature 1994; 371:432-435.
63. Park CH, Mu D, Reardon JT et al. The general transcription-repair factor TFIIH is recruited to the excision repair complex by the XPA protein independent of the TFIIE transcription factor. J Biol Chem 1995; 270:4896-4902.
64. Park CH, Sancar A. Formation of a ternary complex by human XPA, ERCC1 and ERCC4 (XPF) excision repair proteins. Proc Natl Acad Sci USA 1994; 91:5017-5021.
65. Patrick SM, Turchi JJ. Xeroderma pigmentosum complementation group A protein (XPA) modulates RPA-DNA interactions via enhanced complex stability and inhibition of strand separation activity. J Biol Chem 2002; 277:16096-16101.
66. Politi A, Moné MJ, Houtsmuller AB et al. Mathematical modeling of nucleotide excision repair reveals efficiency of sequential assembly strategies. Mol Cell 2005; 19:679-690.

67. Rademakers S, Volker M, Hoogstraten D et al. Xeroderma pigmentosum group A protein loads as a separate factor onto DNA lesions. Mol Cell Biol 2003; 23:5755-5767.

68. Reardon JT, Bessho T, Kung HC et al. In vitro repair of oxidative DNA damage by human nucleotide excision repair system: possible explanation for neurodegeneration in xeroderma pigmentosum patients. Proc Natl Acad Sci USA 1997; 9463-9468.

69. Reardon JT, Nichols AF, Keeney S et al. Comparative analysis of binding of human damaged DNA-binding protein (XPE) and Escherichia coli damage recognition protein (UvrA) to the major ultraviolet photoproducts: T[c,s], T[t,s]T, T[6-4] and T[Dewar]T. J Biol Chem 1993; 268:21301-21308.

70. Reardon JT, Sancar A. Recognition and repair of the cyclobutane thymine dimer, a major cause of skin cancers, by the human excision nuclease. Genes Dev 2003; 17:2539-2551.

71. Reardon JT, Sancar A. Repair of DNA-polypeptide crosslinks by human excision nuclease. Proc Natl Acad Sci USA 2006; 103:4056-4061.

72. Riedl T, Hanaoka F, Egly JM. The comings and goings of nucleotide excision repair factors on damaged DNA. EMBO J 2003; 22:5293-5303.

73. Robins P, Jones CJ, Biggerstaff M et al. Complementation of DNA repair in xeroderma pigmentosum group A cell extracts by a protein with affinity for damaged DNA. EMBO J 1991; 10:3913-3921.

74. Saijo M, Kuraoka I, Masutani C et al. Sequential binding of DNA repair proteins RPA and ERCC1 to XPA in vitro. Nucleic Acids Res 1996; 24:4719-4724.

75. Sancar A. DNA excision repair. Annu Rev Biochem 1996; 65:43-81.

76. Sarker AH, Tsutakawa SE, Kostek S et al. Recognition of RNA polymerase II and transcription bubbles by XPG, CSB and TFIIH: insights for transcription-coupled repair and Cockayne syndrome. Mol Cell 2005; 20:187-19.

77. Satokata I, Iwa K, Matsuda T et al. Genomic characterization of the human DNA excision repair-controlling gene XPAC. Gene 1993; 136:345-348.

78. Shi Q, Thresher R, Sancar A et al. Electron microscopic study of (A)BC excinuclease: DNA is sharply bent in the UvrB-DNA complex. J Mol Biol 1992; 219:425-432.

79. Sijbers AM, de Laat WL, Ariza RR et al. Xeroderma pigmentosum group F caused by a defect in a structure-specific DNA repair endonuclease. Cell 1996; 86:811-822.

80. Sugasawa K, Masutani C, Uchida A et al. hHR23B, a human Rad23 homolog, stimulates XPC protein in nucleotide excision repair in vitro. Mol Cell Biol 1996; 16:4852-4861.

81. Sugasawa K, Ng JM, Masutani C et al. Xeroderma pigmentosum group C protein complex is the initiator of global genome nucleotide excision repair. Mol Cell 1998; 2:223-232.

82. Tanaka K, Miura N, Satokata I et al. Analysis of a human DNA excision repair gene involved in group A xeroderma pigmentosum and containing a zinc-finger domain. Nature 1990; 348:13-14.

83. Tanaka K, Satokata I, Ogita Z et al. Molecular cloning of a mouse DNA repair gene that complements the defect of group-A xeroderma pigmentosum. Proc Natl Acad Sci USA 1989; 86:5512-5516.

84. Thoma BS, Vasquez KM. Critical DNA damage recognition functions of XPC-hHR23B and XPA-RPA in nucleotide excision repair. Mol Carcinog 2003; 38:1-13.

85. Thoma BS, Wakasugi M, Christensen J et al. Human XPC-hHR23B interacts with XPA-RPA in the recognition of triplex-directed psoralen DNA interstrand crosslinks. Nucl Acids Res 2005; 33:2993-3001.

86. Vassylyev DG, Kashiwagi T, Mikami Y et al. Atomic model of a pyrimidine dimer excision repair enzyme complexed with a DNA substrate: structural basis for damaged DNA recognition. Cell 1995; 83:773-782.

87. Volker M, Moné MJ, Karmakar P et al. Sequential assembly of the nucleotide excision repair factors in vivo. Mol Cell 2001; 8:213-224.

88. Wakasugi M, Sancar A. Order of assembly of human DNA repair excision nuclease. J Biol Chem 1999; 274:18759-18768.

89. Wold MS. Replication protein A: a heterotrimeric, single-stranded Dann-binding protein required for eukaryotic DNA metabolism. Annu Rev Biochem 1997; 66:61-66.

90. Wood RD. Nucleotide excision repair in mammalian cells. J Biol Chem 1997; 272:23465-23468.

91. Wu X, Shell SM, Liu Y et al. ATR-dependent checkpoint modulates XPA nuclear import in response to UV irradiation. Oncogene 2006:Epub ahead of print.

92. Wu X, Shell SM, Yang Z et al. Phosphorylation of nucleotide excision repair factor xeroderma pigmentosum group A by ataxia telangiectasia mutated and Rad3-related-dependent checkpoint pathway promotes cell survival in response to UV irradiation. Cancer Res 2006; 66:2997-3005.

93. Yang ZG, Liu Y, Mao LY et al. Dimerization of human XPA and formation of XPA$_2$-RPA protein complex. Biochemistry 2002; 41:13012-13020.

94. You JS, Wang M, Lee SH. Biochemical analysis of the damage recognition process in nucleotide excision repair. J Biol Chem 2003; 278:7476-7485.

CHAPTER 5

XPB and XPD between Transcription and DNA Repair

Brian D. Beck, Dae-Sik Hah and Suk-Hee Lee*

Introduction

Xeroderma pigmentosum group B and D genes (XPB and XPD respectively) are components of the transcription factor IIH (TFIIH), a nine-subunit complex involved in transcription initiation by RNA polymerase II (pol II). Five of these (XPB, p62, p52, p44 and p34) form a tight core subcomplex, while XPD is less tightly associated with the core and mediates the binding of the CAK subcomplex, containing the remaining three subunits, cyclin H, cdk7 and MAT1.[1,2] The TFIIH complex also plays a key role in nucleotide excision repair (NER) by opening duplex DNA at the damage site. Both XPB and XPD possess DNA helicase activities, though XPB functions in a 3' → 5' fashion,[3] while XPD catalyzes unwinding of duplex DNA in the opposite direction.[2] The DNA-dependent helicase activity of XPB and XPD is central to the function of TFIIH in both transcription initiation and NER.

In this chapter, we will discuss the roles of XPB and XPD in transcription and DNA repair and overview the effects of mutations in, or loss of these helicases, both at the molecular and organismal level.

Structure-Function Relationship

XPB, also known as the excision repair cross-complementing rodent repair deficiency group 3 (*ERCC3*), is located at chromosome 2q21. It comprises 15 exons and a transcript length of 2,750 base pairs that encode a protein of 782 amino acids with molecular weight of 89 kDa. *XPD* (*ERCC2*) gene, located at chromosome 19q13.3, comprises 23 exons of about 54,000 base pairs and encodes an 80 kDa protein of 760 amino acids. Because both XPB and XPD proteins are essential for NER, DNA repair-deficient cells, arising from a mutation in these genes, exhibit low unscheduled DNA synthesis and an increased UV sensitivity and direct link to XP, Trichothiodystrophy (TTD), Cockayne syndrome (CS) and cancer-proneness due to a high mutation frequency.

XPB, human homolog of the yeast gene *Rad25* (*SSL2*) belongs to helicase super-family SF2 with seven conserved helicase motifs.[4,5] A recent structural analysis with an Archaeoglobus fulgidus, XPB homolog identified a number of functionally important motifs.6 At the N-terminus, it has damage recognition domain followed by two helicase domains (Fig. 1), a RED motif and a flexible thumb motif. RED motif is conserved among all known XPB homologs and mutations to this motif results in a significant decrease in helicase activity on dsDNA with a 3' overhang. A glycine hinge within the crystal structure suggests a point at which the protein can fold upon itself to bring all of the helicase motifs within close proximity to each other upon DNA binding. A thumb motif within the second helicase domain is structurally similar to T7 and Taq DNA polymerases,[7,8] and has a number of positively charged residues with a possible role in DNA

*Corresponding Author: Suk-Hee Lee—IU Cancer Research Institute (Rm153), 1044 W. Walnut St., Indianapolis, Indiana 46202,USA. Email: slee@iupui.edu

Molecular Mechanisms of Xeroderma Pigmentosum, edited by Shamim I. Ahmad and Fumio Hanaoka. ©2008 Landes Bioscience and Springer Science+Business Media.

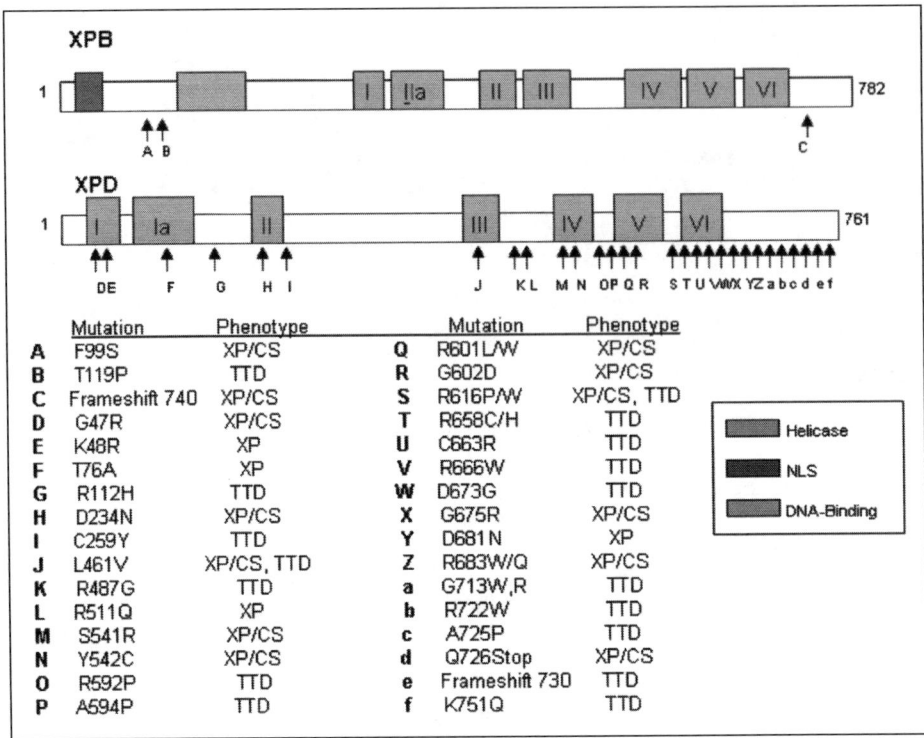

Mutation	Phenotype		Mutation	Phenotype
A	F99S	XP/CS	Q	R601L/W	XP/CS
B	T119P	TTD	R	G602D	XP/CS
C	Frameshift 740	XP/CS	S	R616P/W	XP/CS, TTD
D	G47R	XP/CS	T	R658C/H	TTD
E	K48R	XP	U	C663R	TTD
F	T76A	XP	V	R666W	TTD
G	R112H	TTD	W	D673G	TTD
H	D234N	XP/CS	X	G675R	XP/CS
I	C259Y	TTD	Y	D681N	XP
J	L461V	XP/CS, TTD	Z	R683W/Q	XP/CS
K	R487G	TTD	a	G713W,R	TTD
L	R511Q	XP	b	R722W	TTD
M	S541R	XP/CS	c	A725P	TTD
N	Y542C	XP/CS	d	Q726Stop	XP/CS
O	R592P	TTD	e	Frameshift 730	TTD
P	A594P	TTD	f	K751Q	TTD

Figure 1. Structural motifs of XPB and XPD proteins. Arrows indicate known disease-causing mutations.[9-11,13]

binding. In addition, those positively charged residues facing the groove, formed by the second helicase domain, predicts a likely site for protein-DNA interaction.

XPD, human homolog of the yeast genes *RAD3*[12] shares 15% sequence identity and 62% sequence similarity with UvrB. Both XPD and UvrB belong to the same helicase family (SF2) and exhibit greater similarity in sequence as compared with the three other members of SF helicase family (PcrA, Rep and NS3) whose structures have been determined. The structure of the molecular model developed for the human XPD, based on the X-ray crystal structure of *Bacillus caldotenax*, UvrB[13] shows 7 defined SF2 helicase motifs (Fig. 1). XPD mutations in patients with XP or CS were found in either helicase domains or just before the C-terminal domain (residues 713-761). Mutations at residues 681 and 683, for example, disrupt a region of the protein necessary for interaction with the p44 protein.[14] A comparison of the modeled XPD structures with other solved monomeric helicase structures illustrates a conserved central β-pleated sheet structure with surrounding α-helices common to NTPases.[13] This domain contains the Walker A and B motifs common to NTPases and is involved in ATP binding and hydrolysis. Domain 3 of the XPD model also conserves the central β-pleated sheet structure with surrounding α-helices of the other solved SF1 and SF2 family helicases.

Role in Transcription

RNA pol II-mediated transcription begins with recruitment of basal transcription factors (TFIIF, TFIIB, TFIID and TFIIE) to the promoter region. Following addition of TFIIH, promoter melting and the open complex formation take place, leading to promoter escape, a transition away from promoter into elongation complex. While all 9 subunits of TFIIH are necessary for maximal transcription, only the helicase activity of XPB is required for transcription. Although XPD is required for transcription, its role appears structural rather than catalytic[15] and helps to stimulate the reaction.[16,17]

During transcription initiation, DNA wraps around RNA polymerase II just upstream of the promoter, which begins to destabilize the helix immediately 5' to the transcription initiation site. Entry of TFIIH into the pre-initiation complex induces further unwinding, allowing XPB to bind to single stranded DNA. ATP hydrolysis by XPB is believed to cause a conformational change that pulls the two DNA strands farther apart, providing additional binding sites for RNA Pol II. Mutations from XPB patients showed an inability for TFIIH to properly associate with the pre-initiation complex, leading to deficient promoter melting. A recent study indicated that the two helicase sub-domains of XPB surround the promoter DNA in the pre-initiation complex in a similar manner to that of other DNA-helicase interactions.[18] Promoter opening could still occur in the presence of a disrupted helicase motif, but not in the presence of a mutated ATPase motif, suggesting that only the ATPase activity and not the helicase activity of XPB or XPD is necessary for promoter opening.[19] However, promoter clearance from the transcription complex is blocked in the presence of helicase mutants, suggesting that helicase motif is still required for full transcription activity. The helicase activity may be involved in remodeling of the DNA that is crucial for promoter escape during early transcription.[20] A recent study[21] suggests that in the absence of TFIIH, negative superhelicity is necessary for elongation of a 3 nucleotide RNA transcript, but this requirement ends once the transcript has reached 8 nucleotide that triggers a rate-limiting shift of RNA pol II complexes into elongation, suggesting that the role of XPB and XPD in transcription initiation is completed once the RNA transcript reaches a specific length.

XPB and XPD also play a regulatory role in transcription initiation mediated by effectors such as FUSE (Far UpStream Element)-binding protein (FBP) and FBP-interacting repressor (FIR).[22] For example, expression of c-myc transcription factor is regulated through FBP and FIR, where the opposing transcription factor FBP serves to decrease the time between initiation and promoter escape, while FIR inhibits this passage. Cells from XPB and XPD patients have defective FBP/FIR regulation,[23] where, endogenous c-myc expression was increased 1.5-fold over wild-type cells and this difference grew to 2.7-fold following UV damage[24] due to down-regulation in wild-type cells. On the other hand, a chromatin-immunoprecipitation assay showed very low levels of TFIIH at the c-myc promoter of XPB mutant cells as compared to wild type, suggesting that promoter escape may occur via transcriptional pauses, at which point TFIIH is modulated by effectors such as FBP and FIR. Without functionally active XPB, TFIIH may not be modulated by FIR, which results in increased basal expression of c-myc, suggesting that XPB is a key target of regulation.[16,17]

Role in NER

Assembly of the TFIIH complex at the damaged site occurs immediately following introduction of UV damage and this binds to DNA in a non-ATP-dependent manner after association of XPC to DNA.[25,26] A role for XPB in NER came from an in vitro experiment that cell extracts with a C-terminal deletion mutant of XPB failed to make a 5' incision on the damaged DNA.[27] This occurred despite the presence of functionally active XPF that normally induces a 5' incision.[28] The role for XPB in the 5' incision of damaged DNA is not clear, but it may be linked to remodeling the DNA at the damaged site. It is the ATP-dependent helicase activity of the TFIIH complex that is necessary to remodel the DNA in such a way to allow additional proteins in the NER pathway to associate; a mutation in XPB affecting helicase activity but not DNA binding prevents the association of XPF endonuclease at the damaged sites.

XPB is a phosphoprotein and phosphorylation at serine 751 results in inhibition of NER and a mutant mimicking S751 phosphorylation was unable to complement the NER defective phenotype in cells lacking the protein.[29] Interestingly, transcription and DNA unwinding were unaffected by this mutation, but the ability to make a 5' incision was prevented following phosphorylation. Phosphorylation status did not affect recruitment of XPF to the TFIIH complex, but 5' incision only occurred in the dephosphorylated state. XPB has also been shown to interact with XPG and XPD,[30] although its functional implication in NER is not clear.

Mutations in *XPD* gene that cause a defective helicase appear to affect only the NER pathway and not transcription initiation. On the other hand, XPD-CS mutations show a more complex phenotype that includes defective TCR of oxidative lesions from ionizing radiation and hydrogen peroxide and increased mutations from an 8-oxo-7, 8-dihydroguanine lesions in the transcribed strand in a shuttle vector.[31] Mechanistically, TTD defects have been ascribed to an altered enzyme structure that reduces stability and/or interaction with other TFIIH members, thereby impairing transcription initiation.[32] However, the clinical phenotypes of some *XPD* mutants require a more complex explanation than lack of either helicase activity or native XPD conformation, indicating that XPD patients are more likely to suffer from the transcription repair syndrome rather than DNA repair disorders alone.[33] Additional mutations in XPD have been found in iron-sulfur (Fe-S) cluster domains at the N-terminus of the protein.[34] Three Cys residues (Cys,88 Cys,105 and Cys137) were shown to be essential ligands for Fe-S cluster binding. Mutations in any of these amino acids resulted in normal ssDNA-dependent ATPase activity, but 5' → 3' helicase activity was abolished. In human patients, a XPD variant (R112H) showed a lack of helicase function, despite the mutation occurred outside of the helicase motifs.[13,35] The R112H mutation is a common variant in TTD patients. An equivalent mutation in *Saccharomyces cerevisiae* structurally disrupts the Fe-S cluster binding domain, resulting in loss of helicase activity. Furthermore, point mutations in the Fe-S cluster binding domains in RAD3 of *S. cerevisiae* resulted in heightened UV sensitivity, consistent with an impaired NER pathway.[33]

Separate from NER, XPD may also have a role in interstrand crosslink (ICL) repair mediated by homologous recombination. Over-expression of XPD is positively correlated with increased resistance to cisplatin treatment. This resistance is mediated by elevated homologous repair and sister chromatid exchanges. Glioblastoma cells, stably over-expressing XPD, showed a 2-fold increase in ICL repair, while NER activity remained constant.[36] ICL repair occurs through the homologous recombination (HR) pathway and Rad51 is an essential component of this pathway. Immunohistochemistry and immunoprecipitation analyses, both, showed an interaction between the two proteins and this interaction was increased following cisplatin treatment, suggesting a potential modulating effect of XPD on HR pathway. In keeping with this finding, a recent clinical study showed that advanced squamous cell carcinoma patients, harboring XPD-mutation (Asp312Asn or XPD-Lys751Gln), respond more favorably to cisplatin treatment than do patients expressing wild-type XPD protein.[37]

Other Roles of XPD

XPD also plays a role in regulation of CAK activity through modulation of Cdk7. During the first 13 cell cycles of Drosophila's embryos, transcription is not performed. During the interphase and prophase periods of these cell cycles, XPD is expressed strongly, but the signal decreases dramatically throughout the rest of mitosis, falling to nearly undetectable levels.[38] Lower XPD concentrations result in increased CAK activity, which in turn promotes cell cycle progression, while cells over-expressing XPD show decreased phosphorylation of Cdk1, a downstream target of Cdk7 and blocked cell cycle progression. Increased XPD expression could result in an increased incorporation of CAK into the TFIIH complex, thus fewer free CAK is available to promote cell cycle progression via Cdk1 phosphorylation.[38]

One intriguing thing is that XPB may share with other DNA helicases in protecting cells from p53-mediated apoptosis. Werner syndrome (WS) and Bloom syndrome (BS) are both diseases caused by defects in DNA helicases and patients with either disease are of smaller than average stature. Although the proteins that complement WS and BS are members of the RecQ helicase superfamily and XPB is a member of the SF2 superfamily, all possess 3' → 5' helicase activity. Ectopic expression of any of the three proteins was sufficient to correct the apoptotic phenotype present in BS cells, though not in WS or XPB cells,[39] suggesting that although the proteins are not functionally redundant, they may have similar roles in p53-mediated apoptosis. XPB and XPD may also play a role in the degradation of retroviral cDNA. Host cells defective in XPB or XPD showed more effective transduction of viral cDNA with an increase in total cDNA molecules and integrated provirus.[40] This change was not due to increased apoptosis in XP-defective cells, suggesting that XPB and XPD may have a role in a cellular defense against retroviral infection. Another interesting point is that XPB interacts with BCR through a domain that is conserved when BCR fuses with ABL in chronic myelogenous leukemia (CML).[41,42] In CML, the initial genetic defect is the BCR-ABL fusion protein, but this is followed by blast crisis, an acute period of genomic instability during which multiple mutations occur. Once XPB is bound through BCR-ABL's Dbl-homology domain, the fusion protein was shown to phosphorylate XPB in both C-terminal and N-terminal regions, which resulted in a decrease in both ATPase and the helicase activities. However, in vitro repair activity of UV-sensitive cells complemented with XPB and expressing BCR-ABL was only 5-10% less than that of similar cells not expressing BCR-ABL,[43] indicating that perhaps the mechanism leading to the in vivo results come from phosphorylation-induced transactivation of other TFIIH-interacting proteins.

Lessons from Genetic Variations of XPB/XPD

Unlike most other XP genes, XPB and XPD mutations exhibit variable and complex clinical phenotypes due to the fact that both proteins are involved in transcription and DNA excision repair. Depending on the location of the mutation, one might determine whether it impacts the transcription activity or the repair activity,[44] which would allow for multiple phenotype, as a result of different mutations to a multi-functional protein. For instance, patients with XP are cancer-prone, while CS and TTD patients are not, but a single mutation to XPB could potentially cause any of the three diseases. A mutation resulting in XP probably affects a portion of the protein specific for its NER/transcription function, while one leading to TTD affects transcription only. In support of this rationale, reports have indicated reduced levels of numerous genes in TTD patients, including β-globin[11] in humans and the product of the SPRR skin gene in mice.[45] Regardless of the specific mutations found in patients diagnosed with TTD, those patients invariably display reduced levels of TFIIH expression.[31] Interestingly, however, the relative amount of TFIIH found in the patient appeared to have no impact on the severity of the symptoms. This clinical finding is similar to the report of Satoh et al[11] claiming that cells of XPB/XPD patients exhibit a deficiency in NER and decreased levels of TFIIH, but displayed no change in transcription levels. Together, this suggests that perhaps mutations in one of the XP proteins might lead to improper assembly of the TFIIH complex and that the polypeptide subunits would then be degraded. How a basal level of TFIIH might be maintained in such an occurrence, however, has yet to be explained.

In XPB, a frameshift mutation at the 3' end resulted in inhibition of both repair and transcription function mediated by TFIIH,[46] which was responsible for a combined XP and CS phenotype. As a subunit of the TFIIH complex, XPB is anchored to the core complex by p52. The p52 subunit not only mediates XPB function but also is necessary for TFIIH stability.[47] While both XPB and XPD are necessary for NER, transcription activity may not be affected by defects in either helicase,[48] instead suggesting that problems with cellular function might only be seen under specific conditions. On the other hand, a decreased transcriptional activity with cells from XPB and XPD patients could be overcome by artificial opening of the promoter,[49] suggesting that a mutation of XPD, necessary for ATP hydrolysis, may not be sufficient to alter promoter opening or transcription levels.[50]

Most of the XPD mutations found in XP patients are located in the C-terminal domain that interacts with the p44 protein.[51] Although the mutations at the C-terminus hardly affected XPD's helicase activity, these mutations abolished the interaction of XPD with p44 and the stimulation of helicase activity.[11] Besides point mutations that cause diseases and are found in the homozygous state in patients or on only one XPD allele in asymptomatic parents, at least seven polymorphisms in exons 6, 8, 10, 17, 22 and 23 of the *XPD* gene have been identified in various cancers,[52-58] although the effects of these mutations on gene function are not clear.

Recently, a knock-in mouse model for combined XP/Cockayne syndrome (XPCS) was created to examine the cancer predisposition in the combined disorder.[59] This was accomplished by a single point mutation (G602D) in the XPD gene. Since XP is a cancer-prone disease and CS is a premature aging disease, perhaps this single point mutation can modulate both disease processes. These mice displayed reduced NER activity, residual unscheduled DNA synthesis without damage repair and an elevated cancer-prone phenotype, suggesting that accelerated cell death and cancer predisposition can both result from the same mutation.

Aside from genetic variations in XPB or XPD, virus-encoded proteins also play a role in modulating TFIIH expression/function. For example, a nonstructural protein (NSs) in the nucleus of the Rift Valley Hemorrhagic Fever (RVHF) virus can interact with the p44 protein of the TFIIH complex.[60] The NSs protein outcompetes XPD for p44 binding, sequestering it along with XPB in filamentous structures and preventing the formation of TFIIH. Hepatitis B virus has also been shown to downregulate TFIIH expression, although HBV, unlike RVHF, acts by preventing transcription of XPB and XPD.[44] HBV is a dsDNA virus coding for a protein known as HBx, which has the ability to transactivate cellular genes,[61] and an earlier study showed a novel suppression activity of HBx on XPB and XPD expression in transgenic liver cells.[62] This downregulation may occur through interaction of HBx with transcription factor Sp1, for which both XPB and XPD promoters contain responsive elements. Since chronic Hepatitis B is a risk factor for hepatocellular carcinoma, the authors propose that decreased expression of XPB/XPD might result in decrease NER activity and thus increase genomic instability.

References

1. Drapkin R, Le Roy G, Cho H et al. Human cyclin-dependent kinase-activating kinase exists in three distinct complexes. Proc Natl Acad Sci USA 1996; 93:6488-6493.
2. Reardon J, Ge H, Gibbs E et al. Isolation and characterization of two human transcription factor IIH (TFIIH)-related complexes: ERCC2/CAK and TFIIH. Mol Biol Cell 1996; 10:4231-4246.
3. Schaeffer L, Roy R, Humbert S et al. DNA repair helicase: A component of BTF2 (TFIIH) basic transcription factor. Science 1993; 260:37-38.
4. Park E, Guzder S, Koken M et al. Rad25 (SSL2), the yeast homolog of the human xeroderma pigmentosum group B DNA repair gene, is essential for viability. Proc Natl Acad Sci USA 1992; 89:11416-11420.
5. Weeda G, van Ham R, Vermeulen W et al. A presumed DNA helicase encoded by ERCC-3 is involved in the human repair disorders xeroderma pigmentosum and cockayne's syndrome. Cell 1990; 62:777-791.
6. Fan L, Arvai A, Cooper P et al. Conserved XPB core structure and motifs for DNA unwinding: Implications for pathway selection of transcription of excision repair. Mol Cell 2006; 22:27-37.
7. Doublie S, Tabor S, Long AM et al. Crystal structure of a bacteriophage T7 DNA replication complex at 2.2 A resolution. Nature 1998; 391:251-258.
8. Kim Y, Eom SH, Wang J et al. Crystal structure of Thermus aquaticus DNA polymerase. Nature 1995; 376:612-616.
9. Taylor E, Broughton B, Botta E et al. Xeroderma pigmentosum and trichothiodystrophy are associated with different mutations in the XPD (ERCC2) repair/transcription gene. Proc Natl Acad Sci USA 1997; 94:8658-8663.
10. Bergmann E, Egly JM. Trichothiodystrophy, a transcription syndrome. Trends Genet 2001; 17:279-286.
11. Viprakasit V, Gibbons R, Broughton B et al. Mutations in the general transcription factor TFIIH result in beta-thalassaemia in individuals with trichothiodystrophy. Hum Mol Gen 2001; 10:2797-2802.
12. Sung P, Bailly V, Weber C et al. Human xeroderma pigmentosum group D gene encodes a DNA polymerase. Nature 1993; 365:852-855.
13. Bienstock R, Skorvaga M, Mandavilli B et al. Structural and functional characterization of the human DNA repair helicase XPD by comparative molecular modeling and site-directed mutagenesis of the bacterial repair protein UvrB. J Biol Chem 2003; 278:5309-5316.
14. Coin F, Marinoni J, Rodolfo C et al. Mutations in the XPD helicase gene result in XP and TTD phenotypes, preventing interaction between XPD and the p44 subunit of TFIIH. Nat Genetics 1998; 20:184-188.
15. George J, Salazar E, Vreeswijk M et al. Restoration of nucleotide excision repair in a helicase-deficient XPD mutant from intragenic suppression by a trichothiodystrophy mutation. Mol Cell Biol 2001; 21:7355-7365.

16. Tirode F, Busso D, Coin F et al. Reconstitution of the transcription factor TFIIH: Assignment of functions for the three enzymatic subunits, XPB, XPD and cdk7. Mol Cell 1999; 3:87-95.
17. Moreland R, Tirode F, Yan Q et al. A role for the TFIIH XPB DNA helicase in promoter escape by RNA polymerase II. J Biol Chem 1999; 274:22127-22130.
18. Miller G, Hahn S. A DNA-tethered cleavage probe reveals the path for promoter DNA in the yeast preinitiation complex. Nat Struc Mol Biol 2006; 13:603-610.
19. Lin Y, Choi W, Gralla J. TFIIH XPB mutants suggest a unified bacterial-like mechanism for promoter opening but not escape. Nat Struc Mol Biol 2005; 12:603-607.
20. Goodrich J, Tijan R. Transcription factors IIE and IIH and ATP direct promoter clearance by RNA polymerase II. Cell 1994; 77:145-156.
21. Hieb A, Baran S, Goodrich J et al. An 8nt RNA triggers a rate-limiting shift of RNA polymerase II complexes into elongation. EMBO J 2006; 25:3100-3109.
22. Duncan R, Bazar L, Michelotti G et al. A sequence-specific, single-strand binding protein activates the far upstream element of c-Myc and defines a new DNA-binding motif. Genes Dev 1994; 8:465-480.
23. Liu J, Akoulitchev S, Weber A et al. Defective interplay of activators and repressors with TFIIH in xeroderma pigmentosum. Cell 2001; 104:353-363.
24. Weber A, Liu J, Collins I et al. TFIIH operates through an expanded proximal promotor to fine-tune c-Myc expression. Mol Cell Biol 2005; 25:147-161.
25. Wang Q, Zhu Q, Wani M et al. Tumor supressor p53 dependant recruitment of nucleotide excision repair factors XPC and TFIIH to DNA damage. DNA Repair 2003; 2:483-499.
26. Riedl T, Hanaoka F, Egly JM. The comings and goings of nucleotide excision repair factors on damaged DNA. EMBO J 2003; 22:5293-5303.
27. Weeda G, van Ham R, Vermeulen W et al. A presumed DNA helicase encoded by ERCC-3 is involved in the human repair disorders xeroderma pigmentosum and cockayne's syndrome. Cell 1990; 62:777-791.
28. Evans E, Moggs J, Hwang J et al. Mechanism of open complex and dual incision formation by human nucleotide excision repair factors. EMBO J 1997; 16:6559-6573.
29. Coin F, Auriol J, Tapias A et al. Phosphorylation of XPB helicase regulates TFIIH nucleotide excision repair activity. EMBO J 2004; 23:4835-4846.
30. Iyer N, Reagan M, Wu KJ et al. Interactions involving the human RNA polymerase II transcription/nucleotide excision repair complex TFIIH, the nucleotide excision repair protein XPG and cockayne syndrome group B (CSB) Protein Biochem 1996; 35:2157-2167.
31. Gallego M, Sarasin A. Transcription-coupled repair of 8-oxoguanine in human cells and its deficiency in some DNA repair diseases. Biochemie 2003; 85:1073-1082.
32. Botta E, Nardo T, Lehmann A et al. Reduced level of the repair/transcription factor TFIIH in tricho-thiodystrophy. Hum Mol Gen 2002; 11:2919-2928.
33. Coin F, Bergmann E, Tremeau-Bravard A et al. Mutations in XPB and XPD helicases found in xeroderma pigmentosum patients impair the transcription function of TFIIH. EMBO J 1999; 18:1357-1366.
34. Rudolf J, Makrantoni V, Ingledew WJ et al. The DNA repair helicases XPD and FancJ have essential iron-sulfur domains. Mol Cell 2006; 23:801-808.
35. Queille S, Drougard C, Sarasin A et al. Effects of XPD mutations on ultraviolet-induced apoptosis in relation to skin cancer-proneness in repair deficient syndromes. J Invest Dermatol 2001; 117:1162-1170.
36. Aloyz R, Xu ZY, Bello V et al. Regulation of cisplatin resistance and homologous recombinational repair by the TFIIH subunit XPD. Cancer Res 2002; 62:5457-5462.
37. Quintela-Fandino M, Hitt R, Medina P et al. DNA-repair gene polymorphisms predict favorable clinical outcome among patients with advanced squamous cell carcinoma of the head and neck treated with cisplatin-based induction chemotherapy. J Clin Onc 2006; 24:4333-4339.
38. Chen J, Larochelle S, Li X et al. Xpd/Ercc2 regulates CAK activity and mitotic progression. Nature 2003; 424:228-232.
39. Spillare EA, Wang XW, von Kobbe C et al. Redundancy of DNA helicases in p53-mediated apoptosis. Oncogene 2006; 25:2119-2123.
40. Yoder K, Sarasin A, Kraemer K et al. The DNA repair genes XPB and XPD defend cells from retroviral infection. Proc Natl Acad Sci USA 2006; 103:4622-4627.
41. Maru Y, Kobayashi T, Tanaka K et al. BCR binds to the xeroderma pigmentosum group B protein. Biochem Biophys Res Comm 1999; 260:309-312.
42. Takeda N, Shibuya M, Maru Y. The bcr-abl oncoprotein potentially interacts with xeroderma pigmentosum Group B Protein. Proc Natl Acad Sci USA 1999; 96:203-207.
43. Maru Y, Bergmann E, Coin F et al. TFIIH functions are altered by the P210bcr-Abl oncoprotein produced on the Philadelphia chromosome. Mut Res 2001; 483:82-88.
44. Bootsma D, Hoeijmakers J. Engagement with transcription. Nature 1993; 363:114-115.

45. de Boer J, de Wit J, van Steeg H et al. A mouse model for the basal transcription/DNA repair syndrome trichothiodystrophy. Mol Cell 1998; 1:981-990.
46. Hwang J, Moncollin V, Vermeulen W et al. A 3' → 5' helicase defect in repair/transcription factor TFIIH of xeroderma pigmentosum group B affects both DNA repair and transcription. J Biol Chem 1996; 271:15898-15904.
47. Jawhari A, Laine JP, Dubaele S et al. p52 mediates XPB function within the transcription/repair factor TFIIH. J Biol Chem 2002; 277:31761-31767.
48. Satoh M, Hanawalt P. Competent transcription initiation by RNA polymerase II in cell-free extracts from xeroderma pigmentosum groups B and D in an optimized RNA transcription assay. Biochem Biophys Acta 1997; 1354:241-251.
49. Coin F, Bergmann E, Tremeau-Bravard A et al. Mutations in XPB and XPD Helicases found in xeroderma pigmentosum patients impair the transcription function of TFIIH. EMBO J 1999; 18:1357-1366.
50. Winkler G, Araujo S, Fiedler U et al. TFIIH with inactive XPD helicase functions in transcription initiation but is defective in DNA repair. J Biol Chem 2000; 275:4258-4266.
51. Tirode F, Busso D, Coin F et al. Reconstitution of the transcription factor TFIIH: Assignment of functions for the three enzymatic subunits, XPB, XPD and Cdk7. Mol Cell 1999; 3:87-95.
52. Caggana M, Kilgallen J, Conroy J et al. Associations between ERCC2 polymorphisms and gliomas. Canc Epid Biomarkers Prev 2001; 10:355-60.
53. Tomescu D, Kavanagh G, Ha T et al. Nucleotide excision repair gene XPD polymorphisms and genetic predisposition to melanoma. Carcinogenesis 2001; 22:403-408.
54. Benhamou S, Sarasin A. ERCC2/XPD gene polymorphisms and cancer risk. Mutagenesis 2002; 17:463-469.
55. Schabath M, Delclos G, Grossman H et al. Polymorphisms in XPD exons 10 and 23 and bladder cancer risk. Canc Epidemiol Biomarkers Prev 2005; 14:878-884.
56. Broughton B, Steingrimsdottir H, Lehmann A. Five polymorphisms in the coding sequence of the xeroderma pigmentosum group D gene. Mutat Res 1996; 362:209-211.
57. Shen M, Jones I, Mohrenweiser H. Nonconservative amino acid substitution variants exist at polymorphic frequency in DNA repair genes in healthy humans. Cancer Res 1998; 61:3321-3325.
58. Mohrenweiser H, Xi T, Vazquez-Matias J et al. Identification of 127 amino acid substitution variants in screening 37 DNA repair genes in humans. Canc Epidemiol Biomarkers Prev 2002; 11:1054-1064.
59. Andressoo JO, Mitchell J, de Wit J et al. An Xpd mouse model for the combined xeroderma pigmentosum/cockayne syndrome exhibiting both cancer predisposition and segmental progeria. Cancer Cell 2006; 10:121-132.
60. Le May N, Dubaele S, De Santis L et al. TFIIH transcription factor, a target for the rift valley hemorrhagic fever virus. Cell 2004; 116:541-550.
61. Twu J, Schloemer R. Transcriptional trans-activating function of hepatitis B virus. J Virol 1987; 61:3448-3453.
62. Jaitovich-Groisman I, Benlimame N, Slagle B et al. Transcriptional regulation of the TFIIH transcription repair components XPB and XPD by the hepatitis B virus x protein in liver cells and transgenic liver tissue. J Biol Chem 2001; 276:14124-14132.

CHAPTER 6

XPC:
Its Product and Biological Roles

Kaoru Sugasawa*

Abstract

The XPC protein is a component of a heterotrimeric complex that is essential for damage recognition in a nucleotide excision repair subpathway that operates throughout the genome. Biochemical analyses have revealed that the broad substrate specificity of this repair system is based on the structure-specific DNA binding properties of the XPC complex. Other subunits of this complex, including human Rad23p orthologs and centrin 2, play individual roles in enhancing the damage recognition activity of XPC. Physical interaction with UV-damaged DNA-binding protein is also important for the efficient recruitment of XPC to sites containing DNA damage, particularly UV-induced photolesions. Furthermore, recent studies have suggested that XPC may also be involved in base excision repair and possibly in other cellular functions that may be mediated by posttranslational modifications.

Introduction

The mutation(s) in genetic complementation group C is one of the most common form of autosomal recessive disorder, xeroderma pigmentosum (XP).[1,2] Although XP-C patients exhibit cutaneous hypersensitivity to sunlight exposure and a marked predisposition to skin cancer like patients in other XP groups, they usually do not have neurological abnormalities. Cultured XP-C fibroblasts show relatively mild sensitivity to ultraviolet (UV) irradiation but very limited UV-induced unscheduled DNA synthesis, 10-20% the level of normal fibroblasts, indicating defects in global genome nucleotide excision repair (GG-NER). On the other hand, XP-C cells are proficient in the removal of damage from the transcribed strand of active genes.[3]

The *XPC* gene was first identified by expression cloning as a gene that corrected the repair defect of XP-C fibroblasts.[4] Subsequently, in vitro NER complementation assays enabled purification of a protein factor containing the XPC gene product.[5,6] The human XPC protein is composed of 940 amino acids, the primary sequence of which shares moderate homology with the *Saccharomyces cerevisiae* NER protein Rad4p. XPC exists in vivo as a highly stable heterotrimeric complex with one of the two mammalian homologs of *S. cerevisiae* RAD23p (designated RAD23A and RAD23B) and centrin 2, which is also known as a centrosomal protein.[7] This chapter describes the functions of the XPC complex in the molecular mechanisms of GG-NER, including the roles of each subunit and their possible involvement in other cellular functions beyond NER.

XPC Is a Damage Recognition Factor in GG-NER

The XPC complex was purified from nuclear extracts from NER-proficient human cultured cells as a protein factor that corrected the in vitro NER defects of XP-C cell extracts.[5,6] A key step

*Kaoru Sugasawa—Biosignal Research Center, Organization of Advanced Science and Technology, Kobe University, and SORST, Japan Science and Technology Agency, 1-1 Rokkodai, Nada-ku, Kobe, Hyogo 657-8501, Japan. Email: ksugasawa@garnet.kobe-u.ac.jp

Molecular Mechanisms of Xeroderma Pigmentosum, edited by Shamim I. Ahmad and Fumio Hanaoka. ©2008 Landes Bioscience and Springer Science+Business Media.

of this purification utilized single-stranded DNA-cellulose, indicating that this factor possesses a strong DNA-binding activity. The *XPC* gene encodes a basic protein (theoretical pI ~9.05) that is responsible for its affinity for DNA, whereas neither of the Rad23p homologs nor centrin 2 binds DNA by itself. Electrophoretic mobility shift and DNase I footprinting assays, with DNA substrates containing site-specific lesions, revealed that XPC is capable of binding specifically to various damaged sites, including UV-induced pyrimidine-pyrimidone (6-4) photoproducts (6-4PPs) as well as N-acetoxy 2-acetylaminofluorence (AAF) adducts, whereas it poorly recognizes UV-induced cyclobutane pyrimidine dimers (CPDs) (Fig. 1).[8,9] Scanning force microscopy also demonstrated that XPC binds precisely to damaged sites and induces a ~45° bend in the DNA duplex.[10] Moreover, biochemical analyses indicated that XPC functions in the first stage of cell-free NER reactions. It was demonstrated that, in vitro, NER preferentially takes place on damaged DNA that has been pre-incubated with XPC, as compared to damaged DNA in the same mixture that has been pretreated in its absence.[11] Roles for XPC in damage recognition and in the initiation of repair were also supported by biochemical studies that determined the order of NER complex assembly on damaged DNA substrates immobilized on paramagnetic beads.12 Furthermore, localized UV irradiation experiments, using isopore membrane filters, revealed that the accumulation of XPC in subnuclear domains containing damage does not require any other functional XP gene.[13] This observation is in line with the fact that XP-C cells are deficient only in GG-NER,[3] because primary damage recognition during transcription-coupled repair is thought to be accomplished by RNA polymerase II stalling at lesions on the template strand, a phenomenon that functionally substitutes for lesion recognition by XPC.[14]

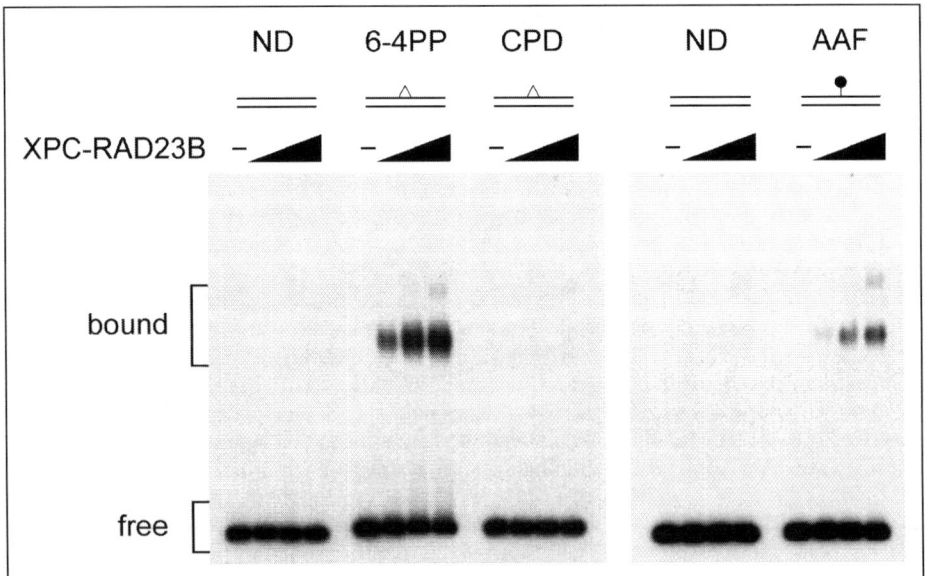

Figure 1. Specificity of DNA binding by the XPC complex. Electrophoretic mobility shift assays were performed using the recombinant human XPC-RAD23B heterodimer (adopted from ref. 9). The radiolabeled substrates used were ~180-bp double-stranded DNAs containing a site-specific lesion (cis-syn CPD or a 6-4PP thymine dimer or an AAF adduct of deoxyguanine). XPC can bind to DNA sites containing lesions like 6-4PP and AAF that induce relatively large DNA helical distortions, whereas CPDs, which only minimally distort the DNA duplex, are poorly recognized. ND: no damage. Reproduced from: Sugasawa K et al. Genes Dev 2001; 15:507-521;[9] ©2001 with permission from Cold Spring Harbor Laboratory Press.

The XPC complex apparently recognizes various helix-distorting base lesions that do not share a common chemical structure. Extensive biochemical studies on its binding specificity revealed that XPC instead recognizes a specific secondary DNA structure (a branched structure containing a junction between double- and single-stranded DNA) rather than lesions themselves.[15] Thus the presence of single-stranded DNA (i.e., one or more nonhydrogen-bonding bases) is a crucial factor for specific binding by XPC, so that its affinity for specific lesions appears to correlate with the extent of helical distortion that they induce. For instance, XPC preferentially binds artificial bubble-like structures, even those without base damage and the extent of affinity parallels bubble size.[9] On the other hand, CPDs induce only minimal helical distortion. Solution structures solved by NMR clearly indicate that all bases, including the two damaged pyrimidine residues themselves, are still hydrogen-bonded with purines on the opposite strand,[16,17] thereby explaining why CPDs are poorly recognized by XPC. Based on these biochemical properties, XPC functions as a versatile damage recognition factor that contributes to the very broad substrate specificity of GG-NER.

It was shown that XPC physically interacts with the multifunctional transcription/NER complex TFIIH, which contains the XPB and XPD helicases.[18,19] The XPB and p62 subunits of TFIIH appear to be responsible for interaction with XPC[19]. Although TFIIH alone does not exhibit significant DNA-binding activity, it appears to be recruited to sites of DNA damage in an XPC-dependent manner both in vitro[18-20] and in vivo.[13] Therefore, one of the mechanistic roles of XPC, after damage recognition, is probably the direct recruitment of TFIIH to the lesion. Since local unwinding of duplex DNA at damaged sites is a key role for TFIIH in NER,[21,22] the conformational change in DNA, induced by XPC,[10] may also play an important role in facilitating entry of the XPB and XPD helicases in DNA.

Several functional domains have been determined in XPC, which include binding domains for DNA, RAD23, centrin 2 and TFIIH (Fig. 2).[23-25] Most of these map near the carboxy terminus and overlap with a domain that shares significant sequence homology with Rad4p,[4] while the functions of the amino terminal region are relatively ambiguous. Numerous pathogenic mutations in the *XPC* gene have been identified from patients; these mutations are distributed throughout the gene with no obvious hotspot.[26-33] The vast majority are either truncating nonsense or frame-shift mutations that cause loss of the functionally important carboxy terminus. In these patients, XPC mRNA levels are also greatly reduced, probably as a result of nonsense-mediated mRNA decay,[28,30,32] so that the expression of truncated XPC proteins has not been detected. Thus most XP-C patients exhibit null phenotypes, including UV sensitivity and reduced levels of UV-induced unscheduled DNA synthesis.

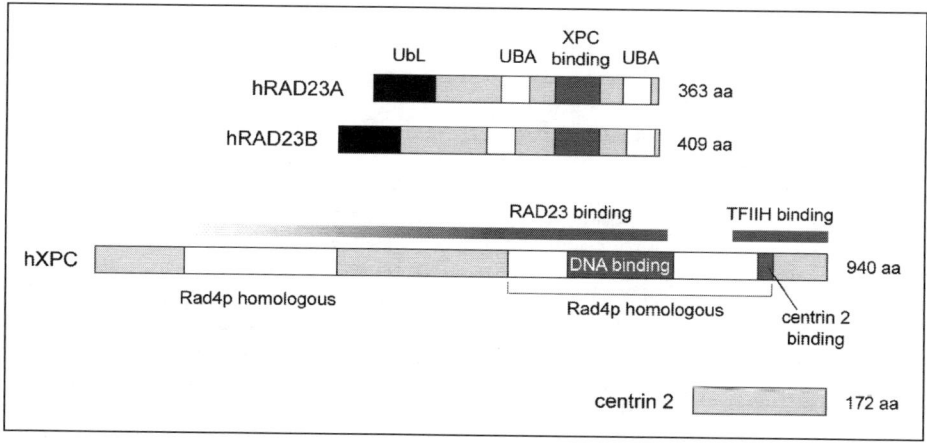

Figure 2. Domain structures of subunits of the human XPC complex.

Roles for Other Subunits

In mammalian cells, most XPC protein forms a complex with RAD23B, while the remaining minor fraction binds to RAD23A.[7] This is probably due to the differential expression levels of the two Rad23p homologs,[34] since the in vitro interactions of recombinant XPC with RAD23A and RAD23B are indistinguishable.[35] The combined cellular level of the two Rad23p homologs appears to be > 10~20 fold higher than that of XPC,[36,37] suggesting that their functions are not restricted to GG-NER. Yeast Rad23p also forms a stable complex with the XPC counterpart Rad4p,[38] showing evolutionary conservation of this interaction and members of the Rad23p family also share unique domain structures (Fig. 2).[39,40] Yeast Rad23p and its homologs possess an amino-terminal ubiquitin-like sequence (UbL), which interacts with the 26S proteasome.[41,42] In mammals, the proteasomal subunit, responsible for this interaction, is S5a (PSMD4), which was identified as a polyubiquitin-chain receptor. In addition, members of the Rad23p family each contain two copies of the ubiquitin-associated (UBA) domain, directly interacts with ubiquitin, particularly lysine 48-linked polyubiquitin chains.[43-46] It has been proposed that RAD23 proteins function as shuttle factors, recruiting polyubiquitinated protein substrates to the 26S proteasome, although they have also been observed to have inhibitory effects on proteolysis under certain experimental conditions.[41,45,47] Their XPC binding domain maps between the two UBA domains.[39]

Targeted disruption of the murine *RAD23B* (*mRad23B*) gene greatly compromises the in vivo stability of XPC, thereby resulting in a marked reduction of its steady-state protein levels.[34,48,49] Simultaneous deletion of *mRad23a* further reduces XPC protein levels, although the *mRad23a* single knockout has only a marginal effect. When XPC is ectopically expressed in mouse embryonic fibroblasts (MEFs), that lack both *mRad23* genes, it is dramatically stabilized by treatment with proteasome inhibitors or DNA damaging agents, strongly suggesting that the Rad23p homologs protect XPC from degradation by the 26S proteasome.[48] Similar phenomena have been observed with yeast Rad4p, which is substantially destabilized in the absence of the *RAD23* gene.[50] Although this is probably also the case for the endogenous XPC protein, its metabolic turnover in normal fibroblasts is so slow that similar stabilization effects are difficult to prove.[34] Importantly, GG-NER activity appears to correlate with steady-state XPC protein levels:[34] *mRad23a/b* double knockout MEFs exhibit specific defects in GG-NER similar to those of XP-C fibroblasts, indicating that Rad23p homologs contribute to the maintenance of cellular repair by stabilizing XPC.[34,48] Intriguingly, *mRad23b* knockout mice exhibit severe phenotypes, including developmental abnormalities and male sterility and the *mRad23a/b* double knockout is lethal.[48,49] These results clearly point to additional important functions of the Rad23p homologs beyond GG-NER, likely in the regulation of proteolysis, because the absence of GG-NER alone does not affect normal development and viability as demonstrated by XP-C patients and *Xpc*⁻/⁻ mice.

The third subunit of the XPC complex is centrin 2, which is a small calmodulin-like calcium-binding protein. Centrin 2 has been identified as a component of the centrosome and knockdown of its expression blocks centriole duplication, thereby arresting the cell cycle.[51] Besides its function in the centrosome, centrin 2 also localizes to the nucleus[52] and is involved in GG-NER as part of the XPC complex and probably in other cellular functions such as homologous recombination.[53] Thus, like the Rad23p homologs, centrin 2 is a multifunctional protein. XPC appears to constitutively bind to centrin 2 rather than inducibly interact with it upon DNA damage.[7,23]

Like other members of the calmodulin family, centrin 2 contains EF-hand motifs and binds to a short putative α-helix in XPC through hydrophobic interactions. The centrin-2-binding site was mapped near the carboxy terminus of XPC (Fig. 2) and in vivo GG-NER activity is partially compromised when several hydrophobic amino acids in this site are replaced by alanine residues.[23] On the other hand, several groups have succeeded in reconstituting cell-free NER reactions with purified protein factors and the XPC-Rad23B heterodimer was routinely utilized in these systems.[54,55] Hence, although centrin 2 is not essential for repair per se, its addition dramatically potentiates in vitro NER activity and this stimulation is due to enhancing both the affinity and damage-specificity of DNA binding by XPC.[23] Therefore, centrin 2 contributes to GG-NER activity in a manner different from that of the RAD23 proteins. Moreover, there is an interesting possibility that centrin 2 mediates crosstalk between DNA repair and other cellular damage response pathways, for instance by regulating cell cycle checkpoints.

Ubiquitination and Interaction with UV-DDB

Recent evidence indicates that the XPC protein undergoes ubiquitination when cells are UV irradiated.[56,57] This modification appears to be reversible and not to induce proteasomal degradation of XPC. Importantly, XPC ubiquitination is dependent on the presence of functional UV-DDB (UV-damaged DNA-binding protein) for which XP-E patients are defective.[56] As described in chapter 7, UV-DDB is a heterodimer consisting of the DDB1 and DDB2 (XPE) proteins, that exhibits much higher damage discrimination than XPC, especially for UV-induced photolesions.[56,58,59] Its affinity and specificity for 6-4PP are particularly notable, while UV-DDB can also recognize CPD.[59-61] XPC and UV-DDB directly interact, suggesting that UV-DDB bound to damaged sites may facilitate the subsequent recruitment of XPC.[56,62,63] This mechanism appears to be especially important for CPD repair, since this lesion is poorly recognized by XPC alone, whereas 6-4PP can be recognized either by XPC directly or through the UV-DDB-mediated pathway (Fig. 3). For chemically-induced lesions, the efficiency and contribution of these two recognition pathways may vary depending on the relative affinities of XPC and UV-DDB for a given lesion. It should be noted that XPC is always required for GG-NER, involving any type of lesion and that UV-DDB cannot substitute for XPC. Furthermore, 6-4PP as well as other chemical adducts are removed efficiently even in the absence of UV-DDB.[64]

UV-DDB is associated in vivo with ubiquitin ligase, which consists of cullin 4A, Roc1 and the COP9 signalosome.[65] Ubiquitin ligase bound to UV-DDB in vitro can polyubiquitinate not only XPC but also DDB2 and cullin 4A.[56,66] The polyubiquitination of DDB2 dramatically alters the damage recognition activities of UV-DDB, which we have proposed to be important for transferring lesions from UV-DDB to XPC and for the subsequent initiation of repair. On the other hand, the precise roles of polyubiquitin chains formed on XPC remain to be elucidated. Although DDB2 has been

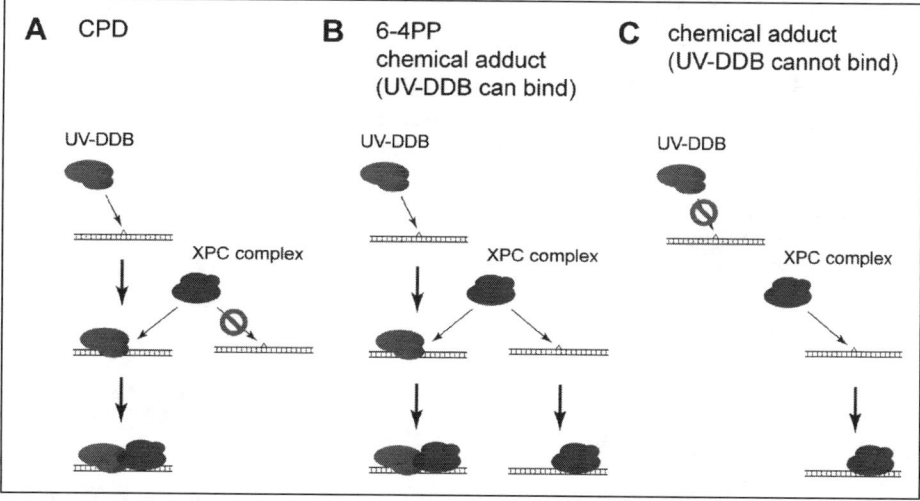

Figure 3. Diversity of damage recognition pathways for GG-NER. A) UV-induced CPDs are poorly recognized by XPC alone and the UV-DDB-dependent pathway has a predominant role in repairing this lesion. B) Since not only UV-DDB but also XPC efficiently recognize 6-4PPs, the two damage recognition pathways operate in parallel. This may also be the case for some helix-distorting chemical adducts that are recognized by UV-DDB. C) Among NER-type lesions, some chemical adducts (e.g., 4-nitroquinoline 1-oxide) are poorly recognized by UV-DDB[83]. Such lesions are repaired mainly through direct recognition by XPC.

reported to be degraded upon UV irradiation of cells,[67,68] ubiquitinated XPC appears to be protected from degradation by the proteasome and the Rad23p homologs may be important for this protection. Moreover, it was reported that XPC may also be sumoylated.[57] In addition to possible roles in mechanisms of GG-NER (e.g., regulation of interactions with other NER factors), these posttranslational modifications of XPC may be involved in specific signaling pathways for cellular damage responses.

Possible Functions of XPC beyond NER

In experiments with a yeast two-hybrid screening system, we discovered that XPC interacts with thymine DNA glycosylase (TDG).[69] TDG is a base excision repair (BER) enzyme that excises uracil and thymine residues from G/U- and G/T-mismatches that arise from the spontaneous deamination of cytosine and 5-methylcytosine, respectively.[70] Thus TDG is thought to contribute to the suppression of spontaneous mutagenesis. In addition to this physical interaction, XPC can stimulate the activity of TDG in vitro.[69] It has been documented that TDG and its *Escherichia coli* homolog, mismatch-specific uracil DNA glycosylase (MUG), remain tightly bound to the abasic sites that result from their activities.[71,72] XPC appears to relieve this product inhibition by facilitating dissociation from abasic sites and enzymatic turnover of TDG (Fig. 4). It has been

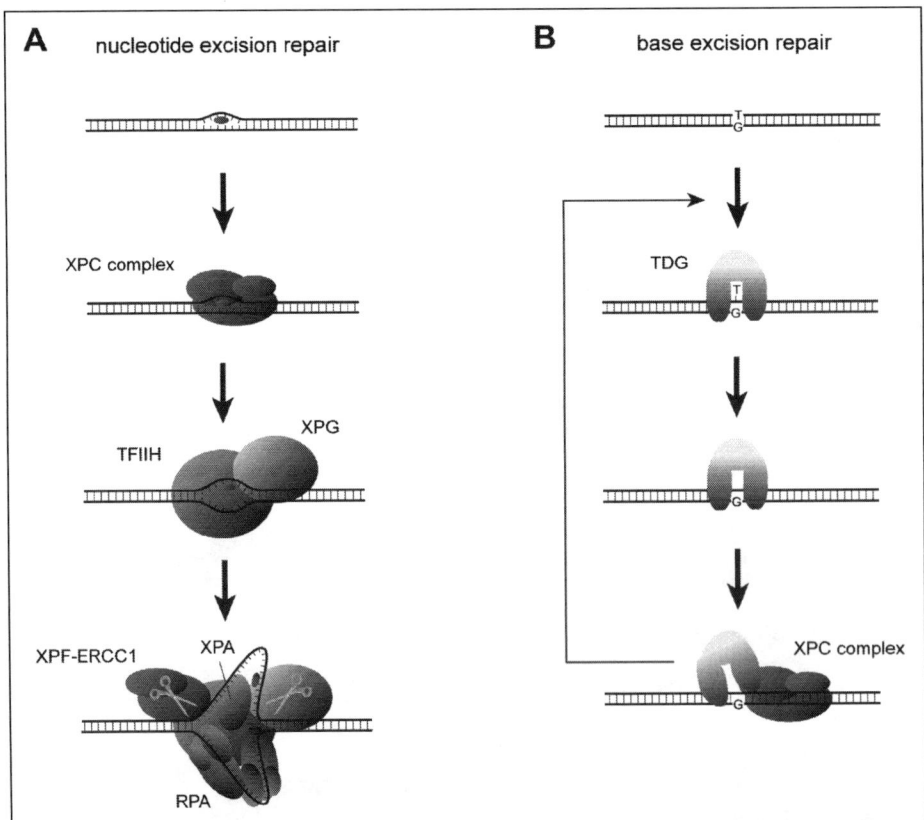

Figure 4. Possible dual functions of XPC in NER and BER. Besides its established function as a damage detector and an initiator of GG-NER (A), XPC may facilitate the enzymatic turnover of TDG (B). In XP-C patients, compromised TDG activity may result in an elevation of the spontaneous mutation frequency, which may further enhance the risk of skin tumors primarily caused by defects in GG-NER.

suggested that AP endonuclease1,[71] and/or sumoylation[73,74] may also be involved in promoting TDG turnover, although the functional relationships between these factors and XPC remain to be elucidated.

Extensive studies have been conducted to identify mutations in the p53 tumor suppressor gene and other genes in UV-induced skin tumors from XP-C patients[75] and *Xpc*[-/-] mice.[76-78] In addition to typical C to T and CC to TT transitions at dipyrimidinic sites, other types of mutations have also been found, for example the frequency of C to T transitions in a specific nondipyrimidinic CpG sequence (affecting the Thr122 residue) of the murine Trp53 gene is considerably increased in *Xpc*[-/-] mice, as compared to skin tumors induced in wild-type and *Xpc* heterozygous mice.[78,79] A possible explanation for this is that the loss of XPC functions may compromise TDG activity, thereby resulting in less efficient correction of deamination-induced G/U and G/T mismatches and a consequent elevation of spontaneous C to T transitions. Also it has been shown that, even in the absence of UV irradiation, *Xpc*[-/-] mice acquire age-dependent spontaneous mutations[80] that cannot be explained by reduced TDG activity and that XP-C fibroblasts may have defects in the repair of oxidative DNA damage.[81,82] These findings suggest that XPC may affect activities of other DNA glycosylases or repair systems in addition to TDG. Indeed it has been very recently reported that XPC stimulates hOGG1, which removes 8-oxoguanine, a major mutagenic oxidative base lesion[81]. Further studies would shed light on the possible composite mechanisms underlying tumorigenesis in XP-C patients.

References

1. Bootsma D, Kraemer KH, Cleaver JE et al. Nucleotide excision repair syndromes: xeroderma pigmentosum, Cockayne syndrome and trichothiodystrophy. In: Scriver CR, Beaudet AL, Sly WS, Valle D, eds. The Metabolic and Molecular Basis of Inherited Disease Vol. 1. New York: McGraw-Hill Book Co, 2001:677-703.
2. Friedberg EC, Walker GC, Siede W et al. DNA Repair and Mutagenesis, 2nd ed. Washington, DC: ASM Press, 2006.
3. Venema J, van Hoffen A, Karcagi V et al. Xeroderma pigmentosum complementation group C cells remove pyrimidine dimers selectively from the transcribed strand of active genes. Mol Cell Biol 1991; 11:4128-4134.
4. Legerski R, Peterson C. Expression cloning of a human DNA repair gene involved in Xeroderma pigmentosum group C. Nature 1992; 359:70-73.
5. Masutani C, Sugasawa K, Yanagisawa J et al. Purification and cloning of a nucleotide excision repair complex involving the Xeroderma pigmentosum group C protein and a human homolog of yeast RAD23. EMBO J 1994; 13:1831-1843.
6. Shivji MKK, Eker APM, Wood RD. DNA repair defect in Xeroderma pigmentosum group C and complementing factor from HeLa cells. J Biol Chem 1994; 269:22749-22757.
7. Araki M, Masutani C, Takemura M et al. Centrosome protein centrin 2/caltractin 1 is part of the xeroderma pigmentosum group C complex that initiates global genome nucleotide excision repair. J Biol Chem 2001; 276:18665-18672.
8. Kusumoto R, Masutani C, Sugasawa K et al. Diversity of the damage recognition step in the global genomic nucleotide excision repair in vitro. Mutat Res 2001; 485:219-227.
9. Sugasawa K, Okamoto T, Shimizu Y et al. A multistep damage recognition mechanism for global genomic nucleotide excision repair. Genes Dev 2001; 15:507-521.
10. Janicijevic A, Sugasawa K, Shimizu Y et al. DNA bending by the human damage recognition complex XPC-HR23B. DNA Repair (Amst) 2003; 2:325-336.
11. Sugasawa K, Ng JMY, Masutani C et al. Xeroderma pigmentosum group C protein complex is the initiator of global genome nucleotide excision repair. Mol Cell 1998; 2:223-232.
12. Riedl T, Hanaoka F, Egly J-M. The comings and goings of nucleotide excision repair factors on damaged DNA. EMBO J 2003; 22:5293-5303.
13. Volker M, Moné MJ, Karmakar P et al. Sequential assembly of the nucleotide excision repair factors in vivo. Mol Cell 2001; 8:213-224.
14. Tornaletti S, Hanawalt PC. Effect of DNA lesions on transcription elongation. Biochimie 1999; 81:139-146.
15. Sugasawa K, Shimizu Y, Iwai S et al. A molecular mechanism for DNA damage recognition by the xeroderma pigmentosum group C protein complex. DNA Repair (Amst) 2002; 1:95-107.
16. Kim J-K, Patel D, Choi B-S. Contrasting structural impacts induced by cis-syn cyclobutane dimer and (6-4) adduct in DNA duplex decamers: implication in mutagenesis and repair activity. Photochem Photobiol 1995; 62:44-50.

17. McAteer K, Jing Y, Kao J et al. Solution-state structure of a DNA dodecamer duplex containing a cis-syn thymine cyclobutane dimer, the major UV photoproduct of DNA. J Mol Biol 1998; 282:1013-1032.
18. Araújo SJ, Nigg EA, Wood RD. Strong functional interactions of TFIIH with XPC and XPG in human DNA nucleotide excision repair, without a preassembled repairosome. Mol Cell Biol 2001; 21:2281-2291.
19. Yokoi M, Masutani C, Maekawa T et al. The Xeroderma pigmentosum group C protein complex XPC-HR23B plays an important role in the recruitment of transcription factor IIH to damaged DNA. J Biol Chem 2000; 275:9870-9875.
20. Li R-Y, Calsou P, Jones CJ et al. Interactions of the transcription/DNA repair factor TFIIH and XP repair proteins with DNA lesions in a cell-free repair assay. J Mol Biol 1998; 281:211-218.
21. Evans E, Moggs JG, Hwang JR et al. Mechanism of open complex and dual incision formation by human nucleotide excision repair factors. EMBO J 1997; 16:6559-6573.
22. Mu D, Wakasugi M, Hsu DS et al. Characterization of reaction intermediates of human excision repair nuclease. J Biol Chem 1997; 272:28971-28979.
23. Nishi R, Okuda Y, Watanabe E et al. Centrin 2 stimulates nucleotide excision repair by interacting with xeroderma pigmentosum group C protein. Mol Cell Biol 2005; 25:5664-5674.
24. Popescu A, Miron S, Blouquit Y et al. Xeroderma pigmentosum group C protein possesses a high affinity binding site to human centrin 2 and calmodulin. J Biol Chem 2003; 278:40252-40261.
25. Uchida A, Sugasawa K, Masutani C et al. The carboxy-terminal domain of the XPC protein plays a crucial role in nucleotide excision repair through interactions with transcription factor IIH. DNA Repair (Amst) 2002; 1:449-461.
26. Chavanne F, Broughton BC, Pietra D et al. Mutations in the XPC gene in families with xeroderma pigmentosum and consequences at the cell, protein and transcript levels. Cancer Res 2000; 60:1974-1982.
27. Emmert S, Wetzig T, Imoto K et al. A Novel Complex Insertion/Deletion Mutation in the XPC DNA Repair Gene Leads to Skin Cancer in an Iraqi Family. J Invest Dermatol 2006; 126:2542-2544.
28. Khan SG, Levy HL, Legerski R et al. Xeroderma pigmentosum group C splice mutation associated with autism and hypoglycinemia. J Invest Dermatol 1998; 111:791-796.
29. Khan SG, Metin A, Gozukara E et al. Two essential splice lariat branchpoint sequences in one intron in a Xeroderma pigmentosum DNA repair gene: mutations result in reduced XPC mRNA levels that correlate with cancer risk. Hum Mol Genet 2004; 13:343-352.
30. Khan SG, Oh KS, Shahlavi T et al. Reduced XPC DNA repair gene mRNA levels in clinically normal parents of Xeroderma pigmentosum patients. Carcinogenesis 2006; 27:84-94.
31. Lam CW, Cheung KK, Luk NM et al. DNA-based diagnosis of xeroderma pigmentosum group C by Whole-genome scan using single-nucleotide polymorphism microarray. J Invest Dermatol 2005; 124:87-91.
32. Li L, Bales ES, Peterson CA et al. Characterization of molecular defects in Xeroderma pigmentosum group C. Nat Genet 1993; 5:413-417.
33. Rivera-Begeman A, McDaniel LD, Schultz RA et al. A novel XPC pathogenic variant detected in archival material from a patient diagnosed with Xeroderma pigmentosum: A case report and review of the genetic variants reported in XPC. DNA Repair (Amst) 2006; (doi:10.1016/j.dnarep.2006.09.008).
34. Okuda Y, Nishi R, Ng JMY et al. Relative levels of the two mammalian Rad23 homologs determine composition and stability of the xeroderma pigmentosum group C protein complex. DNA Repair (Amst) 2004; 3:1285-1295.
35. Sugasawa K, Ng JMY, Masutani C et al. Two human homologs of Rad23 are functionally interchangeable in complex formation and stimulation of XPC repair activity. Mol Cell Biol 1997; 17:6924-6931.
36. Sugasawa K, Masutani C, Uchida A et al. HHR23B, a human Rad23 homolog, stimulates XPC protein in nucleotide excision repair in vitro. Mol Cell Biol 1996; 16:4852-4861.
37. van der Spek PJ, Eker A, Rademakers S et al. XPC and human homologs of RAD23: intracellular localization and relationship to other nucleotide excision repair complexes. Nucleic Acids Res 1996; 24:2551-2559.
38. Guzder SN, Habraken Y, Sung P et al. Reconstitution of yeast nucleotide excision repair with purified Rad proteins, replication protein A and transcription factor TFIIH. J Biol Chem 1995; 270:12973-12976.
39. Masutani C, Araki M, Sugasawa K et al. Identification and characterization of XPC-binding domain of hHR23B. Mol Cell Biol 1997; 17:6915-6923.
40. van der Spek PJ, Visser CE, Hanaoka F et al. Cloning, comparative mapping and RNA expression of the mouse homologues of the Saccharomyces cerevisiae nucleotide excision repair gene RAD23. Genomics 1996; 31:20-27.

41. Hiyama H, Yokoi M, Masutani C et al. Interaction of hHR23 with S5a. The ubiquitin-like domain of hHR23 mediates interaction with S5a subunit of 26 S proteasome. J Biol Chem 1999; 274:28019-28025.

42. Schauber C, Chen L, Tongaonkar P et al. Rad23 links DNA repair to the ubiquitin/proteasome pathway. Nature 1998; 391:715-718.

43. Bertolaet BL, Clarke DJ, Wolff M et al. UBA domains of DNA damage-inducible proteins interact with ubiquitin. Nat Struct Biol 2001; 8:417-422.

44. Chen L, Shinde U, Ortolan TG et al. Ubiquitin-associated (UBA) domains in Rad23 bind ubiquitin and promote inhibition of multi-ubiquitin chain assembly. EMBO Rep 2001; 2:933-938.

45. Raasi S, Pickart CM. Rad23 ubiquitin-associated domains (UBA) inhibit 26 S proteasome-catalyzed proteolysis by sequestering lysine 48-linked polyubiquitin chains. J Biol Chem 2003; 278:8951-8959.

46. Wilkinson CR, Seeger M, Hartmann-Petersen R et al. Proteins containing the UBA domain are able to bind to multi-ubiquitin chains. Nat Cell Biol 2001; 3:939-943.

47. Chen L, Madura K. Rad23 promotes the targeting of proteolytic substrates to the proteasome. Mol Cell Biol 2002; 22:4902-4913.

48. Ng JMY, Vermeulen W, van der Horst GTJ et al. A novel regulation mechanism of DNA repair by damage-induced and RAD23-dependent stabilization of Xeroderma pigmentosum group C protein. Genes Dev 2003; 17:1630-1645.

49. Ng JMY, Vrieling H, Sugasawa K et al. Developmental defects and male sterility in mice lacking the ubiquitin-like DNA repair gene mHR23B. Mol Cell Biol 2002; 22(4):1233-1245.

50. Gillette TG, Yu S, Zhou Z et al. Distinct functions of the ubiquitin-proteasome pathway influence nucleotide excision repair. EMBO J 2006; 25:2529-2538.

51. Salisbury JL, Suino KM, Busby R et al. Centrin-2 is required for centriole duplication in mammalian cells. Curr Biol 2002; 12:1287-1292.

52. Paoletti A, Moudjou M, Paintrand M et al. Most of centrin in animal cells is not centrosome-associated and centrosomal centrin is confined to the distal lumen of centrioles. J Cell Sci 1996; 109:3089-3102.

53. Molinier J, Ramos C, Fritsch O et al. CENTRIN2 modulates homologous recombination and nucleotide excision repair in Arabidopsis. Plant Cell 2004; 16:1633-1643.

54. Araújo SJ, Tirode F, Coin F et al. Nucleotide excision repair of DNA with recombinant human proteins: definition of the minimal set of factors, active forms of TFIIH and modulation by CAK. Genes Dev 2000; 14:349-359.

55. Mu D, Park CH, Matsunaga T et al. Reconstitution of human DNA repair excision nuclease in a highly defined system. J Biol Chem 1995; 270:2415-2418.

56. Sugasawa K, Okuda Y, Saijo M et al. UV-induced ubiquitylation of XPC protein mediated by UV-DDB-ubiquitin ligase complex. Cell 2005; 121:387-400.

57. Wang QE, Zhu Q, Wani G et al. DNA repair factor XPC is modified by SUMO-1 and ubiquitin following UV irradiation. Nucleic Acids Res 2005; 33:4023-4034.

58. Batty D, Rapic'-Otrin V, Levine AS et al. Stable binding of human XPC complex to irradiated DNA confers strong discrimination for damaged sites. J Mol Biol 2000; 300:275-290.

59. Wittschieben BØ, Iwai S, Wood RD. DDB1-DDB2 (Xeroderma pigmentosum group E) protein complex recognizes a cyclobutane pyrimidine dimer, mismatches, apurinic/apyrimidinic sites and compound lesions in DNA. J Biol Chem 2005; 280:39982-39989.

60. Fujiwara Y, Masutani C, Mizukoshi T et al. Characterization of DNA recognition by the human UV-damaged DNA-binding protein. J Biol Chem 1999; 274:20027-20033.

61. Reardon JT, Nichols AF, Keeney S et al. Comparative analysis of binding of human damaged DNA-binding protein (XPE) and Escherichia coli damage recognition protein (UvrA) to the major ultraviolet photoproducts: T[c,s]T, T[t,s]T, T[6-4]T and T[Dewar]T. J Biol Chem 1993; 268:21301-21308.

62. Fitch ME, Nakajima S, Yasui A et al. In vivo recruitment of XPC to UV-induced cyclobutane pyrimidine dimers by the DDB2 gene product. J Biol Chem 2003; 278:46906-46910.

63. Moser J, Volker M, Kool H et al. The UV-damaged DNA binding protein mediates efficient targeting of the nucleotide excision repair complex to UV-induced photo lesions. DNA Repair (Amst) 2005; 4:571-582.

64. Hwang BJ, Ford JM, Hanawalt PC et al. Expression of the p48 Xeroderma pigmentosum gene is p53 dependent and is involved in global genome repair. Proc Natl Acad Sci USA 1999; 96:424-428.

65. Groisman R, Polanowska J, Kuraoka I et al. The ubiquitin ligase activity in the DDB2 and CSA complexes is differentially regulated by the COP9 signalosome in response to DNA damage. Cell 2003; 113:357-367.

66. Matsuda N, Azuma K, Saijo M et al. DDB2, the Xeroderma pigmentosum group E gene product, is directly ubiquitylated by Cullin 4A-based ubiquitin ligase complex. DNA Repair (Amst) 2005; 4:537-545.
67. Fitch ME, Cross IV, Turner SJ et al. The DDB2 nucleotide excision repair gene product p48 enhances global genomic repair in p53 deficient human fibroblasts. DNA Repair (Amst) 2003; 2:819-826.
68. Rapic'-Otrin V, McLenigan MP, Bisi DC et al. Sequential binding of UV DNA damage binding factor and degradation of the p48 subunit as early events after UV irradiation. Nucleic Acids Res 2002; 30:2588-2598.
69. Shimizu Y, Iwai S, Hanaoka F et al. Xeroderma pigmentosum group C protein interacts physically and functionally with thymine DNA glycosylase. EMBO J 2003; 22:164-173.
70. Hardeland U, Bentele M, Lettieri T et al. Thymine DNA glycosylase. Prog Nucleic Acid Res Mol Biol 2001; 68:235-253.
71. Waters TR, Gallinari P, Jiricny J et al. Human thymine DNA glycosylase binds to apurinic sites in DNA but is displaced by human apurinic endonuclease 1. J Biol Chem 1999; 274:67-74.
72. Waters TR, Swann PF. Kinetics of the action of thymine DNA glycosylase. J Biol Chem 1998; 273:20007-20014.
73. Baba D, Maita N, Jee JG et al. Crystal structure of thymine DNA glycosylase conjugated to SUMO-1. Nature 2005; 435:979-982.
74. Steinacher R, Schär P. Functionality of human thymine DNA glycosylase requires SUMO-regulated changes in protein conformation. Curr Biol 2005; 15:616-623.
75. Giglia G, Dumaz N, Drougard C et al. p53 mutations in skin and internal tumors of xeroderma pigmentosum patients belonging to the complementation group C. Cancer Res 1998; 58:4402-4409.
76. Ikehata H, Saito Y, Yanase F et al. Frequent recovery of triplet mutations in UVB-exposed skin epidermis of Xpc-knockout mice. DNA Repair (Amst) 2006; (doi:10.1016/j.dnarep.2006.09.003).
77. Nahari D, McDaniel LD, Task LB et al. Mutations in the Trp53 gene of UV-irradiated Xpc mutant mice suggest a novel Xpc-dependent DNA repair process. DNA Repair (Amst) 2004; 3:379-386.
78. Reis AM, Cheo DL, Meira LB et al. Genotype-specific Trp53 mutational analysis in ultraviolet B radiation-induced skin cancers in Xpc and Xpc Trp53 mutant mice. Cancer Res 2000; 60:1571-1579.
79. Inga A, Nahari D, VelascoMiguel S et al. A novel p53 mutational hotspot in skin tumors from UV-irradiated Xpc mutant mice alters transactivation functions. Oncogene 2002; 21:5704-5715.
80. Wijnhoven SWP, Kool HJM, Mullenders LHF et al. Age-dependent spontaneous mutagenesis in Xpc mice defective in nucleotide excision repair. Oncogene 2000; 19:5034-5037.
81. D'Errico M, Parlanti E, Teson M et al. New functions of XPC in the protection of human skin cells from oxidative damage. EMBO J 2006; 25:4305-4315.
82. Rünger TM, Epe B, Möller K. Repair of ultraviolet B and singlet oxygen-induced DNA damage in xeroderma pigmentosum cells. J Invest Dermatol 1995; 104:68-73.
83. Payne A, Chu G. Xeroderma pigmentosum group E binding factor recognizes a broad spectrum of DNA damage. Mutat Res 1994; 310:89-102.

CHAPTER 7

The XPE Gene of Xeroderma Pigmentosum, Its Product and Biological Roles

Drew Bennett and Toshiki Itoh*

Discovery and Background

Xeroderma Pigmentosum (XP) is an inheritable genetic disorder in which patients become very sensitive to ultraviolet (UV) light exposure and prone to skin cancer. Its genetics are complex and multiallelic. Based on complementation studies, involving UV sensitivity of fused cells, initially XP was classified in 5 subgroups, XP-A to XP-E. Present studies, however, have discovered that there are at least 8 subgroups, XP-A to XP-G and XPV. Studies of these genes have shown that their products play critical roles in nucleotide excision repair (NER) following DNA damage. XP-E subgroup has been considered to be of the mildest XP forms and their exact roles in NER are not yet clear.

From human placenta Feldberg and Grossman (1976) purified a damage-specific DNA binding protein (DDB) which showed the binding ability to UV irradiated DNA. In the same study they showed that this protein did not show any exo- or endonuclease, polymerase or N-glycosidase activity but attached specifically to double stranded irradiated DNA. It did not bind to DNA with pyrimidine dimers suggesting that this protein recognized photoproducts else than pyrimidine dimers. Analysis of XP group A (XP-A), B (XP-B) and C (XP-C) deficient cell lines (only these cell lines were then available) showed that none of these have deficiency in DDB protein.[1]

DDB could not be associated with any XP groups until Chu and Chang (1988) reported that some (and not all) cases of XP-E cell lines were deficient for this protein. Using a broad range of UV-irradiation intensities, the DDB activity; i.e., a specific band shift by gel shift assay, was found in cell extracts to bind to UV-irradiated DNA over most of the range. In addition to UV-irradiation, DNA adducts generated by cisplatin also markedly increased DDB binding. XP-E XP2RO cells did not show wild-type binding activity to UV-irradiated DNA; hence DDB became associated with XP-E pathology. Furthermore, the source of XP-E pathology was found not to be due to a dominant negative factor because complementation with other XP groups to XP-E cells corrected the phenotype and restored damaged DNA binding.[2]

However, other groups failed to identify defects in DDB binding in other XP-E cells.[3,4] Because the defining characteristic of DDB activity was the ability to bind to damaged DNA, understanding more about its function in the cell would help to determine its relationship with XP-E. After purifying DDB from cell extracts using the HeLa human cervical cancer cell line, it was determined that the protein is a heterodimer. It was originally reported as being composed of 124 kDa and 41 kDa polypeptides, which are now referred to as p127^{DDB1} and p48^{DDB2}.[5]

*Corresponding Author: Toshiki Itoh—Department of Pathology, The University of Iowa, Roy J. & Lucille A. Carver College of Medicine, 200 Hawkins Dr., Iowa City, IA 52242, USA. Email: toshiki-itoh@uiowa.edu

Molecular Mechanisms of Xeroderma Pigmentosum, edited by Shamim I. Ahmad and Fumio Hanaoka. ©2008 Landes Bioscience and Springer Science+Business Media.

A critical step forward in XP-E and DDB research was achieved by the cloning of the human DDB subunits. The two subunits were named DDB1 (GeneBank U18299) and DDB2 (GeneBank U18300). Fluorescent in situ hybridization experiment revealed that both subunits are coded by genes are chromosome 11, at 11q12-q13 for DDB1 and 11p11-p12 for DDB2. DDB1 and DDB2 are the only XP genes that are found on chromosome 11.[6] For the DDB2 gene, the presence of WD40 repeats in its sequence suggests that it may be involved in a variety of cellular functions, as other WD40 repeat genes show signal transduction, cell cycle regulation and RNA splicing, transcription and apoptosis.[7] In contrast, DDB1 does not show any sequence homology with any other known, human gene.

Previous studies of XP-E raised the possibility that mutations leading to defect in more than one protein or in different protein domains would lead to the difference in DDB defectiveness in XP-E cells, the sequences of the two DDB genes allowed for analysis of the known XP-E cells in order to determine the link between XP-E and DDB. As the previous data suggest, the XP-E cells, XP2RO and XP82TO, showed mutations in DDB2 while other XP-E cells, which were classified as "Ddb+" XP-E cells were nonmutant. Likewise, there were no XP-E cells which showed mutations in DDB1, implicating DDB2 as the primary site of mutation in the "Ddb-" XP-E cells.[8,9] Finally, additional work by Itoh et al (2000) firmly established that the cause of the XP-E pathology was due to DDB2 mutations alone. The confusion over what role DDB2 mutations had on the XP-E phenotype was clarified by the reclassification of the "Ddb+" XP-E cells into XP-F, XP variant and ultraviolet-sensitive syndrome by genome sequencing and a more refined complementation analysis. Therefore, the correlation between DDB2 mutations and XP-E pathology was confirmed.[10] Currently, there are 8 identified cases of XP-E, all of which possess mutations in the DDB2 gene.[11]

Expression and Regulation of DDB Protein

As a result of DNA damage, p53 becomes phosphorylated and resistant to MDM2-mediated degradation. Then, p53 is available to activate transcription of genes involved in cell cycle arrest, repair and apoptosis. DDB2 RNA and protein have also been found to be upregulated by p53 in response to UV-irradiation. Furthermore, p53 can upregulate the expression of DDB2 outside of UV- and ionizing radiation.[12] Interestingly, DDB2 and p53 seem to show reciprocal regulation. In XP-E cells, DDB2 RNA expression is upregulated by p53 binding to its consensus binding site (CBS) in intron 4 of DDB2 while p53 levels appear to be significantly reduced in XP-E cells. Specifically, a maximum induction of p53 protein was found between approximately 9 and 12 hours after UV-irradiation in normal skin cells. Under the same conditions, DDB2 is induced maximally after 48 hours which implicates p53 as being responsible for its activation. This hypothesis was confirmed by mutation of the intron 4 CBS which did not show an increase in DDB2 expression after UV-irradiation.[13]

The reduced basal levels of p53 (and its downstream target proteins) reached to normal levels only when DDB2 containing the intron 4 p53 CBS was included. This regulation is even more complex because p53 appears to have negative impact on DDB2 levels when the intron 4 p53 CBS is retained. However, the most important aspect of p53 regulation of DDB2 is after UV-irradiation; the XP-E phenotype can be corrected by addition of DDB2 containing intron 4 with the p53 CBS. The correction of the XP-E phenotype was displayed by the ability of the DDB2 construct including intron 4 to restore sensitivity to UV-irradiation by apoptosis.[13] However, the changes in the protein and RNA levels of DDB2 after UV-irradiation do not follow the same time course, implicating posttranscriptional control which has yet to be elucidated.

Exact biochemical function of DDB2 in the overall process of DNA repair is still in the theoretical stage although (i) cell free extracts from DDB deficient cells have normal nucleotide excision repair,[14] (ii) expression levels of DDB2 after UV-irradiation reach their maximum (48 hours) at a time when most NER is complete,[9] and (iii) in vivo DNA repair synthesis is within normal levels in XP-E cells.[9-11] However, XP-E cells are resistant to UV-induced apoptosis, which implies a defect in the apoptotic signal transduction pathway.

Mouse Model

In agreement with cell culture data, XP-E model mice (DDB2-/-) show enhanced skin tumor incidence after UV-exposure over wild-type mice (Figs. 1, 2) but did not develop skin tumors in response to DMBA-exposure. Since other XP model mice are prone to DMBA-induced skin cancer, presumably due to a repair defect of DMBA-induced DNA adducts, DDB2 may have functions beyond NER. The DDB2-/- mice showed relatively normal phenotypes excluding UV-induced skin cancers which give credence to their use as an XP-E model animal. Even DDB2 heterozygous knock-out mice (DDB+/-) showed increased skin tumor incidence in response to UV-irradiation (Fig. 2). Taken together, the DDB2 mouse model has proven useful subjects in understanding the nature of DDB2 in XP-E. Although these results clarify the link between UV sensitivity and DDB2, critical questions remain unanswered about the exact biochemical functions of DDB2 and its role in oncogenesis. Future work is required for understanding the exact role(s) of DDB2 in normal human cells.

Protein Interactions

Either DDB1 or DDB2 individually or the DDB complex have been shown to interact with a variety of proteins. DDB and DDB2 were found to interact with the activation domain of E2F1 and stimulate E2F1-activated transcription.[16] This interaction is interesting because E2F1 is involved in the transcription of genes necessary for DNA replication and cell division. If DDB is directed toward damaged DNA after UV-irradiation, it may no longer be available for E2F1 activation, thereby down-regulating proliferative gene expression.

A number of authors have reported that DDB becomes associated with the cullin 4A (CUL-4A) ubiquitin ligase. It was believed that DDB is a target of CUL-4A since CUL-4A has been found to be upregulated in breast cancers.[17] Over-expression of CUL-4A leads to a decrease in DDB2 while proteasome inhibition increases the level of DDB2.[18] Also, there are XP-E mutations which make DDB2 resistant to CUL-4A ubiquitination. Because most DDB2 mutant proteins in XP-E cells remains undetectable, it is assumed to be degraded,[13] the significance of DDB2 ubiquitination by CUL4A remains unclear. Since CUL-4A is up-regulated in many tumors, its effect on DDB2

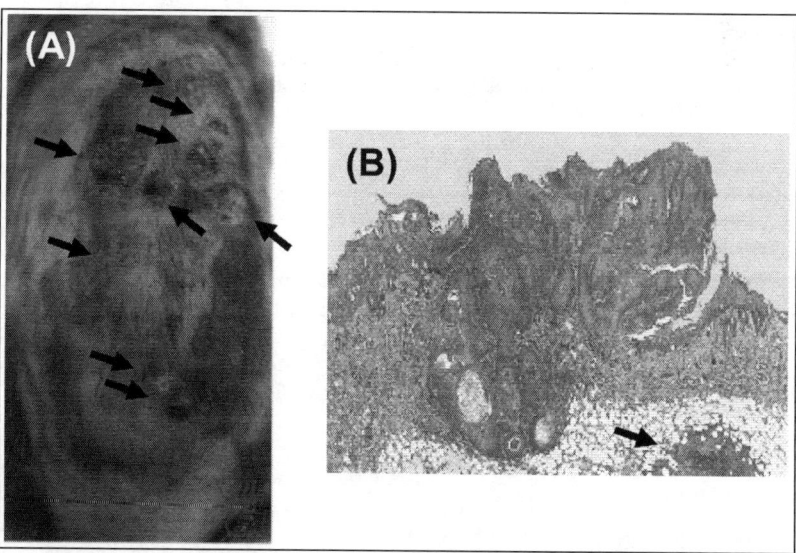

Figure 1. Tumor analysis of DDB2-/- mice after UV-irradiation.[15] A) Arrows indicate tumors appeared on the back skin. B) Arrow points to invasion of atypical squamous cells and mitotic figures. Magnification: X40.

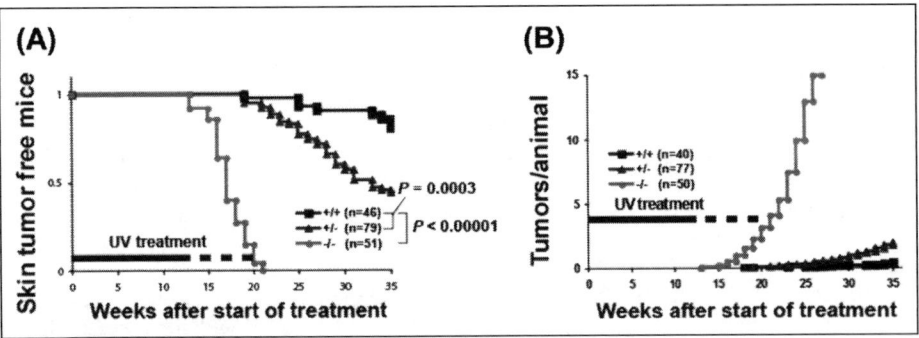

Figure 2. Kaplan-Meier curves (A) and Tumor numbers (B) of UV-treated DDB2 +/+, +/-, and -/- mice.[15] Solid lines indicate exposure to all mice and dotted lines indicate exposure to tumor-free mice only.

may be partly responsible for impaired function of DDB2 in those tumors.[20] Furthermore, some evidence has expanded upon the upstream regulation of DDB2 degradation through CUL-4A by focusing on the actions of c-Abl. The ubiquitination of DDB2 by CUL-4A can be inhibited by CAND1. c-Abl antagonizes CAND1 independently by a kinase activity of c-Abl and leads to degradation of DDB2 through CUL-4A.[21] Current models of NER function divide the pathway into two groups: global genome repair (GGR) and transcription-coupled repair. GGR refers to nontranscribed DNA repair, which is typically less efficient than repair of damaged genes that are actively transcribed. One theory for this observation is that transcribed genes are not condensed into chromatin and are more accessible to the DNA repair machinery. DDB has also been implicated in histone ubiquitination in complex with CUL-4A and associated proteins like Roc1.[22] A possible explanation for the involvement of DDB in histone modification is that after UV-induced damage to DNA, repair enzymes would be able to gain access to condensed chromatin in order to carry out GGR.

It is interesting to note that DDB1 and DDB2 have been found to bind to the hepatitis B viral protein X.[23] The HBV X (HBx) gene encodes a small multi-functional protein which is necessary for the HBV life cycle that trans-activates transcription, stimulates apoptosis and inhibits p53. It is believed that HBx can bind to both DDB1 and DDB2 and that DDB2 can facilitate the nuclear localization of HBx and so contribute to the promotion of HBV life cycle and subsequent pathology. An interesting effect of this binding relationship is that if DDB2 is over-expressed, it competes with HBx for binding to DDB1.[23] In addition, there have been other proteins that have been implicated in binding to DDB which include CBP/p300 and the COP9 signalosome. Recent work implies that DDB may have the potential to associate with CBP or p300 and bind to damaged chromatin. Then, the CBP/p300 would allow other repair enzymes to have access to the damaged site by acetylating the surrounding histones.[24] Furthermore, the data that DDB associates with CUL-4A, the COP9 signalosome has been found to bind to CUL-4A and DDB2 after DNA damage to form a COP9 signalosome complex. This complex is then thought to bind to chromatin and ubiquitinate histones, allowing repair proteins access to damaged sites.[25]

DNA Binding of the DDB Complex

More recent works have shown that DDB can bind to a wide variety of DNA damage with a high affinity to UV-damaged DNA; it can bind to (6-4) photoproducts as well as trans, syn-cyclobutane pyrimidine dimers (CPDs) and T[Dewar]T photoproducts, as well as with low affinity to cis, syn-CPDs.[26] DDB is also able to bind to DNA damaged by a wide variety of damaging agents including cisplatin, 8-methoxypsoralen and cis-diamminedichloroplatinum(II) adducts, N-methyl-N'-nitro-N-nitrosoguanidine and nitrogen mustard (HN^2). It can also bind to single-strand DNA, abasic sites, depurination and base-pair mismatches.[26,28] However, if DDB and/or p48[DDB2] play the same role when binding to all of the various forms of DNA damage that it has affinity to, then it remains to be clearly worked out. Since DDB2 is ubiquitously expressed in DNA damaged cells, it is quite possible that there are still other forms of DNA damage that DDB can recognize and may possess other unknown functions.

Current Models of DDB Function

Sugasawa and colleagues have shown that the XPC protein complex requires functional DDB in order to be ubiquitinated in response to UV-irradiation (Fig. 3).[29] In the absence of UV-irradiation, DDB is not found in complex with the COP9 signalosome and CUL4A. However, after UV-irradiation, the proteins come together to form a functional complex, implying that the ubiquitin ligase activity of the complex is only present after UV-irradiation. A model of DDB function in the context of this complex includes the possibility that this DDB complex has such a high affinity for some forms of damaged DNA, like 6-4 photoproducts, that it must be ubiquitinated and removed before other repair proteins can gain access to the site. One such repair protein is XPC, which also associates with the CUL4A complex after UV-irradiation. In this model, DDB loses its affinity for the DNA lesion when it becomes ubiquitinated while XPC gains affinity after ubiquitination, thereby gaining access to the damaged site.[29] These data are supported by the fact that polyubiquitinated DDB posses a lower affinity for 6-4 photoproducts than unmodified DDB.

Another model was proposed by Wang and colleagues to account for the actual biochemical function of the DDB-CUL4 complex (Fig. 4).[22] However, instead of DDB being replaced by XPC at the damaged site, the DDB complex facilitates the release of the histone associated with the damaged DNA by ubiquitinating it. Then XPC (which has a higher affinity for naked DNA than nucleosomes) along with HR23B would be able to gain access to the damaged site and carry on with the repair process.

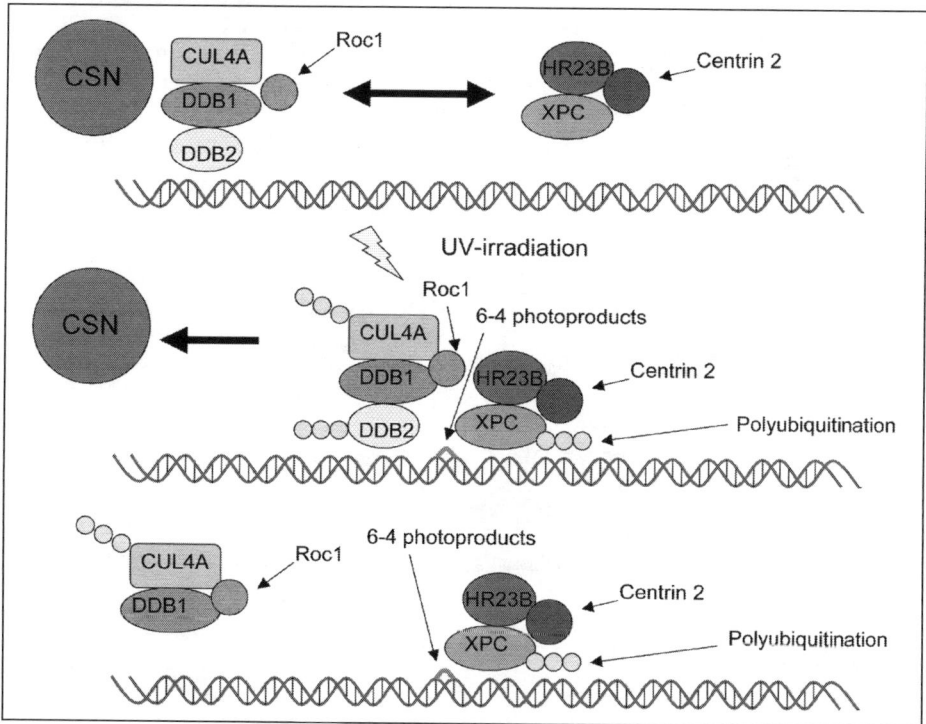

Figure 3. A model of DDB complex function in the NER pathway.[29]

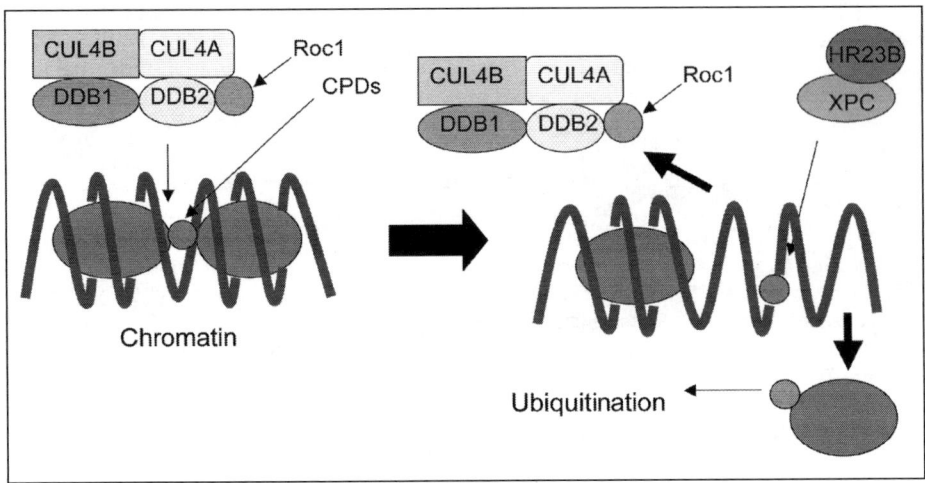

Figure 4. A model of DDB complex function in the NER pathway.[22]

As the previous sections have implied, DDB2 regulation and function in the cell is quite complicated. Therefore, a model which attempts to address some of the incongruous observations surrounding DDB2 function after UV-irradiation was proposed by Kulaksiz et al (Fig. 5).[14] This model, which differs significantly from the previous ones, proposes that DDB2 is not directly involved in NER but in cell cycle regulation, transcription and apoptosis. This model is supported from previous work in that DDB2 has been implicated in the E2F1 transcriptional activation, p53 activation and ubiquitination of a number of proteins in complex with CUL4A and/or the COP9 signalosome. Also, this model does take into account the cell extract studies in which p48DDB2 is excluded from can still carry out in vitro repair of DNA damage as well as XP-E cells being able to carry out DNA repair at near normal levels. This model posits p48DDB2 at the center

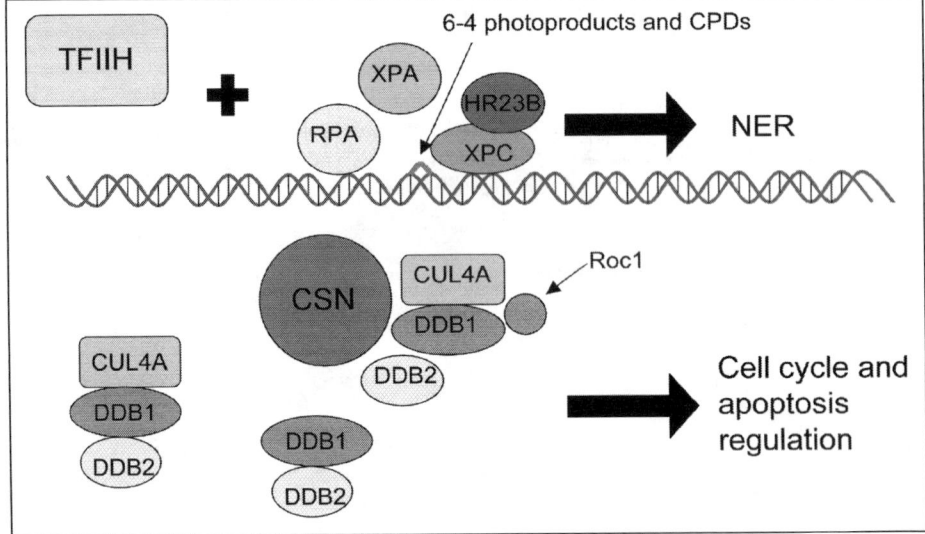

Figure 5. A model of DDB complex function in the NER pathway.[14]

of a variety of cellular activities after UV-irradiation but not as a direct component of the NER pathway. In this model, the cell cycle or apoptotic functions of DDB2 would explain the tumor suppressive effects of DDB2.

Conclusion

Significance advancement has been made in the research and characterization of XP-E and its gene product, p48^{DDB2}. It is now accepted that in all cases XP-E is attributable to mutations in the DDB2 gene. Both, the known human cases and the recent mouse model of DDB2 deficiency, demonstrate that DDB2 has functions as a tumor suppressor in skin cancer. Studies of other tumor types suggest the pathway may be altered in other tumor types. There has been considerable advancement in the understanding of DDB2 transcriptional regulation and protein interactions. However, many questions yet remain unanswered regarding the precise function of DDB2. The exact biochemical function(s) of DDB2 in the context of its various protein complexes require additional research. Continuing research into DDB2 function will provide a better understanding of the pathways that may contribute to developing new treatments and therapies for patients with XP-E and perhaps more generally to other cancer patients with altered DDB2 activity.

Acknowledgements

We are grateful to Drs. Shamim Ahmad and Fumio Hanaoka for giving us an opportunity to write this chapter advancing knowledge of XP-E and we also thank Drs. Sachiyo Iwashita and C. Michael Knudson for their valuable contribution to the manuscript.

References

1. Feldberg RS, Grossman L. A DNA binding protein from human placenta specific for ultraviolet damaged DNA. Biochemistry 1976; 15:2402-2408.
2. Chu G, Chang E. Xeroderma Pigmentosum Group E cells lack a nuclear factor that binds to damaged DNA. Science 1988; 242:564-568.
3. Kataoka H, Fujiwara Y. UV damage-specific DNA-binding protein in xeroderma pigmentosum complementation group E Biochem Biophys Res Commun 1991; 175:1139-1143.
4. Keeney S, Wein H, Linn S. Biochemical heterogeneity in xeroderma pigmentosum complementation group E. Mutat Res 1992; 273:49-56.
5. Keeney S, Chang GJ, Linn S. Characterization of a human DNA damage binding protein implicated in xeroderma pigmentosum E J Biol Chem 1993; 268:21293-21300.
6. Dualan R, Brody T, Keeney S et al. Chromosomal localization and cDNA cloning of the genes (DDB1 and DDB2) for the p127 and p48 subunits of a human damage-specific DNA binding protein. Genomics 1995; 29:62-69.
7. Saeki M, Irie Y, Ni L et al. Monad, a WD40 repeat protein, promotes apoptosis induced by TNF-alpha. Biochem Biophys Res Comm 2006; 342:568-572.
8. Nichols AF, Ong P, Linn S. Mutations specific to the xeroderma pigmentosum group E Ddb- phenotype. J Biol Chem 1996; 271:24317-24320.
9. Nichols AF, Itoh T, Graham JA et al. Human damage-specific DNA-binding protein p48. J Biol Chem 2000; 275:21422-21428.
10. Itoh T, Linn S, Ono T et al. Reinvestigation of the classification of five cell strains of xeroderma pigmentosum group E with reclassification of three of them. J Invest Dermatol 2000; 114:1022-1029.
11. Itoh T. Xeroderma pigmentosum group E and DDB2, a smaller subunit of damage-specific DNA binding protein: proposed classification of xeroderma pigmentosum, Cockayne syndrome and ultra-violet-sensitive syndrome. J Dermatol Sci 2006; 41:87-96.
12. Hwang BJ, Ford JM, Hanawalt PC et al. Expression of the p48 xeroderma pigmentosum gene is p53-dependent and is involved in global genomic repair. Proc Natl Acad Sci USA 1999; 96:424-428.
13. Itoh T, O'Shea C, Linn S. Impaired regulation of tumor suppressor p53 caused by mutations in the xeroderma pigmentosum DDB2 gene: mutual regulatory interactions between p48^{DDB2} and p53. Mol Cell Biol 2003; 23:7540-7553.
14. Kulaksiz G, Reardon JT, Sancar A. Xeroderma pigmentosum complementation group E protein (XPE/DDB2): purification of various complexes of XPE and analyses of their damaged DNA binding and putative DNA Repair Properties. Mol Cell Biol 2005; 25:9784-9792.
15. Itoh T, Cado D, Kamide R et al. DDB2 gene disruption leads to skin tumors and resistance to apoptosis after exposure to ultraviolet light but not a chemical carcinogen. Proc Natl Acad Sci USA 2004; 101:2052-2057.

16. Hayes S, Shiyanov P, Chen X et al. DDB, a putative DNA repair protein, can function as a transcriptional partner of E2F1. Mol Cell Biol 1998; 18:240-249.
17. Shiyanov P, Nag A, Raychaudhuri P. Cullin 4A Associates with the UV-damaged DNA-binding Protein DDB. J Biol Chem 1999; 274:35309-35312.
18. Nag A, Bondar T, Shiv S et al. The xeroderma pigmentosum group E gene product DDB2 is a specific target of Cullin 4A in mammaliam cells. Mol Cell Biol 2001; 21:6738-6747.
19. Matsuda N, Azuma K, Saijo M et al. DDB2, the xeroderma pigmentosum group E gene product, is directly ubiquitylated by Cullin 4A-based ubiquitin ligase complex. DNA Repair 2005; 4:537-545.
20. Chen X, Zhang Y, Douglas L et al. UV-damaged DNA-binding Proteins Are Targets of CUL-4A-mediated Ubiquitination and Degradation. J Biol Chem 2001; 276:48175-48182.
21. Chen X, Zhang J, Lee J et al. A kinase-independent function of c-Abl in promoting proteolytic destruction of damaged DNA binding proteins. Mol Cell 2006; 22:489-499.
22. Wang H, Zhai L, Xu J et al. Histone H3 and H4 ubiquitylation by the CUL4-DDB-ROC1 ubiquitin ligase facilitates cellular response to DNA Damage. Mol Cell 2006; 22:383-394.
23. Nag A, Datta A, Yoo K et al. DDB2 induces nuclear accumulation of the hepatitis B virus X protein independently of binding to DDB1. J Virol 2001; 75:10383-10392.
24. Datta A, Bagchi S, Nag A et al. The p48 subunit of the damaged-DNA binding protein DDB associates with the CBP/p300 family of histone acetyltransferase. Mutat Res 2001; 486:89-97.
25. Groisman R, Polanowska J, Kuraoka I et al. The ubiquitin ligase activity in the DDB2 and CSA complexes is differentially regulated by the COP9 signalosome in response to DNA damage. Cell 2003; 113:357-367.
26. Reardon JT, Nichols AF, Keeney S et al. Comparative analysis of binding of human damaged DNA-binding protein (XPE) and Escherichia coli damage recognition protein (UvrA) to the major ultraviolet photoproducts: T[c,s]T, T[t,s]T, T[6-4]T and T[Dewar]T. J Biol Chem 1993; 268:21301-21308.
27. Payne A, Chu G. Xeroderma pigmentosum group-E binding-factor recognizes a broad spectrum of DNA damage. Mutat Res 1994; 310:89-102.
28. Wittschieben BO, Iwai S, Wood RD. DDB1-DDB2 (xeroderma pigmentosum group E) protein complex recognizes a cyclobutane pyrimidine dimer, mismatches, apurinic/apyrimidinic sites and compound lesions in DNA. J Biol Chem 2005; 280:39982-39989.
29. Sugasawa K, Okuda Y, Saijo M et al. UV-Induced ubiquitylation of XPC protein mediated by UV-DDB-ubiquitin ligase complex. Cell 2005; 121:287-400.

CHAPTER 8

XPF/ERCC4 and ERCC1:
Their Products and Biological Roles

Lisa D. McDaniel* and Roger A. Schultz

Introduction

At the time of writing, a general search of the literature reveals 259 references that specifically refer to XPF/ERCC4. This puts XPF/ERCC4 around the half way point in a ranking for each of the XP groups based on the number of literature citations in which the specific acronym can be found in the title or abstract. Although such a ranking scheme is somewhat contrived, it is interesting to note that the number of citations for XPF/ERCC4 is very close to that of its mechanistic counterpart, XPG (this relationship is discussed more fully below). The large number of references citing XPF/ERCC4 is reflective of the fact that the cellular roles for Nucleotide Excision Repair (NER) proteins are far more complex than at first imagined and the examination of these diverse functions has begun in earnest taking advantage of a collection of high quality reagents that include antibodies, cell lines and purified proteins and protein complexes. In this chapter we review what has been learned about the cellular roles for XPF/ERCC4. The story is an interesting one, but one far from complete with many questions remaining.

Patients in the XP complementation group F have all been found to have pathogenic variations in the *ERCC4* gene[1-3] (Table 1 and Fig. 1), so named as the fourth member of the excision repair cross complementing (ERCC) Chinese hamster cell lines. Given that the *ERCC4* human gene designation was derived from the ability to correct the cellular defects in mutant hamster cells, prior to establishment of an association with XPF patients, there has been some debate over appropriate nomenclature. Similar discrepancies have arisen for a number of human loci where an alias from one scientific field competes with that from another. The HUGO Nomenclature Committee was established and assigned tasks that include the resolution of these discrepancies, determination of a single preferred name and providing such at a readily available database (http://www.gene. ucl.ac.uk/nomenclature/). In the case of *XPF/ERCC4* the committee has chosen *ERCC4* as the accepted name and has withdrawn *XPF*. The reader will find that many journals and books still use the term XPF, particularly in the DNA repair community. However, in support of the aforementioned nomenclature decision and the multidiscipline interest in this gene and the protein it encodes, we chose to use the designation ERCC4. Of course, patients and their cell lines will be referred to by the designation XPF.

XPF Patients, ERCC4 Mutant Cells and Gene Cloning

The clinical features of XP patients are discussed more fully in Chapter 2. However, briefly, patients in the XP-F complementation group have generally mild phenotypes and skin cancer is late onset, between 27 and 47 years of age, compared to other XP groups.[2,4] XPF is more common in Japan[4-13] and few cases have been reported in other parts of the world.[2,14,15] Unscheduled DNA

*Corresponding Author: Lisa D. McDaniel—Department of Pathology, UT Southwestern Medical Center at Dallas, Dallas, Texas TX 75390-9072, USA.
Email: lisa.mcdaniel@utsouthwestern.edu

Molecular Mechanisms of Xeroderma Pigmentosum, edited by Shamim I. Ahmad and Fumio Hanaoka. ©2008 Landes Bioscience and Springer Science+Business Media.

Table 1. XP-F Patients and reported pathogenic variations

Patient	Cell Line	Neu.	Allele 1	Protein	Allele 2	Protein	Ref.
KSP6		N					12
NSP8							12
NSP8							12
MNHN		Y					13
XP1TS		N	r.1809_1829del	p.604_610del	r.1809_1829del	p.604_610del	3
XP2YO	GM04313,	N	r.1969_1971delT	p.Val657GlufsX28	r.1699A > G	p.Thr567Ala	3
XP2YOsv	GM08437	N	r.1469G > A	p.Arg490Gln	r.1823T > C	p.Leu608Pro	3
XP3YO	GM03542	N	r.1504G > A	p.Glu502Lys	r.1586T > C	p.Ile529Thr	3
XP7KA		N	r.1364_1365insA	p.Glu456GlyfsX8	r.1364_1365insA	p.Glu456GlyfsX8	3
XP23OS		N	r.1608_1617del	p.Asn537Argfsx8	r.1360A > T	p.Arg545Trp	3
XP24KY		Y					4
XP25KO		N					4
XP27KO		N					4
XP28KO		N					4
XP38KO		N					4
XP41KO		Y					4
XP42RO		N	r.2395C > T	p.Arg799Trp	r.2395C > T	p.Arg799Trp	2
XP46KO		N					4
XP89TO		N					4
XP90TO		N					4
XP92TO		N					4
XP101OS		N	r.675A > G	p.Ile225Met	r.1537G > A	p.Gly513Arg	3, 6
XP107TO		N					189
XP126LO		N	r.2304_2307del	p.Thr770ProfsX46	r.2395C > T	p.Arg799Trp	1

Neu. = Neurologic symptoms; Ref. = References
Nomenclature has been updated using the sequence NM_005326.1 and the guidelines for nomenclature as outlined by the Human Genome Variation Society (http://www.hgvs.org/mutnomen/). The prefix r. indicates changes found in the RNA and p. the consequences of the changes to the protein. The term used for changes that result in patient phenotypes is pathogenic variation rather than mutation.

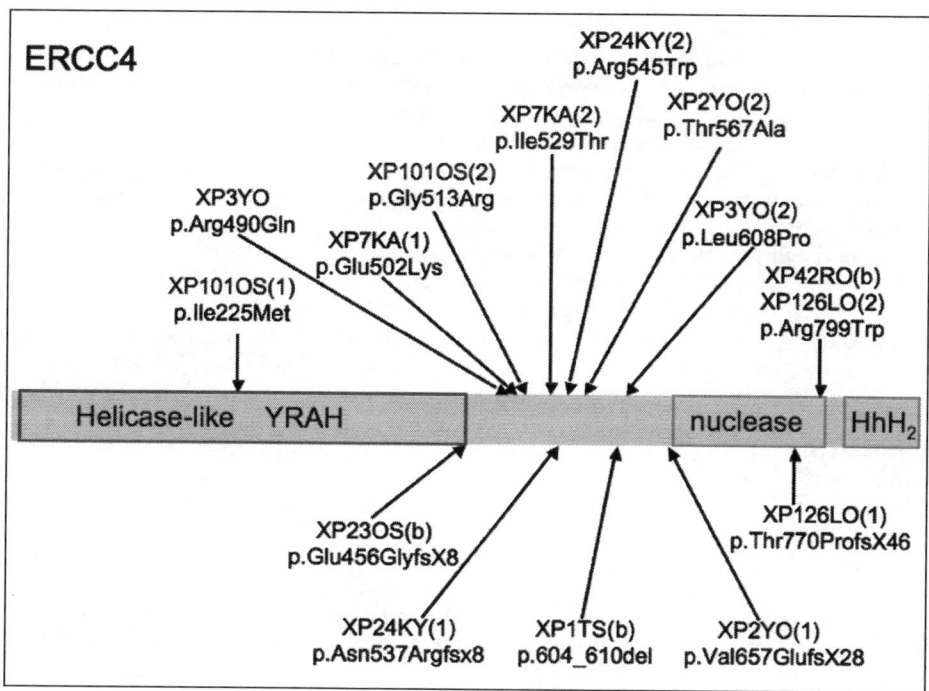

Figure 1. Pathogenic Variations in ERCC4. The ERCC4 protein is indicated by the grey solid line. Helicase-like, Nuclease and Helix-hairpin-Helix (HhH) protein domains are illustrated with boxes in the appropriate region, Missense pathogenic variations are above the gene and are associated with arrows that approximate their location on the protein. Frameshifts and the one deletion identified are below the gene. Nomenclature is as specified by the HGVS (http://www.hgvs.org/mutnomen/) and the reference sequence is NMM_005236.

synthesis is about 10-20% of control cells.[2,4,6,7] UV-survival curves show a sensitivity intermediate between XP-C and control cell lines[2,16] averaging about a 3 to 6-fold increase in sensitivity.[2] The intermediate repair capacity is consistent with the intermediate UV sensitivity. Only a few patients are described that exhibit neurological symptoms and these are reported to be late onset.[2,13] XPF cells are not sensitive to ionizing radiation,[17] but unlike cells representing other NER deficiencies, cells and cell lines from XPF patients are also sensitive to DNA crosslinking agents.[18-20] This biological observation provided an early suggestion that ERCC4 protein might play some role in the repair of DNA crosslinks.

A gene that complements the repair defects in XPF cells was isolated based on identification of cDNAs with homology to yeast *Rad1*.[1] In addition to correction of XPF cells, these authors demonstrated that this cDNA corrected hamster cells representing ERCC groups 4 and 11. Protein purification studies showed that the ERCC4 encoded protein formed a complex with ERCC1, which together comprise a structure-specific endonuclease that cleaves 5′ to the lesion during NER.

One early study suggested that XPF mRNA levels are not reduced in patient cell lines.[3] However, protein levels were dramatically reduced in these same patient cell lines[2,3] suggesting degradation of the protein occurs more rapidly in cells with pathogenic variations. The fact that abundant message is seen in northern blots on RNA from these cells suggest that nonsense-mediated decay (NMD)[21-23] is not occurring in cell lines derived from XP-F patients even when both alleles have a truncating variation.[3] Quantitative RT-PCR studies, examining RNA expression, would be useful in conjunction with western blots to clarify these results.

It was previously reported that truncating pathogenic variations were not found in XPF patients and proposed that complete loss of the protein was lethal.[24] However, XPF patients have now indeed been described that are homozygous for a frameshift resulting in premature termination of the protein. Patient XP23OS is homozygous or hemizygous for p.Glu456GlyfsX8[3] (Table 1). This patient has not been reported to have neurological symptoms or more severe clinical symptoms, including sun sensitivity or tumorigenesis, than patients with missense variations. In addition, cells from this patient were able to perform low level unscheduled DNA synthesis indicating NER is not fully abolished.[3]

Two of the three patients with reported neurologic symptoms have sequence results, one is homozygous for p.Arg799Trp2 and the second is heterozygous for p.Arg799Trp and r.2304_2307del,[3] a truncating variation resulting in the protein p.Thr770ProfsX46. These are the only 2 patients carrying these pathogenic variations. However, the *ERCC4* gene has been sequenced in only nine of the reported 23 patients with 14 alleles identified to date (Table 1). Additionally, as most patients are reported, only at their initial diagnosis, little follow-up has been provided for these individuals. Without further clinical follow-up and sequence analysis it is impossible to examine a genotype-phenotype connection for this disorder. This is an aspect of the disease that warrants further study.

As mentioned above, ERCC4 is a designation given to Chinese hamster cell lines that represent one of the UV sensitive genetic complementation groups totaling eleven. It is therefore relevant to briefly review the cellular phenotypes associated with hamster mutants representing this complementation group. *Ercc4* mutants are 6-fold more sensitive to UV-irradiation and 100-fold more sensitive to mitomycin C (MMC) and other DNA crosslinking agents, once again suggesting a dual role for this protein.[18] The gene was cloned by a combination of functional cloning with human genomic DNA,[25] and additionally by screening of a cDNA library with the complementing clone and subsequently assembling overlapping complementing cDNAs into a 3,845 bp transcript that corrected the UV-sensitivity of UV41 (*Ercc4* deficient hamster cells).[26] The genomic region of the gene contains 11 exons that encompass 28.2 kb on chromosome 16p13.12.[26] The cDNA sequence, NM_005236, is 2751 bp and encodes a protein of 916 amino acids with a mass of 104 kDa.[1,26] There are two leucine zipper motifs at 233-254 and 270-291[26] that are located in the helicase domain.[27] The nuclease domain is between amino acid 656 and 813 with the active site between 670 and 740.[28] There is also a region containing two Helix-hairpin-Helix (HhH2) domains between amino acid 837 and 905. A unique feature of the c-terminal portion of ERCC4 is a striking similarity to ERCC1, a gene that encodes a protein of 297 amino acids with a mass of 32.5 kDa,[29, 30] contains 8 exons and spans 10.1 kb at 19q13.[32] As discussed below, this region of similarity plays a key role in the physical interaction between ERCC4 and ERCC1. The protein is expressed ubiquitously (http://symatlas.gnf.org/SymAtlas/).[31]

It is interesting to note that no 5′ or 3′ untranslated regions (UTRs) have been deposited in Genbank for *ERCC4* cDNA. The average 5′ UTR in humans is 210 bp and the minimum length is 18 bp, the minimum 3′ UTR is 21 bp.[32] In the 86 RNAs that have been sequenced and listed in Unigene (http://www.ncbi.nlm.nih.gov/entrez/query.fcgi?db = unigene), several of which are genomic clones, none contain UTRs. Although in vitro only 1 bp is required for translational start,[33] none of the clones contain even 1 bp before the start codon. The surrounding genomic DNA revealed no ATA, CAAT or GC boxes, indicative of a transcription start site and the 3′ region had no polyadenylation signal.[26] This study also included rapid amplification of cDNA ends (RACE) that achieved 12 to 22 bp of additional sequence, however it is not clear if the sequence matched the adjacent genomic sequence. One can speculate that the 5′ UTR may be toxic to bacteria, allowing expression of the *ERCC4* gene from the cloned sequence in the plasmid.

Mouse Models with Relevance to XPF (*Ercc1* and *Ercc4*)

Mice were generated with a targeted knock-in construct replacing the endogenous exon 8 with a nonsense mutation at codon 445 (*Ercc4^{tm1Fwa}*) the same pathogenic variation in one of the XPF patients with a mild phenotype (XP23OS).[24] The neomycine resistant (neo^r) cassette was removed by cre-mediated recombination leaving a single LoxP site in an intron. An RT-PCR strategy has been devised to distinguish normal and mutant message in MEF cells after restriction digest that differentiates the two cDNAs. However, both, message and protein, were undetectable in the homozygous mutant MEF cells[24] although the patient carrying the same change had abundant message and no detectable protein.[3] Additionally, no mutant message was detected in the heterozygous MEF cells. As expected, MEFs from homozygous mutant mice exhibit sensitivity to UV-irradiation and MMC treatment similar to that seen for XPF patients and the *Ercc4* mutant hamster cell lines.[24] Interestingly, the mutant pups are normal at birth and genotypes occur at near Mendelian ratios. At two weeks of age the mutant pups were smaller than littermates. Mutant mice die by 3 weeks, are small and have abnormal hepatic cells with enlarged nuclei that have 4 to 8N DNA content. The class switch recombination of immunoglobin genes is normal in these mice suggesting that general recombination is normal in these mice and that immunological compromise is not responsible for death.[24]

As briefly mentioned above and elaborated below, ERCC4 physical interaction with ERCC1 is a key aspect of the biochemical function(s) of these proteins. Therefore, not surprisingly the phenotypes of *Ercc1* mutant mouse are somewhat similar to those for *Ercc4*. However, *Ercc1* mutants are recovered at frequencies below the Mendelian expectation and they are runted and short lived with liver, kidney, spleen and skin abnormalities.[34,35] Three mouse models have been developed that lack a functional ERCC1 protein. Two of the mouse models are targeted knock-out mice, *Ercc1*[tm1Dwm34] and *Ercc1*[tm1Jhjb35] and one is a targeted knock-in inserting a stop codon at position 292, *Ercc1*[tm2Jhjb], shortening the protein by 7 amino acids and eliminating 2 amino acids in the essential domain.[35] The specific phenotypes of the mice differ based on the type of mutation and the mouse strain. *Ercc1*[tm1Dwm] and *Ercc1*[tm1Jhjb] both have neomycin resistance gene inserted in exon 5 and 7, respectively. The resulting pups are runted and die by 3 weeks of age in the C57BL/6 × FVB/N and 129P2/OlaHsd × BALB/c mixed background and 6 weeks in 129P2/OlaHsd × FVB/N or 129P2/OlaHsd × C57BL/6 background due to liver failure.[34] *Ercc1*[tm2Jhjb] mice suffer loss of some of the mutant embryos in utero, while mutants that are born survive to about 6 months of age in a mixed 129P2/OlaHsd × C57BL/6 background. Life spans are shorter on a mixed 129P2/OlaHsd × FVB background and mice are runted with spleen, kidney and liver abnormalities including cells with 4,6 and 8N DNA content, which is less pronounced at birth[35] and significantly elevated by 21 days of age.[36] Deficient mice have no subcutaneous fat and are infertile.[35] The size reduction of these mice does not appear to be due to increased apoptosis suggesting it may be due to poor growth and reduced cell proliferation.[35] In fact when proliferation was studied by thymidine uptake and propidium iodide cell cycle analysis, mutant cells were 76% slower to divide than wild type cells.[37] Cell cycle arrest after DNA damage is normal.[35] MEFs from all three mouse models are UV-sensitive and sensitive to crosslinking agents.[34,35] The latter is similar to that seen for XPF cells and *Ercc4* mutant cells, once again supportive of the idea that the ERCC4-ERCC1 complex plays a key role in crosslink repair.

Pathogenic variations in ERCC1 have not been found in the human population leading to the suggestion that such mutations might be incompatible with life.[38] The phenotype of the mice suggests some other role for the ERCC4-ERCC1 complex. As elaborated below, it is known that ERCC4-ERCC1 complex has roles in DNA crosslink repair,[39,40] and some association with homologous recombination.[41] It is also interesting to note that the phenotypes seen in *Ercc1* and *Ercc4* mutant mice are considerably more severe than those associated with other NER-deficient mice. For example, XPA mutant mice are viable and normal appearing exhibiting a UV-sensitivity and cancer predisposition that is similar to that seen in the human counterpart.[42,43]

XPF in Nucleotide Excision Repair

The *Saccharomyces cerevisiae* proteins RAD1 and RAD10 form a stable complex,[44,45] even in the absence of DNA damage,[44,46] that is required for NER and mitotic recombination.[47] The mammalian homologes to these proteins ERCC1 and ERCC4 also form complex and are required for NER.[1,48,49] Extracts from cell lines defective in ERCC1 and ERCC4 do not complement each other in vitro[49-51] although the cell lines are able to complement each other in cell fusion studies.[52] However, when ERCC1 and ERCC4 are produced separately they are able to form a complex in vitro,[53] suggesting that the proteins are unstable in vivo when not in a complex. In fact, the two proteins in the ERCC1/ERCC4 complex are codependent, loss of one component results in a reduction of the other.[1,49,50,54-56] The complex is a structure specific endonuclease,[1,57,58] that interacts with XPA[59] and performs the 5′ incision of a dual incision event during NER[57,59-61] after unwinding of the region surrounding the lesion by other components of the NER machinery. In addition, in vitro studies have shown that the ERCC1/ERCC4 complex may also cut duplex DNA at the site of 3′ single-stranded flaps, bubble structures, stem loops, splayed arms and single-strand protruding structures in vitro.[57,62-65] The interaction between XPA and ERCC1/ERCC4 is essential for recruitment of ERCC1/ERCC4 to the NER site and for the dual incision.[59,66-69]

The 3′ incision is made by ERCC5 (encoded by the gene defective in XPG) and binding of ERCC5 precedes[70] and is necessary for the recruitment of and the 5′ incision made by ERCC1/ERCC4. However, the 3′ incision made by ERCC5 is not required for the incision by the ERCC1/ERCC4 complex.[71] The excised oligonucleotide is 24-32 bases in length.[60,72] Both the binding of the ERCC1/ERCC4 complex and the incision are enhanced by RPA[64]. Addition of RPA to in vitro reactions stops the nonspecific endonuclease activity of ERCC1/ERCC4 enabling the complex to hydrolyze the fourth phosphodiester bond before the junction in dsDNA on the strand entering the bubble in the 5′ to 3′ direction.[48,64] The RPA trimer[73] binds to the undamaged ssDNA strand using four OB (oligonucleotide/oligosaccharide-binding)-fold motifs that can bind either 8 or about 30 nucleotides, consistent with the size of the open bubble.[74] RPA is required for the dual incision step.[75-77]

Two helix-hairpin-helix motifs at the C-terminal end of the two subunits mediate the ERCC1/ERCC4 interaction[78] and the complex binds DNA in a sequence independent fashion.[79] The presence of ERCC1 in the heterodimer is required for NER, although the actual nuclease domain is in the ERCC4 protein.[28] ERCC4 contains the conserved GDX_nERKX_3D motif involved in metal-dependant nuclease activity.[28,79] Mutations in the conserved domain display little enzymatic activity.[28] Site-directed mutation studies have shown that the mutations in many of the acidic and basic amino acids near the metal (Mg^+) binding site reduce the enzymatic activity of the protein but do not affect the DNA binding function.[28] Only mutations in the acidic amino acids reduced the metal-induced cleavage of DNA by the ERCC1/ERCC4 complex.[28]

The c-Fos protein and a member of the Jun family, ATF1, form a heterodimeric activator protein AP-1[80,81], a transcription factor that binds to AP-1 sites stimulating a broad spectrum of genes that are immediate-early inducible after transcriptional activation by UV light[82] and other agents.[83-86] MEFs, deficient in c-Fos, are sensitive to UV-irradiation, a result that was originally explained as impaired recovery of the cells from the UVC induced block of DNA replication.[87] It has recently been shown that the levels of *Ercc4* mRNA do not recover in *c-fos*-/- MEFs after DNA damage and the damage is not removed after UV-irradiation, when compared to wild type cells suggesting that resynthesis of DNA repair proteins is required after damage.[88] The ERCC5 protein is also affected. However, ERCC5 protein appears more stable than ERCC4 as the 3′ incision occurs in *c-fos*-/- MEFs after DNA damage.[88]

The ERCC1/ERCC4 complex is not required for bubble formation around the sites of DNA damage.[89] The endonuclease and nucleotide binding domains reside in the N-terminal 378 amino acids while the ERCC1 interacting domain resides in the c-terminal end of the protein from 667-854 amino acids.[78,90] Incision products have a 3′ OH group allowing gap filling to start at the 5′ incision site with no further modifications of the DNA.[60] Purified ERCC1/ERCC4 complex is capable of cleaving Y structures[62] however a free 3′ end is not required for cleavage.[91]

Sulfolobus solfataricus from the Archaea group of bacteria has a short ERCC4 homolog containing a C-terminal nuclease domain and two HhH domains, but lacking the N-terminal helicase domain. It also contains a PCNA-interacting region required for cleavage.[92] *Pyrococcus furiosus*, contain a longer XPF homolog termed Hef (helicase associated endonuclease for forked structure).[27] Hef contains the functional helicase region motif in the N-terminal region, the nuclease region and 2 HhH domains.[27] Both archaeal proteins are homodimers, cleave bubble and 3′-flap structures and the catalytic and HhH domains form homodimers independently.[27,92] When the homodimer is bound to DNA one subunit or protomer is bound to the DNA through the nuclease and HhH domains and the HhH domain is rotated by 95° from the unbound structure closing around the DNA.[93] The linker between the two domains acts as a hinge for this rotation and the second protomer has very little contact with the DNA.[93] It has been shown that only one active site participates in the cleavage while both HhH structures are required.[94]

As a heterodimer the ERCC1/ERCC4 complex differs from the homodimer with one nuclease domain in ERCC4 and HhH domains in both proteins. The other region in ERCC1 is the central domain and it does not bind independently to the nuclease domain of ERCC4.[95] The use of a truncated XPF, lacking the helicase-like domain, was shown to have the activity of the full-length protein and was capable of binding with a ssDNA-dsDNA junction. The ERCC1/ERCC4 complex had 6-fold affinity for ssDNA over dsDNA structures, suggesting that the HhH domains bind the two ssDNA regions allowing the nuclease domain of ERCC4 to cleave near the junction of dsDNA.[95] However, the ERCC1/ERCC4 complex appears to release the excised ssDNA strand as this fragment has been shown to be free of additional NER proteins.[96]

Expression has been examined using a green fluorescent protein tagged ERCC1. Tagged protein interacts with ERCC4 and is homogeneous in the nucleus although about 25% of the nuclei in the study had 1-3 bright fluorescent spots and the ERCC1/ERCC4 complex was able to move freely through the nucleus.[54] After UV-irradiation the number of cells with areas of intense GFP spots decreased to 9% unlike other repair proteins[97-101] and 35% of the protein became immobilized and remained so for 4 minutes[54] suggesting that the complex is engaged in NER for that period.

ERCC4 in Immunoglobin Switching

Immunoglobin switching utilizes two mechanisms; (i) Somatic hypermutation (SHM) which introduces mutations in the variable region genes and with antigen selection, results in antibodies with increased diversity (ii) Class switch recombination (CSR); this allows B-cells to maintain the same antigen-binding domain while diversifying the constant region resulting in varied effector function. Isotype switching during B-cell development has been shown to utilize mismatch repair proteins, with ERCC1 interacting with Msh2.[102] ERCC1/ERCC4 may also be utilized in CSR in a small percent of cases where microhomology mediates switching, as the ERCC1/ERCC4 complex can cleave R-loops in synthetic substrates used in those studies.[103]

It has been shown that isolated splenocytes from 15-day of age *Ercc4* or *Ercc1* mutant mice that were stimulated with cytokines to undergo class switch recombination had normal levels of different antibody isotypes comparable to control mice.[24] However, class switching analyzed in another study with cells that had divided six to nine times in all genotypes indicated that switching was reduced in all *Ercc1-/-* cell populations when compared to wild type cells. Therefore, the moderate but consistent reduction in class switching, observed in these experiments, with *Ercc1-/-* B-cells is not due to reduced cell division.[37] The mutation rates were not significantly different between *Ercc1-/-* and wild type cells, but the mutations in the mutants were located in a discrete region while the wild type were more diffused. These data suggest that ERCC1/ERCC4 participates in the error-prone mechanisms that repair DNA lesions initiated in the Sμ segment via the essential protein activation-induced cytidine deaminase (AID)[104,105] for conversion of dC to dU in variable and S regions.[106-108] Resolution of the dU residues in an error-prone manner is likely the mechanism for introduction of mutations in these regions.[109,110] However, since the mutation rate is not increased in mutant cells, the ERCC1/ERCC4 complex may participate in CSR, but not in a required fashion. This is an interesting area of study that illustrates the possible diversity of biological functions employing ERCC1/ERCC4 and an additional area of investigation that warrants further study.

ERCC4 in Crosslink Repair

Certain DNA damaging agents are capable of forming interstrand crosslinks (ICL), including bi-functional alkylating agents, platinum compounds and 8-methoxypsoralen in the presence of UVA (PUVA). It is estimated that 20 ICL provide lethal blocks to DNA replication in cells that are unable to repair this damage.[111,112] Traditional NER and base excision repair (BER) require an undamaged template strand for DNA synthesis and so are not directly useful in repairing ICL. The mutations induced by these agents are deletions,[113,114] or rearrangements[115] consistent with defects in double strand break (DSB) repair. Crosslinking agents are also strong inducers of mitotic recombination, once again consistent with a suggestion and now supported with additional studies that a recombinational repair process is involved in the repair of these lesions.[116-119] Additionally, there are reports suggesting that the processing of ICL utilizes diverse nonrecombinational pathways to repair this DNA damage including lesion bypass,[120,121] postreplication repair,[122,123] transcription,[124,125] and a process involving some mismatch proteins.[126] However, the interactive and independent roles of these alternative pathways, particularly their relative importance in the repair of ICL, have yet to be revealed.

Given the well-established and fundamental role for ERCC4 in NER, one might expect that mutations in this gene or its genetic homologues would not exhibit phenotypes relevant to crosslink repair. However, exactly the opposite has been observed and a critical role for ERCC1 in ICL repair is currently accepted. In yeast the RAD1/RAD10 complex and additional NER genes have been shown to be crucial for ICL repair.[127,128] The retention of crosslinks in DNA as measured by treatment with alkali or heat demonstrates defective repair in NER mutants *rad1, rad2, rad3, rad4* and *rad10*. Interestingly, such yeast mutants generate DSB as efficiently as control cells following exposure to crosslinking agents. These observations suggest that strand breaks arise due to replication fork arrest at an ICL and that the DSB is a key intermediate in the repair of ICL damage. The RAD1/RAD10 complex has been shown to have roles in intrachromosomal recombination distinct from the RAD52 recombination pathway.[47,129-132] RAD1 and RAD10 mutants have reduced targeted homologous integration of plasmids into chromosomal loci[47,132,133] and both proteins are required for removal of the 3'-end tails from single-strand annealing (SSA) during repair of the DSB that occur due to ICL and removing nonhomologous end-blocking sequences from invading homologous strands during DSB repair.[131,134-138] These studies show that RAD1/RAD10 complex is required for removing end-blocking nonhomologies of greater than 20 bases in length. Shorter nonhomologies can be removed by independent pathways.

The repair of ICL damage in mammalian cells and the role(s) for ERCC4 in this process can be inferred from the results in yeast. Furthermore, as in yeast, DSB are also generated in mammalian cells as a result of ICL.[126,139,140] Rodent cells lacking ERCC1 or ERCC4 are reported to be deficient in DSB formation in response to PUVA-induced ICL[126,141] but not to nitrogen mustard or mitomycin C.[139,140] Phosphorylated histone H2AFX (formerly H2AX) designated γ-H2AX is used as a marker for the presence of DSB[142,143] and as a maker for processing of ICL.[144,145] DSB as a result of ICL as measured by γ-H2AX formation, requires ERCC1/ERCC4 in mammalian cells.[146] It is probable that some psoralen-induced ICL are repaired in normal cells without the introduction of DSB through a mechanism that requires XPA,[146] perhaps through a process that involves damage recognition and strand cleavage, with subsequent bypass replication.

The data presented above and additional studies implicate homologous recombination (HR) as one of the pathways in cells associated with DNA crosslink repair. In eukaryotes, most ICL damage leads to DSB. In yeast, most double strand DNA breaks are repaired by products of the genes in the *RAD52* epistasis group,[147,148] and mutants in these genes are sensitive to ICL damaging agents.[149,150] As mentioned, DSB arising in response to crosslinking agents is thought to occur at the replication fork when an ICL is encountered and the fork collapses.[150] Consistent with this notion, the induction of ICL in nondividing cells produces few DSB.[150] Mammalian cells with mutations in genetic homologues of RAD52, XRCC2 and XRCC3, are also sensitive to ICL damage once again suggesting a role for some aspect(s) of HR in ICL repair.[151-153] Additionally, there is evidence that RAD52 interacts directly with ERCC1/ERCC4 and this interaction stimulates the endonuclease activity of ERCC1/ERCC4. This interaction reduces the strand annealing activity of RAD52.[154]

Studies have looked at frequencies of HR between extrachromosomal substrates,[155] plasmid-plasmid or plasmid-chromosomal events[156] and intrachromosomal recombination between direct repeats.[157] No differences in recombination rates were found between *Ercc1+/+* and *Ercc1-/-* mouse cells. However, recombinant-dependant deletions and rearrangements involving duplicated sequences are reduced in *Ercc1+/+* cells over *Ercc1-/-* cells.[157] These mutations that occurred at the APRT locus required diploidy for the locus and proved dependant on flanking direct repeats.[158,157] The results are consistent with the roles of a number of repair genes in strand processing and rearrangement in yeast as discussed above. In support of that notion, recombination requiring the removal of long nonhomologous termini are reduced in *Ercc1-/-* cells.[159] Interestingly, one report indicated that ERCC1 is required for HR between genomic DNA and gene targeting constructs in mouse ES cells.[41]

Summarizing evidence for the role of ERCC1/ERCC4 in ICL repair, XPF and ERCC1 mutant mammalian cell lines are very sensitive to crosslinking agents.[139,160] It is instructive to compare the sensitivity of these mutant cells to that of XPG mutant cells. While all of the aforementioned mutants exhibit comparable UV sensitivity, XPG mutant cells are not broadly hypersensitive to crosslinking agents, but rather exhibit sensitivity to a subgroup of these agents (including PUVA).[18,139,160-162] This has led to the suggestion that the repair pathways utilized are agent specific (for review see McHugh[163]). For the purposes of the current review, it is clear that both ERCC1 and XPF mutant cell lines are hypersensitive to at least some crosslink-induced damage. The precise role for ERCC1 and ERCC4 proteins in ICL repair may involve two independent pathways (see Fig. 2) and this is a focus of ongoing investigations. It is possible that repair occurs by a modified

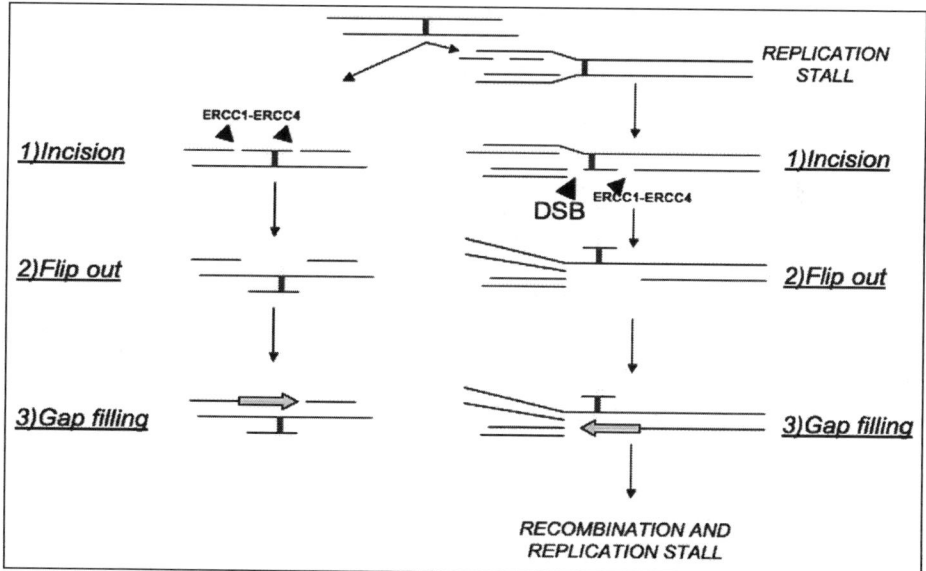

Figure 2. ERCC1/ERCC4 in the repair of interstrand crosslinks. Two models are presented. On the right, the nuclease complex cleaves the DNA strand to permit the lesion to "flip out" of the way. Gap filling might occur through replication with a bypass polymerase or through strand exchange with a homologue. The flipped out lesion resides on a single strand and might be repaired by NER. On the left, S-phase replication stalls when the lesion is encountered. Cleavage leads to a double strand break. ERCC1/ERCC4 cleaved on the other side once again providing a means for the lesion to flip out. New gap filling synthesis coupled with recombinational repair is required to achieve replication restart.

excision repair reaction focused principally on incision of only one strand of DNA, an event that would be dependent on the ERCC1/ERCC4 complex. Additionally or alternatively, as in yeast, the trimming of a DNA strand end in a recombination intermediate may require ERCC1/ERCC4. The latter presumes that DSB occurs in mammalian cells following treatment with crosslinking agents, which has been well-documented. For example, DSB are seen in Chinese hamster ovary (CHO) cells, treated with nitrogen mustard,[139] are observed only in dividing cells and are not dependent on an intact NER pathway.[139] Another study has shown that ICL damage is repaired primarily, if not exclusively, during S-phase and is associated with the appearance of DSB.[164]

One final note, DNA-protein crosslinks (DPC) are another class of DNA damage that inhibits DNA replication and transcription and is toxic to cells (reviewed in 163 and 165). ERCC4-deficient cells are capable of removing DPC at near normal rates indicating that ERCC4 is not required for DPC removal.[17] However, in certain conditions such as gamma irradiation and under hypoxic stress, XPF cells and ERCC1-deficient cells are more sensitive than wild-type cells, a phenotype considered to reflect a deficiency in the repair of DPC.[166]

ERCC4 in Telomeres

Telomeres in mammalian cells are composed of TTAGGG repeats including a 3′ ssDNA overhang that prevent degradation of the ends of chromosomes.[167] The ends of chromosomes are protected by invasion of the single-stranded DNA into the dsDNA forming the T-loop[168] and proteins bound to the telomeres including the 3′ overhang strand[167] prevent the ssDNA from being recognized as DNA damage. The proteins form a complex that is specific to telomeres[169] and vital to telomere protection and maintenance (reviewed in 170 and 171). The cellular dilemma of replication of the DNA at the ends of the chromosome requires telomerase to synthesize these repeats in a nontemplate directed manner.[172] TERF2 (formerly TRF2) is a component of the complex[169] which, when lost, results in degradation of the 3′ G-rich single strand overhangs. Studies suggest that this occurs mainly by the recognition of the telomere by the DNA repair machinery triggering ATM activation and p53-dependant cell cycle arrest.[173,174] Most of the telomere loss in a TERF2 deficient cell occurs in progressive cell cycles as repeats are not added to the end of the telomere.[175] However some of the losses occurring in cells, not progressing through DNA synthesis,[176] imply a nuclease function associated with repeat loss.

TERF2 has been shown to interact with ERCC1/ERCC4 and the complex localizes to the telomeres.[177] Overexpression of TERF2 results in telomere shortening in human fibroblasts,[178] in telomerase expressing cancer cells[179] and in mice.[180] It appeared that ERCC1/ERCC4 was the endonuclease responsible for the cleavage, next to a 3′ overhang just inside the neighboring duplex DNA,[177] since it does not occur in cell lines deficient in ERCC1/ERCC4.[177] Degradation of the 3′ overhang results in telomere loss, leading to telomere fusion, when the telomeres have become too short,[175,176,181] an event that does not occur in ERCC4 deficient cells.[177] End-trimming reactions such as this are dependent on the NHEJ machinery,[181] but how and if the NHEJ machinery recruits ERCC1/ERCC4 is unknown.

Recently, however, it has been demonstrated that cells with nuclease inactive ERCC4 are capable of forming complexes with ERCC1 and are defective in their DNA repair functions but are able to rescue the TERF2 mediated telomere shortening.[180,182] These studies reveal a nuclease-independent function for ERCC1/ERCC4 in TERF2-mediated telomere shortening, that the authors suggest, may be due to a structural role for ERCC1/ERCC4 in the binding of the telomeric proteins.[180] This study brings into question whether ERCC1/ERCC4 is actually responsible for removal of the 3′ overhang. Indeed a study now suggests that DCLRE1B (DNA cross-link repair 1B, termed Apollo by the authors) is a novel TERF2-interacting factor related to Artemis (DCLRE1C), a factor involved in V(D)J recombination and DNA repair[183] and belonging to the β-CASP metallo-β-lactamase family of DNA caretaker proteins.[184,185] DCLRE1B is mainly localized at telomeres in a TERF2-dependent manner. Reduced levels of DCLRE1B increase the sensitivity of cells to TERF2 inhibition, resulting in growth defects and an increased telomere-induced DNA-damage foci and telomere fusions. DCLRE1B exhibits a 5′ to 3′ DNA exonuclease activity, suggesting that DCLRE1B is a component of the human telomeric complex and works with TERF2 to protect chromosome termini from being recognized and processed as DNA damage.

TERF2 has recently been shown to bind to DSB genome-wide, suggesting it has a broader role in DSB damage processing.[186] The rapid binding of TERF2 to induced DSB may act to protect these regions from the action of exonucleases such as ERCC1/ERCC4[177] or to stabilize DSB for processing by DNA repair proteins such as WRN and BLM, whose in vitro activities on nontelomeric DNA substrates are enhanced by TERF2.[187,188] However, the precise role of ERCC1/ERCC4 in this type of DNA damage event remains to be determined.

There may also be a form of HR between a telomere and interstitial telomeric DNA. Chromosomal internal telomeric DNA is present infrequently in human cells, but in many other vertebrates these sequences are abundant throughout the chromosomes. During recombination between telomeres and these elements generation of terminal deletions, extrachromosomal fragments, inversions and translocations could occur. This type of recombination appears to take place in mouse cells lacking ERCC1, which generate large extrachromosomal elements that contain a single stretch of telomeric DNA, presumably at a chromosome internal site.[177] These elements, referred to as Telomeric DNA-containing Double Minute chromosomes (TDMs) could be formed by recombination between a telomere and interstitial telomeric DNA on the same chromosome.

Summary

ERCC4 is the gene mutated in XPF cells and also in rodent cells representing the mutant complementation groups ERCC4 and ERCC11. The protein functions principally as a complex with ERCC1 in a diversity of biological pathways that include NER, ICL repair, telomere maintenance and immunoglobulin switching. Sorting out these roles is an exciting and challenging problem and many important questions remain to be answered. The ERCC1/ERCC4 complex is conserved across most species presenting an opportunity to examine some functions in model organisms where mutants can be more readily generated and phenotypes more quickly assessed.

References

1. Sijbers AM, de Laat WL, Ariza RR et al. Xeroderma pigmentosum group F caused by a defect in a structure-specific DNA repair endonuclease. Cell 1996; 86(5):811-822.
2. Sijbers AM, van Voorst Vader PC, Snoek JW et al. Homozygous R788W point mutation in the XPF gene of a patient with Xeroderma pigmentosum and late-onset neurologic disease. J Invest Dermatol 1998; 110(5):832-836.
3. Matsumura YC, Nishigori T, Yagi S et al. Characterization of molecular defects in xeroderma pigmentosum group F in relation to its clinically mild symptoms. Hum Mol Genet 1998; 7(6):969-974.
4. Kondo S, Mamada A, Miyamoto C et al. Late onset of skin cancers in 2 xeroderma pigmentosum group F siblings and a review of 30 Japanese Xeroderma pigmentosum patients in groups D, E and F. Photodermatol 1989; 6(2):89-95.
5. Arase S, Kozuka T, Tanaka K et al. A sixth complementation group in Xeroderma pigmentosum. Mutat Res 1979; 59(1):143-146.
6. Fujiwara Y, Ichihashi M, Uehara Y et al. Xeroderma pigmentosum groups C and F: additional assignments and a review of the subjects in Japan. J Radiat Res (Tokyo) 1985; 26(4):443-449.
7. Fujiwara Y, Uehara Y, Ichihashi M et al. Xeroderma pigmentosum complementation group F: more assignments and repair characteristics. Photochem Photobiol 1985; 41(5):629-634.
8. Hayakawa H, Ishizaki K, Inoue M et al. Repair of ultraviolet radiation damage in Xeroderma pigmentosum cells belonging to complementation group F. Mut Res 1981; 80(2):381-388.
9. Nishigori C, Ishizaki K, Takebe H et al. A case of Xeroderma pigmentosum group F with late onset of clinical symptoms. Arch Dermatol 1986; 122(5):510-511.
10. Takebe H, Nishigori C, Satoh Y. Genetics and skin cancer of Xeroderma pigmentosum in Japan. Jpn J Cancer Res 1987; 78(11):1135-1143.
11. Yamamura K, Ichihashi M, Hiramoto T et al. Clinical and photobiological characteristics of Xeroderma pigmentosum complementation group F: a review of cases from Japan. Br J Dermatol 1989; 121(4):471-480.
12. Itoh T, Watanabe H, Yamaizumi M et al. A young woman with Xeroderma pigmentosum complementation group F and a morphoeic basal cell carcinoma. Br J Dermatol 1995; 132(1):122-127.

13. Moriwaki S, Nishigori C, Imamura S et al. A case of Xeroderma pigmentosum complementation group F with neurological abnormalities. Br J Dermatol 1993; 128(1):91-94.
14. Norris PG, Hawk JL, Avery JA et al. Xeroderma pigmentosum complementation group F in a nonJapanese patient. J Am Acad Dermatol 1988; 18(5 Pt 2):1185-1188.
15. Zghal M, Fazaa B, Zghal A et al. A whole family affected by Xeroderma pigmentosum: clinical and genetic particularities. Ann Dermatol Venereol 2003; 130(1 Pt 1):31-36.
16. Friedberg EC, Walker GC, Siede W et al. DNA repair and mutagenesis. 2nd ed. Washington DC: ASM Press, 2005.
17. Murray D, Vallee-Lucic L, Rosenberg E et al. Sensitivity of nucleotide excision repair-deficient human cells to ionizing radiation and cyclophosphamide. Anticancer Res 2002; 22(1A):21-26.
18. Hoy CA, Thompson LH, Mooney CL et al. Defective DNA cross-link removal in Chinese hamster cell mutants hypersensitive to bifunctional alkylating agents. Cancer Res 1985; 45(4):1737-1743.
19. Collins AR. Mutant rodent cell lines sensitive to ultraviolet light, ionizing radiation and cross-linking agents: a comprehensive survey of genetic and biochemical characteristics. Mut Res 1993; 293(2):99-118.
20. Busch DB, van Vuuren H, de Wit J et al. Phenotypic heterogeneity in nucleotide excision repair mutants of rodent complementation groups 1 and 4. Mut Res 1997; 383(2):91-106.
21. Baumann B, Potash MJ, Kohler G. Consequences of frameshift mutations at the immunoglobulin heavy chain locus of the mouse. EMBO J 1985; 4(2):351-359.
22. Leeds P, Peltz SW, Jacobson A et al. The product of the yeast UPF1 gene is required for rapid turnover of mRNAs containing a premature translational termination codon. Genes Dev 1991; 5(12A):2303-2314.
23. Nilsson G, Belasco JG, Cohen SN et al. Effect of premature termination of translation on mRNA stability depends on the site of ribosome release. Proc Natl Acad Sci USA 1987; 84(14):4890-4904.
24. Tian M, Shinkura R, Shinkura N et al. Growth retardation, early death and DNA repair defects in mice deficient for the nucleotide excision repair enzyme XPF. Mol Cell Biol, 2004; 24(3):1200-1205.
25. Thompson LH, Brookman KW, Weber CA et al. Molecular cloning of the human nucleotide-excision-repair gene ERCC4. Proc Natl Acad Sci USA 1994; 91(15):6855-6859.
26. Brookman KW, Lamerdin JE, Thelen MP et al. ERCC4 (XPF) encodes a human nucleotide excision repair protein with eukaryotic recombination homologs. Mol Cell Biol 1996; 16(11):6553-6662.
27. Komori K, Fujikane R, Shinagawa H et al. Novel endonuclease in Archaea cleaving DNA with various branched structure. Genes Genet Syst 2002; 77(4):227-241.
28. Enzlin JH, Scharer OD. The active site of the DNA repair endonuclease XPF-ERCC1 forms a highly conserved nuclease motif. EMBO J 2002; 21(8):2045-2053.
29. van Duin M, de Wit J, Odijk H et al. Molecular characterization of the human excision repair gene ERCC-1: cDNA cloning and amino acid homology with the yeast DNA repair gene RAD10. Cell 1986; 44(6):913-923.
30. Westerveld A, Hoeijmakers JH, van Duin M et al. Molecular cloning of a human DNA repair gene. Nature 1984; 310(5976):425-429.
31. Su AI, Cooke MP, Ching KA et al. Large-scale analysis of the human and mouse transcriptomes. Proc Natl Acad Sci USA 2002; 99(7):4465-4470.
32. Mignone F, Gissi C, Liuni S et al. Untranslated regions of mRNAs. Genome Biol 2002; 3(3): REVIEWS0004.
33. Hughes MJ, Andrews DW. A single nucleotide is a sufficient 5′ untranslated region for translation in an eukaryotic in vitro system. FEBS Letts 1997; 414(1):19-22.
34. McWhir J, Selfridge J, Harrison DJ et al. Mice with DNA repair gene (ERCC-1) deficiency have elevated levels of p53 liver nuclear abnormalities and die before weaning. Nat Genet 1993; 5(3):217-224.
35. Weeda G, Donker I, de Wit J et al. Disruption of mouse ERCC1 results in a novel repair syndrome with growth failure, nuclear abnormalities and senescence. Curr Biol 1997; 7(6):427-439.
36. Kirschner K, Singh R, Prost S et al. Characterisation of Ercc1 deficiency in the liver and (Amst) 2006;in conditional Ercc1-deficient primary hepatocytes in vitro. DNA Repair.
37. Schrader CE, Vardo J, Linehan E et al. Deletion of the nucleotide excision repair gene Ercc1 reduces immunoglobulin class switching and alters mutations near switch recombination junctions. J Exp Med 2004; 200(3):321-330.
38. Van Duin M, Hoeijmakers JH. Cloning of human repair genes by genomic DNA transfection. Ann Ist Super Sanita 1989; 25(1):131-142.
39. Quievryn G, Zhitkovich A. Loss of DNA-protein crosslinks from formaldehyde-exposed cells occurs through spontaneous hydrolysis and an active repair process linked to proteosome function. Carcinogenesis 2000; 21(8):1573-1580.
40. Zhang N, Zhang X, Peterson C et al. Differential processing of UV mimetic and interstrand crosslink damage by XPF cell extracts. Nucleic Acids Res 2000; 28(23):4800-4804.

41. Niedernhofer LJ, Essers J, Weeda G et al. The structure-specific endonuclease Ercc1-Xpf is required for targeted gene replacement in embryonic stem cells. EMBO J 2001; 20(22):6540-6549.
42. de Vries A, van Oostrom CT, Hofhuis FM et al. Increased susceptibility to ultraviolet-B and carcinogens of mice lacking the DNA excision repair gene XPA. Nature 1995; 377(6545):169-173.
43. Nakane H, Takeuchi S, Yuba S et al. High incidence of ultraviolet-B-or chemical-carcinogen-induced skin tumours in mice lacking the Xeroderma pigmentosum group A gene. Nature 1995; 377(6545):165-168.
44. Bailly V, Sommers CH, Sung P et al. Specific complex formation between proteins encoded by the yeast DNA repair and recombination genes RAD1 and RAD10. Proc Natl Acad Sci USA 1992; 89(17):8273-8277.
45. Bardwell AJ, Bardwell L, Tomkinson AE et al. Specific cleavage of model recombination and repair intermediates by the yeast Rad1-Rad10 DNA endonuclease. Science 1994; 265(5181):2082-2085.
46. Bardwell L, Cooper AJ, Friedberg EC. Stable and specific association between the yeast recombination and DNA repair proteins RAD1 and RAD10 in vitro. Mol Cell Biol 1992; 12(7):3041-3049.
47. Schiestl RH, Prakash S. RAD10, an excision repair gene of Saccharomyces cerevisiae, is involved in the RAD1 pathway of mitotic recombination. Mol Cell Biol 1990; 10(6):2485-2491.
48. Park CH, Bessho T, Matsunaga T et al. Purification and characterization of the XPF-ERCC1 complex of human DNA repair excision nuclease. J Biol Chem 1995; 270(39):22657-22660.
49. van Vuuren AJ, Appeldoorn E, Odijk H et al. Evidence for a repair enzyme complex involving ERCC1 and complementing activities of ERCC4, ERCC11 and Xeroderma pigmentosum group F. EMBO J 1993; 12(9):3693-3701.
50. Biggerstaff M, Szymkowski DE, Wood RD. cocorrection of the ERCC1, ERCC4 and xeroderma pigmentosum group F DNA repair defects in vitro. EMBO J 1993; 12(9):3685-3692.
51. Reardon JT, Thompson LH, Sancar A. Excision repair in man and the molecular basis of Xeroderma pigmentosum syndrome. Cold Spring Harb Symp Quant Biol 1993; 58:605-617.
52. Busch D, Greiner C, Lewis K et al. Summary of complementation groups of UV-sensitive CHO cell mutants isolated by large-scale screening. Mutagenesis 1989; 4(5):349-354.
53. Gaillard PH, Wood RD. Activity of individual ERCC1 and XPF subunits in DNA nucleotide excision repair. Nucleic Acids Res 2001; 29(4):872-829.
54. Houtsmuller AB, Rademakers S, Nigg AL et al. Action of DNA repair endonuclease ERCC1/XPF in living cells. Science 1999; 284(5416):958-961.
55. Yagi T, Matsumura Y, Sato M et al. Complete restoration of normal DNA repair characteristics in group F Xeroderma pigmentosum cells by over-expression of transfected XPF cDNA. Carcinogenesis 1998; 19(1):55-60.
56. Yagi T, Wood RD, Takebe H. A low content of ERCC1 and a 120 kDa protein is a frequent feature of group F Xeroderma pigmentosum fibroblast cells. Mutagenesis 1997; 12(1):41-44.
57. Bessho T, Sancar A, Thompson LH et al. Reconstitution of human excision nuclease with recombinant XPF-ERCC1 complex. J Biol Chem 1997; 272(6):3833-3837.
58. Damia G, Guidi G, D'Incalci M. Expression of genes involved in nucleotide excision repair and sensitivity to cisplatin and melphalan in human cancer cell lines. Eur J Cancer 1998; 34(11):1783-1788.
59. Park CH, Sancar A. Formation of a ternary complex by human XPA, ERCC1 and ERCC4(XPF) excision repair proteins. Proc Natl Acad Sci USA 1994; 91(11):5017-21.
60. Moggs JG, Yarema KJ, Essigmann JM et al. Analysis of incision sites produced by human cell extracts and purified proteins during nucleotide excision repair of a 1,3-intrastrand d(GpTpG)-cisplatin adduct. J Biol Chem 1996; 271(12):7177-7186.
61. Aboussekhra A, Biggerstaff M, Shivji MK et al. Mammalian DNA nucleotide excision repair reconstituted with purified protein components. Cell 1995; 80(6):859-868.
62. de Laat WL, Appeldoorn E, Jaspers NG et al. DNA structural elements required for ERCC1-XPF endonuclease activity. J Biol Chem 1998; 273(14):7835-7842.
63. Evans E, Moggs JG, Hwang JR et al. Mechanism of open complex and dual incision formation by human nucleotide excision repair factors. EMBO J 1997; 16(21):6559-6573.
64. Matsunaga T, Park CH, Bessho T et al. Replication protein A confers structure-specific endonuclease activities to the XPF-ERCC1 and XPG subunits of human DNA repair excision nuclease. J Biol Chem 1996; 271(19):11047-11050.
65. O'Donovan A, Davies AA, Moggs JG et al. XPG endonuclease makes the 3' incision in human DNA nucleotide excision repair. Nature 1994; 371(6496):432-435.
66. Li L, Elledge SJ, Peterson CA et al. Specific association between the human DNA repair proteins XPA and ERCC1. Proc Natl Acad Sci USA 1994; 91(11):5012-5016.
67. Li L, Peterson CA, Lu X et al. Mutations in XPA that prevent association with ERCC1 are defective in nucleotide excision repair. Mol Cell Biol 1995; 15(4):1993-1998.

68. Nagai A, Saijo M, Kuraoka I et al. Enhancement of damage-specific DNA binding of XPA by interaction with the ERCC1 DNA repair protein. Biochem Biophys Res Commun 1995; 211(3):960-966.
69. Saijo M, Kuraoka I, Masutani C et al. Sequential binding of DNA repair proteins RPA and ERCC1 to XPA in vitro. Nucleic Acids Res 1996; 24(23):4719-4724.
70. Mu D, Hsu DS, Sancar A. Reaction mechanism of human DNA repair excision nuclease. J Biol Chem 1996; 271(14):8285-8294.
71. Wakasugi M, Reardon JT, Sancar A. The noncatalytic function of XPG protein during dual incision in human nucleotide excision repair. J Biol Chem 1997; 272(25):16030-16034.
72. Huang JC, Svoboda DL, Reardon JT et al. Human nucleotide excision nuclease removes thymine dimers from DNA by incising the 22nd phosphodiester bond 5′ and the 6th phosphodiester bond 3′ to the photodimer. Proc Natl Acad Sci USA 1992; 89(8):3664-3668.
73. Henricksen LA, Umbricht CB, Wold MS. Recombinant replication protein A: expression, complex formation and functional characterization. J Biol Chem 1994; 269(15):11121-11132.
74. Bochkarev A, Bochkareva E. From RPA to BRCA2: lessons from single-stranded DNA binding by the OB-fold. Curr Opin Struct Biol 2004; 14(1):36-42.
75. Coverley D, Kenny MK, Lane DP et al. A role for the human single-stranded DNA binding protein HSSB/RPA in an early stage of nucleotide excision repair. Nucleic Acids Res 1992; 20(15):3873-3880.
76. Guzder SN, Habraken Y, Sung P et al. Reconstitution of yeast nucleotide excision repair with purified Rad proteins, replication protein A and transcription factor TFIIH. J Biol Chem 1995; 270(22):12973-12976.
77. Mu D, Park CH, Matsunaga T et al. Reconstitution of human DNA repair excision nuclease in a highly defined system. J Biol Chem 1995; 270(6):2415-2418.
78. de Laat WL, Sijbers AM, Odijk H et al. Mapping of interaction domains between human repair proteins ERCC1 and XPF. Nucleic Acids Res 1998; 26(18):4146-4152.
79. Nishino T, Komori K, Ishino Y et al. X-ray and biochemical anatomy of an archaeal XPF/Rad1/Mus81 family nuclease: similarity between its endonuclease domain and restriction enzymes. Structure 2003; 11(4):445-457.
80. Chiu R, Boyle WJ, Meek J et al. The c-Fos protein interacts with c-Jun/AP-1 to stimulate transcription of AP-1 responsive genes. Cell 1988; 54(4):541-552.
81. Rauscher FJ 3rd, Sambucetti LC, Curran T et al Common DNA binding site for Fos protein complexes and transcription factor AP-1. Cell 1988; 52(3):471-480.
82. Buscher M, Rahmsdorf HJ, Litfin M et al. Activation of the c-fos gene by UV and phorbol ester: different signal transduction pathways converge to the same enhancer element. Oncogene 1988; 3(3):301-311.
83. Dosch J, Kaina B. Induction of c-fos, c-jun, junB and junD mRNA and AP-1 by alkylating mutagens in cells deficient and proficient for the DNA repair protein O6-methylguanine-DNA methyltransferase (MGMT) and its relationship to cell death, mutation induction and chromosomal instability. Oncogene 1996; 13(9):1927-1935.
84. Gubits RM, Fairhurst JL. c-fos mRNA levels are increased by the cellular stressors, heat shock and sodium arsenite. Oncogene 1988; 3(2):163-168.
85. Hollander MC, Fornace AJ Jr. Induction of fos RNA by DNA-damaging agents. Cancer Res 1989; 49(7):1687-1692.
86. Muller R, Bravo R, Burckhardt J et al. Induction of c-fos gene and protein by growth factors precedes activation of c-myc. Nature 1984; 312(5996):716-720.
87. Haas S, Kaina B. c-Fos is involved in the cellular defence against the genotoxic effect of UV radiation. Carcinogenesis 1995; 16(5):985-991.
88. Christmann M, Tomicic MT, Origer J et al. c-Fos is required for excision repair of UV-light induced DNA lesions by triggering the resynthesis of XPF. Nuc Acids Res, 2006.
89. Mu D, Wakasugi M, Hsu DS et al. Characterization of reaction intermediates of human excision repair nuclease. J Biol Chem 1997; 272(46):28971-28979.
90. McCutchen-Maloney SL, Giannecchini CA, Hwang MH et al. Domain mapping of the DNA binding, endonuclease and ERCC1 binding properties of the human DNA repair protein XPF. Biochemistry 1999; 38(29):9417-9425.
91. Kuraoka I, Kobertz WR, Ariza RR et al. Repair of an interstrand DNA cross-link initiated by ERCC1-XPF repair/recombination nuclease. J Biol Chem 2000; 275(34):26632-26636.
92. Roberts JA, Bell SD, White MF. An archaeal XPF repair endonuclease dependent on a heterotrimeric PCNA. Mol Microbiol 2003; 48(2):361-371.
93. Newman M, Murray-Rust J, Lally J et al. Structure of an XPF endonuclease with and without DNA suggests a model for substrate recognition. EMBO J 2005; 24(5):895-905.

94. Nishin T, Komori K, Ishino Y et al. Structural and functional analyses of an archaeal XPF/Rad1/Mus81 nuclease: asymmetric DNA binding and cleavage mechanisms. Structure (Camb) 2005; 1(8):1183-1192.

95. Tsodikov OV, Enzlin JH, Scharer OD et al. Crystal structure and DNA binding functions of ERCC1, a subunit of the DNA structure-specific endonuclease XPF-ERCC1. Proc Natl Acad Sci USA 2005; 102(32):11236-11241..

96. Riedl T, Hanaoka F, Egly JM. The comings and goings of nucleotide excision repair factors on damaged DNA. EMBO J 2003; 22(19):5293-5303.

97. Harless J. Hewitt RR. Intranuclear localization of UV-induced DNA repair in human VA13 cells. Mutat Res 1987; 183(2):177-184.

98. Jackson DA, Balajee AS, Mullenders L et al. Sites in human nuclei where DNA damaged by ultraviolet light is repaired: visualization and localization relative to the nucleoskeleton. J Cell Sci 1994; 107(7):1745-1752.

99. McCready SJ, Cook PR. Lesions induced in DNA by ultraviolet light are repaired at the nuclear cage. J Cell Sci 1984; 70:189-196.

100. Mullenders LH, van Kesteren van Leeuwen AC, van Zeeland AA et al. Nuclear matrix associated DNA is preferentially repaired in normal human fibroblasts, exposed to a low dose of ultraviolet light but not in Cockayne's syndrome fibroblasts. Nucleic Acids Res 1988; 16(22):10607-10622.

101. Park MS, Knauf JA, Pendergrass SH et al.Ultraviolet-induced movement of the human DNA repair protein, Xeroderma pigmentosum type G, in the nucleus. Proc Natl Acad Sci USA, 1996; 93(16):8368-8373.

102. Lan L, Hayashi T, Rabeya RM et al. Functional and physical interactions between ERCC1 and MSH2 complexes for resistance to cis-diamminedichloroplatinum(II) in mammalian cells. DNA Repair (Amst) 2004; 3(2):135-143.

103. Tian M, Alt FW. Transcription-induced cleavage of immunoglobulin switch regions by nucleotide excision repair nucleases in vitro. J Biol Chem 2000; 275(31):24163-24172.

104. Muramatsu M, Kinoshita K, Fagarasan S et al. Class switch recombination and hypermutation require activation-induced cytidine deaminase (AID), a potential RNA editing enzyme. Cell 2000; 102(5):553-563.

105. Revy P, Muto T, Levy Y et al. Activation-induced cytidine deaminase (AID) deficiency causes the autosomal recessive form of the Hyper-IgM syndrome (HIGM2). Cell 2000; 12(5):565-575.

106. Bransteitter R, Pham P, Scharff MD et al. Activation-induced cytidine deaminase deaminates deoxycytidine on single-stranded DNA but requires the action of RNase. Proc Natl Acad Sci USA 2003; 100(7):4102-4107.

107. Chaudhuri J, Tian M, Khuong C et al. Transcription-targeted DNA deamination by the AID antibody diversification enzyme. Nature 2003; 422(6933):726-730.

108. Di Noia J, Neuberger MS. Altering the pathway of immunoglobulin hypermutation by inhibiting uracil-DNA glycosylase. Nature 2002; 419(6902):43-48.

109. Petersen-Mahrt SK, Harris RS, Neuberger MS. AID mutates E. coli suggesting a DNA deamination mechanism for antibody diversification. Nature 2002; 418(6893):99-103.

110. Storb U, Stavnezer J. Immunoglobulin genes: generating diversity with AID and UNG. Curr Biol 2002; 12(21):R725-727.

111. Lawley PD, Phillips DH. DNA adducts from chemotherapeutic agents. Mutat Res 1996; 355(1-2):13-40.

112. Murnane JP, Byfield JE. Irreparable DNA cross-links and mammalian cell lethality with bifunctional alkylating agents. Chem Biol Interact 1981; 38(1):75-86.

113. Wijen JP, Nivard MJ, Vogel EW. The in vivo genetic activity profile of the monofunctional nitrogen mustard 2-chloroethylamine differs drastically from its bifunctional counterpart mechlorethamine. Carcinogenesis 2000; 21(10):1859-1867.

114. Yaghi BM, Turner PM, Denny WA et al. Comparative mutational spectra of the nitrogen mustard chlorambucil and its half-mustard analogue in Chinese hamster AS52 cells. Mutat Res 1998; 401(1-2):153-164.

115. Vogel EW, Nivard MJ, Ballering LA et al. DNA damage and repair in mutagenesis and carcinogenesis: implications of structure-activity relationships for cross-species extrapolation. Mutat Res 1996; 353(1-2):177-218.

116. Bodell WJ, Aida T, Berger MS et al. Repair of O6-(2-chloroethyl)guanine mediates the biological effects of chloroethylnitrosoureas. Environ Health Perspect 1985; 62:119-126.

117. Bodell WJ, Aida T, Rasmussen J. Comparison of sister-chromatid exchange induction caused by nitrosoureas that alkylate or alkylate and crosslink DNA. Mutat Res 1985; 149(1):95-100.

118. Tokuda K, Bodell WJ. Cytotoxicity and sister chromatid exchanges in 9L cells treated with monofunctional and bifunctional nitrogen mustards. Carcinogenesis 1987; 8(11):1697-1701.

119. Vogel EW, Nivard MJ. Performance of 181 chemicals in a Drosophila assay predominantly monitoring interchromosomal mitotic recombination. Mutagenesis 1993; 8(1):57-81.

120. McHugh PJ, Sarkar S. DNA interstrand cross-link repair in the cell cycle: a critical role for polymerase zeta in G1 phase. Cell Cycle 2006; 5(10):1044-1047.

121. Sarkar S, Davies AA, Ulrich HD et al. DNA interstrand crosslink repair during G1 involves nucleotide excision repair and DNA polymerase zeta. EMBO J 2006; 25(6):1285-1294.

122. Cheng S, Van Houten B, Gamper HB et al. Use of psoralen-modified oligonucleotides to trap three-stranded RecA-DNA complexes and repair of these cross-linked complexes by ABC excinuclease. J Biol Chem 1988; 263(29):15110-15117.

123. Saffran WA, Ahmed S, Bellevue S et al. DNA repair defects channel interstrand DNA cross-links into alternate recombinational and error-prone repair pathways. J Biol Chem 2004; 279(35):36462-36469.

124. Islas AL, Baker FJ, Hanawalt PC. Transcription-coupled repair of psoralen cross-links but not monoadducts in Chinese hamster ovary cells. Biochemistry 1994; 3(35):10794-10799.

125. Islas AL, Vos JM, Hanawalt PC. Differential introduction and repair of psoralen photoadducts to DNA in specific human genes. Cancer Res 1991; 51(11):2867-2873.

126. Zhang N, Lu X, Zhang X et al. hMutSbeta is required for the recognition and uncoupling of psoralen interstrand cross-links in vitro. Mol Cell Biol 2002; 22(7):2388-2397.

127. Meniel V, Magana-Schwencke N, Averbeck D. Preferential repair in Saccharomyces cerevisiae rad mutants after induction of interstrand cross-links by 8-methoxypsoralen plus UVA. Mutagenesis 1995; 10(6):543-548.

128. Miller RD, Prakash L, Prakash S. Genetic control of excision of Saccharomyces cerevisiae interstrand DNA cross-links induced by psoralen plus near-UV light. Mol Cell Biol 1982; 2(8):939-948.

129. Klein HL. Different types of recombination events are controlled by the RAD1 and RAD52 genes of Saccharomyces cerevisiae. Genetics 1988; 120(2):367-377.

130. Klein HL. Genetic control of intrachromosomal recombination. BioEssays 1995; 17(2):147-159.

131. Prado F, Aguilera A. Role of reciprocal exchange, one-ended invasion crossover and single-strand annealing on inverted and direct repeat recombination in yeast: different requirements for the RAD1, RAD10 and RAD52 genes. Genetics 1995; 139(1):109-123.

132. Schiestl RH, Prakash S. RAD1, an excision repair gene of Saccharomyces cerevisiae, is also involved in recombination. Mol Cell Biol 1988; 8(9):3619-3626.

133. Saffran WA, Greenberg RB, Thaler-Scheer MS et al. Single strand and double strand DNA damage-induced reciprocal recombination in yeast. Dependence on nucleotide excision repair and RAD1 recombination. Nucleic Acids Res 1994; 22(14):2823-2829.

134. Colaiacovo MP, Paques F, Haber JE. Removal of one nonhomologous DNA end during gene conversion by a RAD1- and MSH2-independent pathway. Genetics 1999; 151(4):1409-1423.

135. Fishman-Lobell J, Rudin N, Haber JE. Two alternative pathways of double-strand break repair that are kinetically separable and independently modulated. Mol Cell Biol 1992; 12(3):1292-1303.

136. Ivanov EL, Haber JE. RAD1 and RAD10, but not other excision repair genes, are required for double-strand break-induced recombination in Saccharomyces cerevisiae. Mol Cell Biol 1995; 15(4):2245-2251.

137. Paques F, Haber JE. Two pathways for removal of nonhomologous DNA ends during double-strand break repair in Saccharomyces cerevisiae. Mol Cell Biol 1997; 17(11):6765-6771.

138. Sugawara N, Paques F, Colaiacovo M et al. Role of Saccharomyces cerevisiae Msh2 and Msh3 repair proteins in double-strand break-induced recombination. Proc Natl Acad Sci USA 1997; 94(17):9214-9219.

139. De Silva IU, McHugh PJ, Clingen PH et al. Defining the roles of nucleotide excision repair and recombination in the repair of DNA interstrand cross-links in mammalian cells. Mol Cell Biol 2000; 20(21):7980-7990.

140. Niedernhofer LJ, Odijk H, Budzowska M et al. The structure-specific endonuclease Ercc1-Xpf is required to resolve DNA interstrand cross-link-induced double-strand breaks. Mol Cell Biol 2004; 24(13):5776-5787.

141. Rothfuss A, Grompe M. Repair kinetics of genomic interstrand DNA cross-links: evidence for DNA double-strand break-dependent activation of the Fanconi anemia/BRCA pathway Mol Cell Biol 2004; 24(1):123-134.

142. Redon C, Pilch D, Rogakou E et al.Histone H2A variants H2AX and H2AZ. Curr Opin Genet Dev 2002; 12(2):162-169.

143. Thiriet C, Hayes JJ. Chromatin in need of a fix: phosphorylation of H2AX connects chromatin to DNA repair. Mol Cell 2005; 18(6):617-622.

144. Royer-Pokora B, Peterson WD Jr, Haseltine WA. Biological and biochemical characterization of an SV40-transformed Xeroderma pigmentosum cell line. Exp Cell Res,1984; 151 (2):408-420.

145. Yagi T, Takebe H. Establishment by SV40 transformation and characteristics of a cell line of Xeroderma pigmentosum belonging to complementation group F. Mutat Res 1983; 112(1):59-66.
146. Mogi S, Oh DH. Gamma-H2AX formation in response to interstrand crosslinks requires XPF in human cells. DNA Repair (Amst) 2006; 5(6):731-740.
147. Krejci L, Chen L, Van Komen S et al. Mending the break: two DNA double-strand break repair machines in eukaryotes. Prog Nucleic Acid Res Mol Biol 2003; 74:159-201.
148. Symington LS. Role of RAD52 epistasis group genes in homologous recombination and double-strand break repair. Microbiol Mol Biol Rev 2002; 66(4):630-670, table of contents.
149. Jachymczyk WJ, von Borstel RC, Mowat MR et al. Repair of interstrand cross-links in DNA of Saccharomyces cerevisiae requires two systems for DNA repair: the RAD3 system and the RAD51 system. Mol Gen Genet 1981; 182(2):196-205.
150. McHugh PJ, Sones WR, Hartley JA. Repair of intermediate structures produced at DNA interstrand cross-links in Saccharomyces cerevisiae. Mol Cell Biol 2000; 20(10):3425-3433.
151. Liu N, Lamerdin JE, Tebbs RS et al. XRCC2 and XRCC3, new human Rad51-family members, promote chromosome stability and protect against DNA cross-links and other damages. Mol Cell 1998; 1(6):783-793.
152. Rijkers T, Van Den Ouweland J, Morolli B et al. Targeted inactivation of mouse RAD52 reduces homologous recombination but not resistance to ionizing radiation. Mol Cell Biol 1998; 18(11):6423-6429.
153. Yamaguchi-Iwai Y, Sonoda E, Buerstedde JM et al. Homologous recombination, but not DNA repair, is reduced in vertebrate cells deficient in RAD52. Mol Cell Biol 1998; 18(11):6430-6435.
154. Motycka TA, Bessho T, Post SM et al. Physical and functional interaction between the XPF/ERCC1 endonuclease and hRad52. J Biol Chem 2004; 279(14):13634-13639.
155. Nairn RS, Adair GM, Christmann CB et al. Ultraviolet stimulation of intermolecular homologous recombination in Chinese hamster ovary cells. Mol Carcinog 1991; 4(6):519-526.
156. Melton DW, Ketchen AM, Nunez F et al. Cells from ERCC1-deficient mice show increased genome instability and a reduced frequency of S-phase-dependent illegitimate chromosome exchange but a normal frequency of homologous recombination. J Cell Sci 1998; 111(3):395-404.
157. Sargent RG, Rolig RL, Kilburn AE et al. Recombination-dependent deletion formation in mammalian cells deficient in the nucleotide excision repair gene ERCC1. Proc Natl Acad Sci USA 1997; 94(24):13122-13127.
158. Sargent RG, Meservy JL, Perkins BD et al. Role of the nucleotide excision repair gene ERCC1 in formation of recombination-dependent rearrangements in mammalian cells. Nucleic Acids Res 2000; 28(19):3771-3778.
159. Adair GM, Rolig RL, Moore-Faver D et al. Role of ERCC1 in removal of long nonhomologous tails during targeted homologous recombination. EMBO J 2000; 19(20):5552-5561.
160. Damia G, Imperatori L, Stefanini M et al. Sensitivity of CHO mutant cell lines with specific defects in nucleotide excision repair to different anticancer agents. Int J Cancer 1996; 66(6):779-783.
161. Kaye J, Smith CA, Hanawalt PC. DNA repair in human cells containing photoadducts of 8-methoxypsoralen or angelicin. Cancer Res 1980; 40(3):696-702.
162. Vuksanovic L, Cleaver JE. Unique cross-link and monoadduct repair characteristics of a Xeroderma pigmentosum revertant cell line. Mutat Res 1987; 184(3):255-263.
163. McHugh PJ, Spanswick VJ, Hartley JA. Repair of DNA interstrand crosslinks: molecular mechanisms and clinical relevance. Lancet Oncol 2001; 2(8):483-490.
164. Akkari YM, Bateman RL, Reifsteck CA et al. DNA replication is required To elicit cellular responses to psoralen-induced DNA interstrand cross-links. Mol Cell Biol 2000; 20(21):8283-8289.
165. Barker S, Weinfeld M, Murray D. DNA-protein crosslinks: their induction, repair and biological consequences. Mutat Res 2005; 589(2):111-135.
166. Murray D, Rosenberg E. The importance of the ERCC1/ERCC4[XPF] complex for hypoxic-cell radioresistance does not appear to derive from its participation in the nucleotide excision repair pathway. Mutat Res 1996; 364(3):217-226.
167. Henderson ER, Blackburn EH. An overhanging 3′ terminus is a conserved feature of telomeres. Mol Cell Biol 1989; 9(1):345-348.
168. Griffith JD, Comeau L, Rosenfield S et al. Mammalian telomeres end in a large duplex loop. Cell 1999; 97(4):503-514.
169. De Lange T. Telomere-related genome instability in cancer. Cold Spring Harb Symp Quant Biol 2005; 70:197-204.
170. de Lange T. Shelterin: the protein complex that shapes and safeguards human telomeres. Genes Dev 2005; 19(18):2100-2110.
171. Munoz P, Blanco R, Blasco MA. Role of the TRF2 telomeric protein in cancer and ageing. Cell Cycle 2006; 5(7):718-721.

172. Greider CW, Blackburn EH. The telomere terminal transferase of Tetrahymena is a ribonucleoprotein enzyme with two kinds of primer specificity. Cell 1987; 51(6):887-898.
173. Chin L, Artandi SE, Shen Q et al. p53 deficiency rescues the adverse effects of telomere loss and co-operates with telomere dysfunction to accelerate carcinogenesis. Cell 1999; 97(4):527-538.
174. Karlseder J, Broccoli D, Dai Y et al. p53- and ATM-dependent apoptosis induced by telomeres lacking TRF2. Science 1999; 283(5406):1321-1325.
175. van Steensel B, Smogorzewska A, de Lange T. TRF2 protects human telomeres from end-to-end fusions. Cell 1998; 92(3):401-413.
176. Smogorzewska A, Karlseder J, Holtgreve-Grez H et al. DNA ligase IV-dependent NHEJ of deprotected mammalian telomeres in G1 and G2. Curr Biol 2002; 12(19):1635-1644.
177. Zhu XD, Niedernhofer L, Kuster B et al. ERCC1/XPF removes the 3′ overhang from uncapped telomeres and represses formation of telomeric DNA-containing double minute chromosomes. Mol Cell 2003; 12(6):1489-1498.
178. Karlseder J, Smogorzewska A, de Lange T. Senescence induced by altered telomere state, not telomere loss. Science 2002; 295(5564):2446-2449.
179. Smogorzewska A, van Steensel B, Bianchi A et al. Control of human telomere length by TRF1 and TRF2. Mol Cell Biol 2000; 20(5):1659-1668.
180. Munoz P, Blanco R, Flores JM et al. XPF nuclease-dependent telomere loss and increased DNA damage in mice overexpressing TRF2 result in premature aging and cancer. Nat Genet 2005; 37(10):1063-1071.
181. Celli GB, de Lange T. DNA processing is not required for ATM-mediated telomere damage response after TRF2 deletion. Nat Cell Biol 2005; 7(7):712-718.
182. Wu Y, Zacal NJ, Rainbow AJ et al. XPF with mutations in its conserved nuclease domain is defective in DNA repair but functions in TRF2-mediated telomere shortening DNA Repair (Amst), 2007. doi:10.1016/j.dnarep.2006.09.005.
183. Le Deist F, Poinsignon C, Moshous D et al. Artemis sheds new light on V(D)J recombination. Immunol Rev 2004; 200:142-155.
184. Bonatto D, Revers LF, Brendel M et al. The eukaryotic Pso2/Snm1/Artemis proteins and their function as genomic and cellular caretakers. Braz J Med Biol Res 2005; 8(3):321-334.
185. Callebaut I, Moshous D, Mornon JP et al. Metallo-beta-lactamase fold within nucleic acids processing enzymes: the beta-CASP family. Nucleic Acids Res 2002; 30(16):3592-3601.
186. Bradshaw PS, Stavropoulos DJ, Meyn MS. Human telomeric protein TRF2 associates with genomic double-strand breaks as an early response to DNA damage. Nat Genet 2005; 37(2):193-197.
187. Opresko PL, von Kobbe C, Laine JP et al. Telomere-binding protein TRF2 binds to and stimulates the Werner and Bloom syndrome helicases. J Biol Chem 2002; 277(43):41110-9.
188. Stavropoulos DJ, Bradshaw PS, Li X et al. The Bloom syndrome helicase BLM interacts with TRF2 in ALT cells and promotes telomeric DNA synthesis. Hum Mol Genet 2002; 11(25):3135-3144.
189. Nishigori C, Fujisawa H, Uyeno K et al. Xeroderma pigmentosum patients belonging to complementation group F and efficient liquid-holding recovery of ultraviolet damage. Photodermatol Photoimmunol Photomed 1991; 8(4):146-150.

CHAPTER 9

XPG:
Its Products and Biological Roles

Orlando D. Schärer*

Abstract

Xeroderma pigmetosum patients of the complementation group G are rare. One group of XP-G patients displays a rather mild and typical XP phenotype. Mutations in these patients interfere with the function of XPG in the nucleotide excision repair, where it has a structural role in the assembly of the preincision complex and a catalytic role in making the incision 3' to the damaged site in DNA. Another set of XP-G patient is much more severely affected, displaying combined symptoms of xeroderma pigmentosum and Cockayne syndrome, referred to as XP/CS complex. Although the molecular basis leading to the XP/CS complex has not yet been fully established, current evidence suggests that these patients suffer from a mild defect in transcription in addition to a repair defect. Here, the history of how the *XPG* gene was discovered, the biochemical properties of the XPG protein and the molecular defects found in XP-G patients and mouse models are reviewed.

Introduction

Numerous important discoveries have resulted from studies related to xeroderma pigmentosum (XP) since the initial demonstration by Jim Cleaver that this disorder is caused by a cellular defect of repairing UV lesions in DNA.[1] Several complementation groups were discovered, the genes defective in XP cells were cloned, the biochemical basis for the nucleotide excision repair (NER) pathway, which is defective in XP cells, was uncovered and links between XP, cancer and premature aging were explored.[2] One of the fascinating aspects of these studies has been that although all of the eight genes that cause XP (*XPA-XPG* and *XP-V* or variant) with the exception of XP-V work in the same biochemical pathway, many of the genes have their own characteristic mutant patient alleles that can result in unique phenotypic manifestations in addition to the XP symptoms of varying severity. Furthermore, each of the XP genes (except XP-V) codes for proteins with a unique set of biochemical activities that contributes to the overall NER pathway in a distinct way. In this chapter the unique properties of the *XPG* gene will be explored.[3]

Discovery and Cloning of *XPG*

In the early 1970s the genetic heterogeneity of XP was established and XP patients were assigned to the complementation groups with an approach using cell fusion and complementation developed by Bootsma and colleagues.[4] In 1979, the first patient was assigned to the complementation group G,[5] marking the birth of XPG research. The first and second XP patient both exhibited some features not commonly seen in XP patients, such as bird-like faces, mental retardation and dental

*Orlando D. Schärer—Department of Pharmacological Sciences and Chemistry, Graduate Building 619, Stony Brook University, Stony Brook, NY 11974-3400, USA.
Email: orlando@pharm.stonybrook.edu

Molecular Mechanisms of Xeroderma Pigmentosum, edited by Shamim I. Ahmad and Fumio Hanaoka. ©2008 Landes Bioscience and Springer Science+Business Media.

caries.[6,7] We now know that those are features of Cockayne syndrome (CS) and that mutations in *XPG* can lead to both XP and combined XP/CS (see below).[8]

Just like all the XP patients and complementation groups have their unique properties, each one also has a unique story of how the corresponding gene was discovered and cloned. In the case of XPG, serendipity played an important role. While studying tRNA transcription, the laboratory of Stuart Clarkson isolated cDNAs encoding frog and human homologs of the yeast *RAD2* gene (the *XPG* equivalent in *Saccharomyces cerevisiae*) and demonstrated that the human cDNA was able to complement the UV sensitivity of XP-G cells.[9] XP-G cells were subsequently found to be equivalent to Chinese hamster ovary (CHO) cells from complementation group ERCC5[10-12] and the *XPG* gene was mapped at chromosome 13q32-33.[13,14]

Biochemical Properties of the XPG Protein

Analysis of the primary sequence of XPG quickly revealed that it harbors two nuclease domains placing it in the Fen1 family of structure-specific endonucleases.[9,15] Consistent with this observation XPG was shown to be a structure specific endonuclease that cleaves substrates with the polarity that would be required for the incision 3' to the lesion in NER.[15,16] XPG cleaves a variety of substrates that contain a ss/dsDNA junction, including splayed arm and bubble substrates, which contain a 5' single-stranded overhang.[17,18] At the primary sequence level, XPG shares two conserved regions, the N and I regions, with other nucleases (Fig. 1). A first glance of the salient features of the active site of XPG was provided by crystal structures of the exonuclease domains of T4 RNase H and T5 5'-exonuclease.[19,20] These studies showed a number of conserved acidic residues that coordinate metal ions. Sequence alignments of XPG with these structures revealed that several conserved acidic residues, including Glu77, Glu791 and Asp812, are poised to be part of the active site. Biochemical analysis confirmed a role of these residues in mediating XPG cleavage activity since mutation of these three residues to alanine abolishes the catalytic activity of the protein.[21,22]

While XPG shares its catalytic core with other nucleases, notably Fen1, the flap endonuclease with roles in replication and other aspects of DNA metabolism,[23] it contains additional unique domains that are responsible for mediating its NER specific functions. The roughly 600 amino acids separating the N and I region in XPG are commonly referred to as the spacer region.[9] This region is highly acidic and does not contain any known structural or functional motifs. Parts of the spacer region have been shown to interact with other proteins. The interaction of the N-terminal part of the spacer region with TFIIH has been shown to be important for the NER reaction.[24-26] An interaction with RPA has also been reported, but the functional significance of this interaction remains to be established.[27] The 300 amino acids at the C-terminal end of the protein, beyond the I region, are also engaged in protein-protein interactions; this region contains interaction sites with TFIIH and CSB and a PIP-box motif that mediates an interaction with PCNA.[24,28,29] A second interaction site with TFIIH is consistent with a strong functional interaction between XPG and TFIIH.[30] The functional significance of the interaction of XPG with CSB and PCNA has not yet been established. The C-terminus also contains two putative nuclear localization signals at residues 1051-1084 and 1169-1186.[31,32]

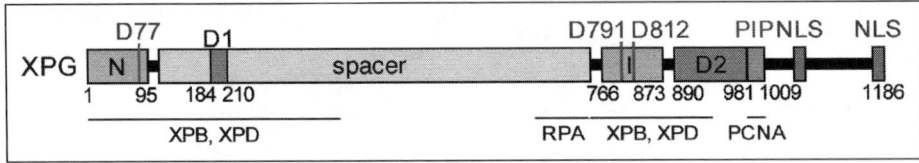

Figure 1. Functional domains of the XPG protein. The N and I domain (blue) make up the active site of XPG and are separated by the 600 amino acid spacer region. The D1 and D2 boxes are conserved in higher eukaryotes. The PCNA binding motif and nuclear localization signals in the C-terminus are indicated in orange and green, respectively. Interaction regions with TFIIH (XPB, XPD), RPA and PCNA are indicated.

The Role of XPG in Nucleotide Excision Repair

NER operates by two distinct pathways, global genome repair (GG-NER) and transcription-coupled repair (TC-NER) and XPG plays a key role in both of them. GG-NER is the better-understood pathway of the two and involves the removal of lesions from all sites of the genome, while TC-NER refers to the preferential repair of lesion from the transcribed strands of active genes.[33] GG-NER operates by the sequential assembly of all the proteins involved at sites of DNA damage.[34-36] Accordingly, NER proteins diffuse freely trough the cell and are recruited in a defined order to sites of DNA damage and are released again from NER complexes once the damage has been repaired. Current evidence suggests that the XPC/HR23B/centrin-2 protein complex is the initial damage recognition factor in GG-NER (Fig. 2).[35,37,38] XPC then recruits TFIIH to sites of UV lesion and two helicase subunits XPB and XPD of TFIIH initiate the opening of the DNA

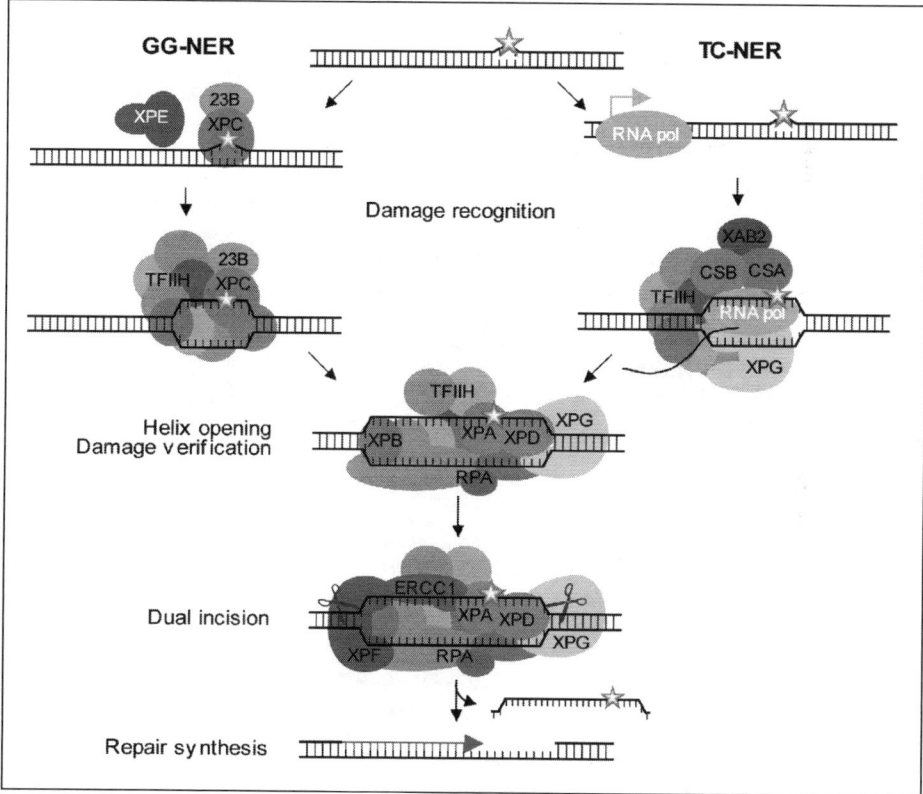

Figure 2. The role of XPG in nucleotide excision repair. In GG-NER, the XPC-HR23B protein is the initial damage recognition factor for helix distorting lesions. The DDB1/2 (XPE) proteins and associated factors play a role in recognizing lesions in the context of chromatin. XPC-HR23B recruits TFIIH to the site of the lesion and the two helicase subunits XPB and XPD open the DNA around the lesion. RPA, XPA and XPG are then recruited to assemble the "preincision" complex and verify the damage. XPG does not appear to be catalytically active at this stage. ERCC1-XPF finally joins the complex and the dual incision (5' by ERCC1-XPF and 3' by XPG) occurs. The resulting gap is filled in by the replication machinery and DNA ligase I seals the nick. TC-NER is less well characterized and is believed to be initiated by an RNA polymerase stalled at a damaged site. The CSB, CSA, XAB2, TFIIH and XPG proteins appear to play a role early in TC-NER, while the damage verification, dual incision and repair synthesis step are the same as in GG-NER.

around the lesion. XPA, RPA and XPG subsequently join the complex to form a stable open structure and recruitment of ERCC1-XPF by XPA initiates the dual incision of the DNA 5' and 3' to the lesion by ERCC1-XPF and XPG, respectively. The replication machinery subsequently fills the gap and seals the nick.

XPG has a structural as well as a catalytic role in GG-NER. Upon arrival at the NER complex it forms a complex at damaged sites with TFIIH, XPA and RPA (Fig. 2). This complex is stable enough to be subjected to band shift and footprinting assays.[39,40] XPG appears to have none or at least very limited nuclease activity at this stage and the full catalytic activity of XPG is only revealed once ERCC1-XPF has made the 5' incision. While efficient 3' incision by XPG appears to require the presence and catalytic activity of ERCC1-XPF, the 5' incision by ERCC1-XPF is efficiently carried out in the presence of catalytically inactive XPG.[21,22,40] One possibility therefore is that the 3' incision is triggered by a conformational change brought about in the NER complex by the incision activity of ERCC1-XPF. Consistent with this notion, XPG shows distinct requirements for binding and cleaving its DNA substrates.[18]

Protein-protein interactions are essential in mediating the various steps in NER. XPG interacts tightly with TFIIH and it is the interaction with TFIIH that recruits XPG to NER complexes at sites of UV damage.[30,35,41] Although investigation of the intracellular behavior of XPG suggests that the proteins has a mobility consistent with it being present in monomeric form,[41] some studies have found a tight and perhaps constitutive association of XPG with TFIIH.[42,50] It is currently believed that this constitutive interaction between XPG and TFIIH is important for the additional roles of the protein outside of NER (see below).

The role of XPG in TC-NER is less well understood. TC-NER is initiated by a RNA polymerase, stalled at the site of the lesion.[43] TC-NER involves almost all of the GG-NER factors with the exception of XPC and additionally requires the CSA, CSB and XAB2 proteins. Current evidence suggests that the presence of CSB at a site of a stalled RNA polymerase is required for recruitment of the core NER factors and enabling the dual incision reaction.[44,45] According to these studies, XPG is recruited as one of the core NER factors, playing a similar role as it does in GG-NER. Another study has suggested that XPG interacts directly with RNA polymerase II pointing to an earlier role of the protein in TC-NER in cooperation with the CSB protein.[46] The detailed mechanism of TC-NER and the role of XPG in this process are still not fully resolved and continue to be an active area of research.[47]

Roles of XPG Outside of Nucleotide Excision Repair

The severe phenotype displayed by the XP/CS patient suggested early on that XPG might have important roles outside of NER. Despite the phenotypic connection with the CSA and CSB genes, a defect in TC-NER alone cannot account for this observation, as *XPG* alleles that lead exclusively to XP symptoms are also deficient in TC-NER. It has been suggested that CS does not only result in a defect in TC-NER, but is rather results in a mild transcription defect.[48] This transcription defect can of course not be absolute, as this would be incompatible with viability. Although a role for XPG in transcription has not yet been clearly demonstrated in mammals, data from *S. cerevisiae* studies support a role for Rad2 (the XPG homolog) in transcription.[49] Importantly, this transcription function is independent of the nuclease activity of XPG, as a nuclease-deficient variant of the protein containing E794A mutation (equivalent to E791A in human XPG) retains full transcriptional activity. By contrast, a truncated version of RAD2, lacking 200 C-terminal amino acids but with intact nuclease activity, did not support the transcriptional activity. The inability of the truncated Rad2 protein to support transcription correlates well with the XP/CS phenotype displayed by XPG patients that have truncated *XPG* alleles (see below). By contrast, *XPG* alleles that form full-length protein, but have nuclease deficiency display the normal XP phenotype. This observation provides strong evidence that a defect in RNA polymerase II transcription is indeed responsible for the XP/CS phenotype. A very recent study, however, has suggested that at least a fraction of XPG resides in a tight complex with TFIIH.[50] Interestingly, *XPG* alleles, with truncations in the C-terminus that lead to XP/CS phenotype, fail to associate with TFIIH and this lack of association leads to a dissociation of the CAK subunit from TFIIH. As a consequence and in accordance with what had previously been observed for XPD alleles with C-terminal truncations with defective anchoring of the CAK subunit, phosphorylation and transactivation of nuclear receptors are reduced in cells from XP-G/CS patients. These findings provide further evidence for the notion that XP/CS patients suffer from some degree of impairment in transcription.

XPG has additional activities that are unique among the XP proteins that point to a role of the protein in the repair of oxidative damage. XPG has been shown to stimulate the activity of the DNA glycosylase NTH1, an enzyme involved in the base excision repair pathway that removes oxidized bases from DNA, independent of its nuclease activity.[51,52] Although this observation suggests a possible role in the removal of oxidative DNA lesions in a NER-independent pathway, initial in vivo studies supporting this suggestion have remained unconfirmed.[3,53,54] Thus although it is attractive to speculate that the nuclease-independent stimulation of the repair of oxidative lesions might contribute to the XP/CS phenotype, conclusive evidence to support this notion is currently lacking.

XP-G Patients and Their Mutant Alleles

XP-G patients are rare and only 14 XP cases have been assigned to this complementation group to date.[25,55] XP-G patients can be subdivided into three categories that have been found to display typical XP symptoms (4 patients; XP124LO, XP125LO, XP65BE, XP31KO), XP with late onset CS symptoms (3 patients; XP3BR, XP2BI, XPCS1BD), or XP with severe CS symptoms (7 patients; XPCS1LV, XPCS2LV, XPCS1RO, XP82DC, XP96TA, XP20BE, XPCS4RO), respectively (Fig. 3). In addition to the classical XP symptoms, this last group of patients exhibit developmental retardation, dwarfism, severe neurological abnormalities, bird-like faces and early death that are characteristic of CS. In 13 cases the molecular defects that account for the defects in the XPG genes and clinical symptoms have been characterized (Fig. 3). Based on this analysis a clear distinction can be made in the nature of the mutations that lead to the XP and XP/CS phenotypes.[56]

XP Group G Patients without CS

The patients of the XPG only group exhibit a mild XP phenotype with sensitivity to UV light, abnormal pigmentation, increased risk of skin cancer and no neurological abnormalities. All the patients of the XP class expressed at least one full-length allele with a point mutation or frame-shift induced sequence change. In the case of the XP124LO and XP125LO sibling patients, the only stably expressed allele contains an alanine to valine mutation at residue 792,[56,57] immediately adjacent to the highly conserved glutamic acid 791, which is essential for catalysis. Although the A792V allele stably expresses full length XPG protein, this protein has severely reduced nuclease and NER activities.[22] The mild XP phenotype of these two patients (no skin cancer at ~20 years of age) can be explained by the residual activity and UV resistance conferred by the A792A allele in vivo.

The full-length allele of patient XP65BE, containing a point mutation of Ala 874 to Thr, also resulted in the expression of full-length protein and conferred significant residual repair activity on XP-G/CS cells.[55] The patient accordingly exhibited a mild XP phenotype at age 14 without skin cancer, as proper care was taken to avoid exposure to sunlight.

The mutation in the XP-G patient, XP31KO, has not been established but this patient exhibited a very mild form of XP due a residual level of unscheduled DNA synthesis of about 25%.[58]

Patients with Severe XP and CS Symptoms

All of the XP-G patients, belonging to XP/CS complex class, displayed very severe phenotype including severe developmental retardation, dwarfism and neurological abnormalities in addition to the XP symptoms and suffered from early death (at or before six years of age).[55] Clarkson et al first reported in 1997 a common mutational pattern that distinguishes XP-G and XP-G/CS patients.[56] Analysis of the mutations in three XP-G/CS patients (XPCS1LV, XPCS2LV and XPCS1RO)

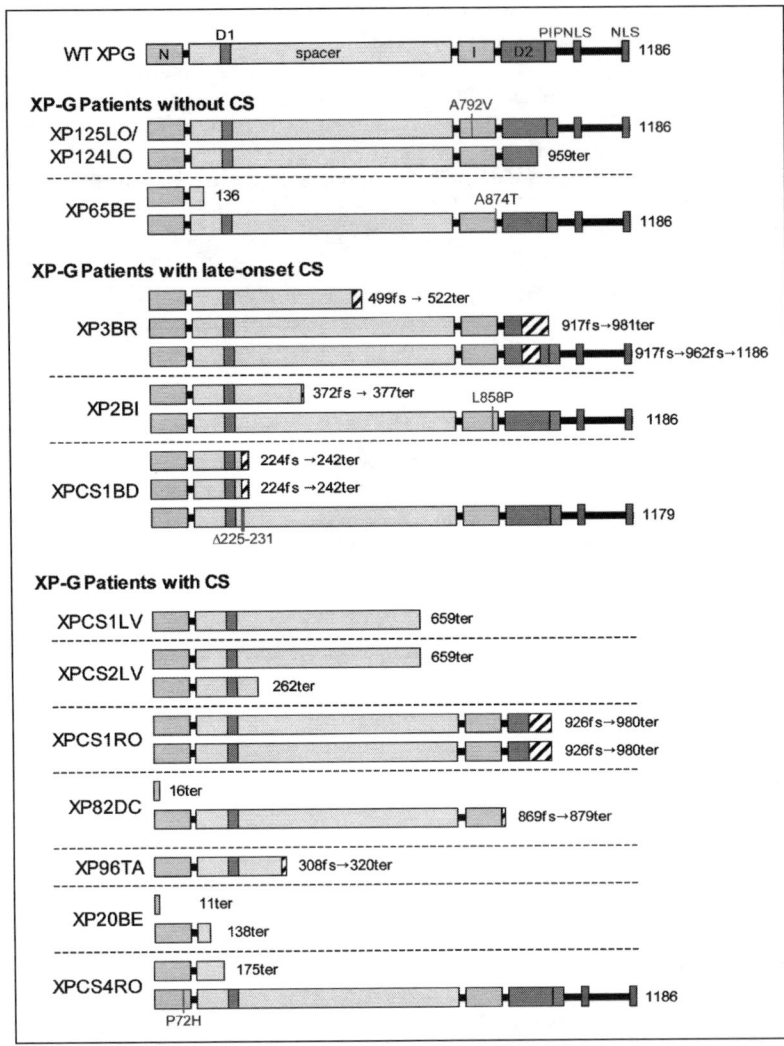

Figure 3. Known alleles of XP-G patients. The characterized alleles from XPG patients are divided into those from patients without CS, late onset CS and severe CS symptoms. The wild type XPG sequence is shown for reference. The positions of point mutations, frameshift mutations (fs) and premature stop codons (ter) are indicated. Striped boxes indicate coding sequences that are foreign to XPG.

revealed that each of them contained mutations that prevented the expression of full-length protein. The patient XPCS1LV expressed only one allele with a stop codon introduced by a single A deletion resulting in a protein of 659 amino acids in length. Patient XPCS2LV, who like XPCS1LV, is of Flemish origin, expressed the same allele and additionally a protein variant spanning the first 262 amino acids of the protein. The protein expressed in the cells of patient XPCS1RO (formerly known as 94RD27) is considerably longer. This patient is homozygous for an allele that contains a frameshift in the codon for amino acid 926 and expresses an XPG variant of 980 amino acids, of which the last 55 amino acids are unrelated to XPG.[57] Although this protein contains both, the N- and I-region that make up the nuclease active site, it does not confer any UV resistance to

XP-G cells and does not avert the symptoms of the XP/CS complex. This observation suggested that the C-terminal region of XPG plays an essential role in mediating functions of XPG within and outside of NER, while a localized defect in the nuclease activity only affects the NER specific functions. The first detailed neuropathological study of the XP/CS complex syndrome was carried out on XP-G patient XP20BE.[59,60] This individual expresses two very short XPG proteins of 10 and 137 amino acids, respectively.[59,61] These studies revealed that, consistent with the observed overt phenotype, the neuropathology is equivalent to that found in CS rather than the one found in severe XP patients with neurological abnormalities.

Two more recently characterized XP-G/CS patients, XP82DC, which expresses XPG proteins of 16 and 869 amino acids and XP96TA, which is homozygous for a mutation that leads to a frameshift in amino acid 308 resulting in a protein of 320 amino acids of which the last 12 amino acids are unrelated to XPG confirm the observations that truncation of the XPG protein results in an XP/CS phenotype.[55]

The last XP-G/CS patient shows a different genetic make up. XPCS4RO is one of the most severe XP/CS patients known and she died at 11 months of age. One of her *XPG* alleles encoded a truncated protein of 175 amino acids. The other allele encoded a full-length protein with a single amino acid substitution of the highly conserved proline 72 to histidine.[62] This allele does not confer any resistance to UV irradiation and the protein has been found to be highly unstable in overexpressed form, although the stability of this protein has not been assessed in cells of the patient directly. Based on these observation it is likely that the P72H mutation renders XPG completely unstable, resulting effectively in a null mutation.

XP-G Patients with Late-Onset CS Symptoms

The XP2BI and XP3BR patients both express full length XPG protein from one allele.[63] XP2BI contains a mutation of Leu 858 to Pro, a conserved residue in the I region while XP3BR expresses two splice variants from one allele, one of which expresses full length protein but with an internal stretch of 44 amino acids (917-962) unrelated to XPG. Both of these XPG variants are completely deficient in nuclease activity and their expression does not confer any UV resistance to XP-G/CS cells. The XP2BI and XP3BR patients both suffered from XP symptoms and a mild, late onset version of CS, clearly distinct from the typical symptoms of the XP/CS complex.

Perhaps the most unusual patient at the molecular level is XPCS1DB. This patient is homozygous for deletion of a nucleotide in the splice acceptor site in the last nucleotide of intron 6, which leads to a frameshift mutation in exon 7 resulting in the expression of a protein with 224 amino acids of XPG and 18 unrelated amino acids.[25] Based on the considerations discussed previously this would place him in the severe XP/CS category. However this patient displayed only mild late-onset CS symptoms with XP symptoms. Upon closer examination a second minor splice product was detected that results in the formation of a full-length protein with a deletion of amino acids 225-231. A cDNA expressing the XPG Δ225-231 did not confer any UV resistance to XP-G/CS cells indicating that this allele was completely NER deficient. The corresponding protein had nuclease activity and the cause of the NER deficiency could be attributed to a defect in the interaction with TFIIH.[25] Interestingly the XPGΔ225-231 protein was still recruited to NER complexes in living cells, but failed to mediate appropriate processing of substrates, presumably because it failed to stably associate with the NER complexes.

As these examples demonstrate, the study of the various alleles found in XP-G patients has not only revealed insight into the molecular basis underlying the XP and XP/CS complex symptoms, but also into the mechanisms by which XPG exerts its roles in NER and other pathways.

Mouse Models with XPG Deficiency (see also Chapter 17)

The consequences of XPG deficiency have also been studied in mouse models. These studies are in broad agreement with observations made in the analyses of the various alleles found in XP-G patients. A mouse in which the wild-type allele was exchanged for one that expresses XPG containing an E791A mutation that abolishes nuclease activity displayed the classical XP phenotype.[64] These mice displayed hypersensitivity to UV irradiation, but did not display any of the developmental defects characteristic of the XP/CS complex. By contrast, mice in which exon 3 corresponding to nucleotides 264 to 380 of XPG was deleted, lacked any functional XPG protein and displayed a phenotype much more reminiscent of XP-G/CS patients.[65] These mice showed severe growth retardation and died prematurely before weaning at about three weeks of age. Cells from these mice underwent premature replicative senescence and readily underwent transformation, suggestive of genetic instability. Like the E791A mutation, cells from the knock out mice were hypersensitive to UV irradiation.

The generation of additional XPG mutant mice provided insight into the nature of the mutations in the gene that cause XP and combined XP/CS phenotypes. Consistent with observations made in patients, mice in which the last 360 amino acids in the C-terminus of XPG were deleted displayed the expected symptoms associated with the XP/CS complex.[3,66,67] By contrast deletion of only 183 C-terminal amino acids (exon 15) resulted only in partial UV sensitivity and none of the CS characteristics. However, when this 183 amino acid deletion in XPG was combined with a deletion in XPA, it resulted in CS-like phenotypes reminiscent of XPG/CS, indicating that this deletion has serious consequences only in an NER deficient background. This observation underscores the complexity of genotype/phenotype relationships and warrants further exploration.

Conclusion

Research of the XP group G patients has uncovered the genetic bases of why some patients display CS in addition to XP symptoms, while others do not. We now understand the role of the XPG protein in preventing XP symptoms as a structure-specific endonuclease in nucleotide excision repair in some detail. The mechanisms by which XPG prevents CS syndromes are less well understood, but there is increasing recent evidence that a XPG plays a role in transcription by interacting with TFIIH. To what extent a defect in the suggested role of XPG in the NER-independent repair of oxidative damage contributes to the CS phenotype is still in need of further clarification. Without doubt, current and future research efforts will clarify this and other questions regarding the function of XPG.

Acknowledgements

I thank Dr. Stuart G. Clarkson for insightful comments on the manuscript and Dr. Kiyoji Tanaka for communicating data prior to publication.

References

1. Cleaver JE. Defective Repair Replication of DNA in Xeroderma Pigmentosum. Nature 1968; 218:652-656.
2. Friedberg EC, Walker GC, Siede W et al. DNA Repair and Mutagenesis. 2nd edition ed. ASM Press, Washington DC, 2005.
3. Clarkson SG. The XPG story. Biochimie 2003; 85:1113-1121.
4. De Weerd-Kastelein EA, Keijzer Bootsma D. Genetic heterogeneity of xeroderma pigmentosum demonstrated by somatic cell hybridization. Nat New Biol 1972; 238:80-83.
5. Keijzer W, Jaspers NG, Abrahams PJ et al. A seventh complementation group in excision-deficient xeroderma pigmentosum. Mutat Res 1979; 62:183-190.
6. Cheesbrough MJ. Xeroderma pigmentosum—a unique variant with neurological involvement. Br J Dermatol 1978; 99(suppl 16):61.
7. Arlett CF, Harcourt SA, Lehmann AR et al. Studies on a new case of xeroderma pigmentosum (XP3BR) from complementation group G with cellular sensitivity to ionizing radiation. Carcinogenesis 1980; 1:745-751.
8. Vermeulen W, Jaeken J, Jaspers NG et al. Xeroderma pigmentosum complementation group G associated with Cockayne syndrome. Am J Hum Genet 1993; 53:185-192.
9. Scherly D, Nouspikel T, Corlet J et al. Complementation of the DNA repair defect in xeroderma pigmentosum group G cells by a human cDNA related to yeast RAD2. Nature 1993; 363:182-185.
10. Mudgett JS, MacInnes MA. Isolation of the functional human excision repair gene ERCC5 by inter-cosmid recombination. Genomics 1990; 8:623-633.
11. O'Donovan A, Wood RD. Identical defects in DNA repair in xeroderma pigmentosum group G and rodent ERCC group 5. Nature 1993; 363:185-188.

12. MacInnes MA, Dickson JA, Hernandez RR et al. Human ERCC5 cDNA-cosmid complementation for excision repair and bipartite amino acid domains conserved with RAD proteins of Saccharomyces cerevisiae and Schizosaccharomyces pombe. Mol Cell Biol 1993; 13:6393-6402.

13. Takahashi E, Shiomi N, Shiomi T. Precise localization of the excision repair gene, ERCC5, to human chromosome 13q32.3-q33.1 by direct R-banding fluorescence in situ hybridization. Jpn J Cancer Res 1992; 83:1117-1119.

14. Samec S, Jones TA, Corlet J et al. The human gene for xeroderma pigmentosum complementation group G (XPG) maps to 13q33 by fluorescence in situ hybridization. Genomics 1994; 21:283-285.

15. Harrington JJ, Lieber MR. Functional domains within FEN-1 and RAD2 define a family of structure-specific endonucleases:implications for nucleotide excision repair. Genes Dev 1994; 8:1344-1355.

16. O'Donovan A, Davies AA, Moggs JG et al. XPG endonuclease makes the 3' incision in human DNA nucleotide excision repair. Nature 1994; 371:432-435.

17. Evans E, Fellows J, Coffer A et al. Open complex formation around a lesion during nucleotide excision repair provides a structure for cleavage by human XPG protein. EMBO J 1997; 16:625-638.

18. Hohl M, Thorel F, Clarkson SG et al. Structural Determinants for Substrate Binding and Catalysis by the Structure-specific Endonuclease XPG. J Biol Chem 2003; 278:19500-19508.

19. Mueser TC, Nossal NG, Hyde CC. Structure of bacteriophage T4 RNase H, a 5' to 3' RNA-DNA and DNA-DNA exonuclease with sequence similarity to the RAD2 family of eukaryotic proteins. Cell 1996; 85:1101-1112.

20. Ceska TA, Sayers JR, Stier G et al. A helical arch allowing single-stranded DNA to thread through T5 5'- exonuclease. Nature 1996; 382:90-93.

21. Wakasugi M, Reardon JT, Sancar A. The noncatalytic function of XPG protein during dual incision in human nucleotide excision repair. J Biol Chem 1997; 272:16030-16034.

22. Constantinou A, Gunz D, Evans E et al. Conserved residues of human XPG protein important for nuclease activity and function in nucleotide excision repair. J Biol Chem 1999; 274:5637-5648.

23. Liu Y, Kao HI, Bambara RA. Flap endonuclease 1:a central component of DNA metabolism. Annu Rev Biochem 2004; 73:589-615.

24. Iyer N, Reagan MS, Wu KJ et al. Interactions involving the human RNA polymerase II transcription/ nucleotide excision repair complex TFIIH, the nucleotide excision repair protein XPG and Cockayne syndrome group B (CSB) protein. Biochemistry 1996; 35:2157-2167.

25. Thorel F, Constantinou A, Dunand-Sauthier I et al. Definition of a Short Region of XPG Necessary for TFIIH Interaction and Stable Recruitment to Sites of UV Damage. Mol Cell Biol 2004; 24:10670-10680.

26. Dunand-Sauthier I, Hohl M, Thorel F et al. The spacer region of XPG mediates recruitment to nucleotide excision repair complexes and determines substrate specificity. J Biol Chem 2005; 280:7030-7037.

27. He Z, Henricksen LA, Wold MS et al. RPA involvement in the damage-recognition and incision steps of nucleotide excision repair. Nature 1995; 374:566-569.

28. Gary R, Ludwig DL, Cornelius HL et al. The DNA repair endonuclease XPG binds to proliferating cell nuclear antigen (PCNA) and shares sequence elements with the PCNA-binding regions of FEN-1 and cyclin-dependent kinase inhibitor p21. J Biol Chem 1997; 272:24522-24529.

29. Warbrick E. The puzzle of PCNA's many partners. BIOESSAYS, 2000; 22:997-1006.

30. Araujo SJ, Nigg EA, Wood RD. Strong functional interactions of TFIIH with XPC and XPG in human DNA nucleotide excision repair, without a preassembled repairosome. Mol Cell Biol 2001; 21:2281-2291.

31. Knauf JA, Pendergrass SH, Marrone BL et al. Multiple nuclear localization signals in XPG nuclease. Mutat Res 1996; 363:67-75.

32. Park MS, Knauf JA, Pendergrass SH et al. Ultraviolet-induced movement of the human DNA repair protein, Xeroderma pigmentosum type G, in the nucleus. Proc Natl Acad Sci USA 1996; 93:8368-8373.

33. Hoeijmakers JH. Genome maintenance mechanisms for preventing cancer. Nature 2001; 411:366-374.

34. Houtsmuller AB, Rademakers S, Nigg AL et al. Action of DNA repair endonuclease ERCC1/XPF in living cells. Science 1999; 284:958-961.

35. Volker M, Mone MJ, Karmakar P et al. Sequential assembly of the nucleotide excision repair factors in vivo. Mol Cell 2001; 8:213-224.

36. Riedl T, Hanaoka F, Egly JM. The comings and goings of nucleotide excision repair factors on damaged DNA. EMBO J 2003; 22:5293-5303.

37. Sugasawa K, Ng JM, Masutani C et al. Xeroderma pigmentosum group C protein complex is the initiator of global genome nucleotide excision repair. Mol Cell 1998; 2:223-232.

38. Gillet LC, Schärer OD. Molecular mechanisms of Mammalian global genome nucleotide excision repair. Chem Rev 2006; 106:253-276.

39. Wakasugi M, Sancar A. Assembly, subunit composition and footprint of human DNA repair excision nuclease. Proc Natl Acad Sci USA 1998; 95:6669-6674.

40. Tapias A, Auriol J, Forget D et al. Ordered conformational changes in damaged DNA induced by nucleotide excision repair factors. J Biol Chem 2004; 279:19074-19083.

41. Zotter A, Luijsterburg MS, Warmerdam DO et al. Recruitment of the Nucleotide Excision Repair Endonuclease XPG to Sites of UV-induced DNA Damage Depends on Functional TFIIH. Mol Cell Biol 2006; 26:8868-8869.

42. Mu D, Park CH, Matsunaga T. Reconstitution of human DNA repair excision nuclease in a highly defined system. J Biol Chem 1995; 270:2415-2418.

43. Svejstrup JQ. Mechanisms of transcription-coupled DNA repair. Nat Rev Mol Cell Biol 2002; 3:21-29.

44. Fousteri M, Vermeulen W, van Zeeland AA et al. Cockayne syndrome A and B proteins differentially regulate recruitment of chromatin remodeling and repair factors to stalled RNA polymerase II in vivo. Mol Cell 2006; 23:471-482.

45. Laine JP, Egly JM. Initiation of DNA repair mediated by a stalled RNA polymerase IIO. EMBO J 2006; 25:387-397.

46. Sarker AH, Tsutakawa SE, Kostek S et al. Recognition of RNA polymerase II and transcription bubbles by XPG, CSB and TFIIH: insights for transcription-coupled repair and Cockayne Syndrome. Mol Cell 2005; 20:187-198.

47. Sarasin A, Stary A. New insights for understanding the transcription-coupled repair pathway. DNA Repair (Amst) 2007; 6:265-269.

48. van Gool AJ, van der Horst GT, Citterio E et al. Cockayne syndrome: defective repair of transcription? EMBO J 1997; 16:4155-4162.

49. Lee SK, Yu SL, Prakash L et al. Requirement of yeast RAD2, a homolog of human XPG gene, for efficient RNA polymerase II transcription. implications for Cockayne syndrome. Cell 2002; 109:823-834.

50. Ito S, Kuraoka I, Chymkowitch P et al. XPG stabilizes TFIIH allowing transactivation of nuclear receptors: Implications for Cockayne syndrome in XP-G/CS patients. Mol Cell 2007; 26:231-243.

51. Bessho T. Nucleotide excision repair 3' endonuclease XPG stimulates the activity of base excision repairenzyme thymine glycol DNA glycosylase. Nucleic Acids Res 1999; 27:979-983.

52. Klungland A, Hoss M, Gunz D et al. Base excision repair of oxidative DNA damage activated by XPG protein. Mol Cell 1999; 3:33-42.

53. Cooper PK, Nouspikel T, Clarkson SG et al. Defective transcription-coupled repair of oxidative base damage in Cockayne syndrome patients from XP group G. Science 1997; 275:990-993.

54. Cooper PK, Nouspikel T, Clarkson SG. Retraction. Science 2005; 308:1740.

55. Emmert S, Slor H, Busch DB et al. Relationship of neurologic degeneration to genotype in three xeroderma pigmentosum group G patients. J Invest Dermatol 2002; 118:972-982.

56. Nouspikel T, Lalle P, Leadon SA et al. A common mutational pattern in Cockayne syndrome patients from xeroderma pigmentosum group G: implications for a second XPG function. Proc Natl Acad Sci USA 1997; 94:3116-3121.

57. Nouspikel T, Clarkson SG. Mutations that disable the DNA repair gene XPG in a xeroderma pigmentosum group G patient. Hum Mol Genet 1994; 3:963-967.

58. Ichihashi M, Fujiwara Y, Uehara Y et al. A mild form of xeroderma pigmentosum assigned to complementation group G and its repair heterogeneity. J Invest Dermatol 1985; 85:284-287.

59. Moriwaki S, Stefanini M, Lehmann AR et al. DNA repair and ultraviolet mutagenesis in cells from a new patient with xeroderma pigmentosum group G and cockayne syndrome resemble xeroderma pigmentosum cells. J Invest Dermatol 1996; 107:647-653.

60. Lindenbaum Y, Dickson D, Rosenbaum P et al. Xeroderma pigmentosum/cockayne syndrome complex: first neuropathological study and review of eight other cases. Eur J Paediatr Neurol 2001; 5:225-242.

61. Okinaka RT, Perez-Castro AV, Sena A et al. Heritable genetic alterations in a xeroderma pigmentosum group G/Cockayne syndrome pedigree. Mutat Res 1997; 385:107-114.

62. Zafeiriou DI, Thorel F, Andreou A et al. Xeroderma pigmentosum group G with severe neurological involvement and features of Cockayne syndrome in infancy. Pediatr Res 2001; 49:407-412.

63. Lalle P, Nouspikel T, Constantinou A et al. The founding members of xeroderma pigmentosum group G produce XPG protein with severely impaired endonuclease activity. J Invest Dermatol 2002; 118:344-351.

64. Tian M, Jones DA, Smith M et al. Deficiency in the nuclease activity of xeroderma pigmentosum G in mice leads to hypersensitivity to UV irradiation. Mol Cell Biol 2004; 24:2237-2242.

65. Harada YN, Shiomi N, Koike M et al. Postnatal growth failure, short life span and early onset of cellular senescence and subsequent immortalization in mice lacking the xeroderma pigmentosum group G gene. Mol Cell Biol 1999; 19:2366-2372.

66. Shiomi N, Kito S, Oyama M et al. Identification of the XPG region that causes the onset of Cockayne syndrome by using Xpg mutant mice generated by the cDNA-mediated knock-in method. Mol Cell Biol 2004; 24:3712-3719.

67. Shiomi N, Mori M, Kito S et al. Severe growth retardation and short life span of double-mutant mice lacking Xpa and exon 15 of Xpg. DNA Repair (Amst) 2005; 4:351-357.

CHAPTER 10

Xeroderma Pigmentosum Variant, XP-V:
Its Product and Biological Roles

Chikahide Masutani, Fumio Hanaoka and Shamim I. Ahmad*

Introduction

Xeroderma pigmentosum (XP) is a rare autosomal recessive genetic disorder first reported in 1874 by Hebra and Kaposi[1] and now known to involve a number of phenotypic characteristics, including photophobia, early onset of freckling and neoplastic alterations on sun exposed areas of body. So far, eight complementation groups of XP have been identified including XP-A through -G and XP-V (XP variant). About 80% of XP patients belong to XP-A to XP-G groups and has deficiency in Nucleotide Excision Repair (NER). The remaining 20% of XP patients are of the XP-V type. They have normal NER but are deficient in Translesion DNA synthesis (TLS). In addition, there are XP cases where mutations in XP genes have effects on other disorders such as Cockayne syndrome, trichothiodystrophy and progeroid syndrome (see Chapter 14). This chapter will mainly focus on XP-V.

Human DNA Polymerases

Cells from XP-V patients lack a specific DNA repair polymerase and since its activity is usually associated with other polymerases, it will be useful to give a brief description of other polymerases in human cells. Fourteen different human DNA dependent DNA polymerases have been identified. These have been classified into four families: DNA polymerase γ (gamma), θ (theta) and ν (nu) belong to the A family. Polymerase γ is the only mitochondrial enzyme and this and all polymerases of A family having homology with *Escherichia coli* DNA polymerase-I. DNA polymerases α (alpha), δ (delta), ε (epsilon) and ζ (zeta) make up group B; the former three polymerases are involved in general replication (and repair) of nuclear DNA and Pol ζ has been found to play a role in mutagenesis. DNA polymerases β (beta), λ (lambda) and μ (mu) are in the X family and polymerase β has been shown to play a significant role in Base Excision Repair (BER). Polymerase η (eta), ι (iota), κ (kappa) and Rev1 make up the Y family and their primary roles are in TLS. All XP-V patients have been found to be deficient in pol η activity.[2-4]

XP-V Gene and Its Homologues

In humans *POLH* (also known as *hRad30A*) codes for pol η. Its homologues have been found in mouse (pol η)[5], Drosophila (pol η)[6] and *Arabidopsis thaliana* (pol η).[7] Homology analysis of the Y family of polymerases in different organisms can be presented on the basis of evolution. Homologues of *din*B seem to have been conserved from *E. coli* through *Saccharomyces pombe* to humans. Pol κ is a homologue of *din*B in humans. In *Saccharomyces cerevisiae* a homologue,

*Corresponding Author: Shamim Ahmad—Nottingham Trent University, Clifton Lane, Nottingham NG11 8NS, England. Email: shamim.ahmad@ntu.ac.uk

Molecular Mechanisms of Xeroderma Pigmentosum, edited by Shamim I. Ahmad and Fumio Hanaoka. ©2008 Landes Bioscience and Springer Science+Business Media.

Rad30, has been found and its next evolutionary homologue in *S. pombe* is Eso1.[8-10] In humans pol η and pol ι are homologues of Rad30. Other members of the Y family are polV in *E. coli*, Dpo4 in *Sulfolobus solfataricus* P2[8] and Rev1 in *S. cerevisiae*, *S. pombe* and humans.[11,12]

Structure and Activities of Polymerase η

From in vitro DNA replication experiments, in cell-free extracts from XP-V patients, it has been shown that XP-V cells are defective in replicating cyclobutane pyrimidine dimer (CPD) on the template DNA, whereas extracts from normal cells can proficiently bypass it.[13] The pol η protein, able to complement the defect of XP-V repair in cell-free extracts, has been purified from HeLa cell and shown that it can bypass CPDs as efficiently as undamaged DNA, but stops at (6-4) photoproducts.[13,14] The latter type of damage is more efficiently repaired by NER compared to the former type. Hence the two types of damages are efficiently tolerated by a combined action of these two processes, TLS and NER.

POLH is located on human chromosome 6p21.1-6p12 and comprises 11 exons spanning 40 kilobases.[15,16] Human pol η is 713 amino acids long and the N-terminal 511 amino acids contain conserved motifs among Y family polymerases. These 511 amino acids have been shown to be adequate for the DNA polymerase activity.[13,14] The C-terminal portion is important for cellular localization of the protein including its nuclear localization and foci formation with the replication machinery.[17] Pol η is mostly localized uniformly in the nucleus and is associated with replication foci during S phase. The same study group identified an XP-V mutant patient whose pol η polymerase motif was intact but who had lost the relocalization domain in the C-terminal portion.[17]

The C-terminal portion of pol η also contains interacting residues with other proteins such as pol ι, Rev1, Rad18, proliferating cell nuclear antigen (PCNA) and ubiquitin, suggesting that this part of the protein plays a regulatory role.[17-21]

Pol η has low processivity and lacks proof-reading exonucleolytic activity. Structural analysis of this protein showed that it has more opened active sites than most replicative DNA polymerases such as T7 DNA polymerase.[22] Studies have shown that, except for passing through CPD, pol η is an error-prone enzyme and in fact pol η can mis-insert wrong nucleotides with a frequency of 10^{-2} to 10^{-3} if it passes through undamaged DNA.[23,24] However, pol η has also been shown to bind more stably to DNA having thymine-thymine type cyclobutane pyrimidine dimer (TT-CPD) if it incorporates the correct nucleotides, and can usually carry out accurate DNA synthesis by incorporating 2 As opposite TT-CPD and in that case the chain elongation can move further two or more nucleotides and then the enzyme dissociate from the DNA (for further steps of DNA synthesis see below). Nonetheless, if a G is incorporated opposite the CPD the elongation process stops at this site.[25,26]

In addition to TLS, pol η participates in somatic hypermutation of immunoglobulin genes. Analysis of mutation spectra of immunoglobulin variable gene VH6, obtained from peripheral blood lymphocytes of 3 XP-V patients showed that although the mutation frequency was normal, nevertheless the types of base change were different. There was a decrease in mutation frequency at A and T and a concomitant rise in mutation in G and C. It was proposed that more than one polymerase contributes to hypermutation and that pol η is involved in causing errors predominantly at A and T sites and other polymerase at G and C.[27,28] Pol η has also been reported to participate in homologous recombinational repair; for example it can extend DNA synthesis from the D loop recombination intermediate in which an invading strand serves as the primer.[29] On the other hand, in the same assay, mutant XP-V cell extracts failed to extend the primer. It was also shown that RAD51 recombinase interacts with pol η and this protein stimulates pol η-mediated D loop extension. Pol η mutants display a significant decrease in the frequency of both *Ig* gene conversion and double-strand break-induced homologous recombination in chicken DT40 cells.[30]

Bypassing of Unusual Nucleotides by Pol η

In the last decade a number of papers have appeared showing the bypassing of pol η (and of, pol ι, pol κ and Rev1) through a number of different unusual nucleotides and abasic sites. Studies have shown that pol η can bypass 8-oxoguanine and O[6]-methylguanine lesions efficiently and C and C/T respectively are inserted opposite the lesions.[31] Interestingly, this enzyme cannot insert a nucleotide opposite an abasic site, which means that it essentially requires a template for its activity. The enzyme can also catalyze relatively accurately TLS past thymine glycol lesions by preferentially inserting A opposite thymine glycol.[32] Choi's group have carried out a number of studies and have shown that pol η can effectively bypass N2-methyl(ME)G, N2-ethyl(Et)G, N2-isobutyl(Ib)G, N2-benzyl(BZ)G and N2-CH2(2-nzyl)G but was severely blocked by N2-CH2(9-anthracenyl)G, (N2-SnthG) and N2-CH2(6-benzo[a]pyrenylG (N2BPG).[33-36]

Pol η can also bypass N2-isopropyle G adduct lesions with better efficiency and accuracy compared to the un-adducted G.[37] On the other hand, this enzyme can bypass N6-isopropyl A with only modest efficiency and accuracy. The kinetics of insertion of nucleotides in the polymerizing chains was also studied and shown to be variable with different nucleotides in the parental strand. Pol η can also bypass the DNA adduct of cisplatin causing intra-strand crosslinks, inserting mainly C in the complementary strand.[38,39] (To note that cisplatin can also cause a variety of other types of DNA damage including mono-adducts, inter-strand DNA cross-links, protein to DNA cross-links and glutathione DNA cross-links). XP-V cells are hyper-sensitive to cisplatin, carbo-platin and oxaliplatin[40] which indicates that pol η probably plays a repair role in damage induced by these agents. The same sensitivity was found in XP-V mutants. Interestingly XP-A cells are also sensitive to these agents, but an important difference between XP-A and XP-V is that the absence of pol η expression results in a reduced ability to overcome cisplatin induced S phase arrest. Pol η also plays an important role in modulating cellular sensitivity to agents that are used to target DNA as anticancer drugs;[41] for example XP-V patients were found to be 3 fold more sensitive to β-d-arabinofuranocyl cytosine, gemcitabine and cisplatin used individually. Combination therapies such as gemcitabine+cisplatin gave 10-fold higher sensitivity. These results may signify important roles of pol η in developing and employing anti-cancer drugs.

Yasui et al[42] studied the importance of the interaction of tamoxifen (TAM, a widely used chemotherapeutic agent for breast cancer and also attributed to mutagenecity) with DNA. This agent causes the formation of α-(N2-deoxyguanosinyl) tamoxifen (dG-N2-TAM) adducts in DNA. They produced and studied site-specifically modified oligodeoxynucleotides containing a single diastereoisomer of trans or cis forms of dG-N2-TAM. The results of primer extension reactions showed that, although pol η bypassed the modified nucleotide, inserting dCMP opposite the adduct, pol κ extended this more efficiently. Thus the properties of pol η and pol κ are consistent with the mutagenic event attributed to the TAM-DNA adduct.

Shibutani's group carried out studies on the risk involved in hormone replacement therapy (HRT).[43,44] This therapy increases the risk of developing breast, ovarian and endometrial cancers. Two major components of the HRT drug are equilin and equilenin. 4-Hydroxyequilenin (4-OHEN) is a major metabolite of equilin and equilenin which promotes 4-OHEN-modified dC, dA and dG adducts. These adducts are present in breast and other tumors of patients receiving HRT. TLS studies by pol η showed that the 4-OHEN-dC DNA adduct is a highly mutagenic lesion generating C → T transition and C → G transversion. However, opposite 4-OHEN-dA lesions pol η (and pol κ) are able to introduce the correct nucleotide, dTMP. In another study the activity of pol η for bypassing 5-methylcytosine (5MC) was analysed and showed that the catalytic core of the enzyme was inserting dGMP opposite the 5MC of the CPD with about 120:1 selectivity relative to dAMP. Furthermore dTTP or dCMP were not inserted opposite 5MC.[45]

Other studies show that, during a single cycle of processive DNA synthesis, Dpo4 (13-30% efficiency) and pol η (10-13% efficiency) bypass synthetic AP (apurinic or apyrimidinic) sites.[46] Furthermore, the bypass is nearly 100% mutagenic for the AP site, lacking A or G and inserting dAMP. This dAMP insertion occurs with between 70-80% efficiency opposite the AP site. Thus AP site bypass could be a source of substitution or frameshift mutation.

In contrast to several abnormal and modified nucleotides where pol η can effectively or semi-effectively carry out TLS, Barone et al[47] have shown that this enzyme bypassed 2-hydroxy-adenine very inefficiently whereas Dpo4 did it efficiently.

In addition to the bypass ability of damaged nucleotides in the template DNA, pol η can incorporate oxidized DNA precursors during DNA synthesis.[48] Thus pol η can incorporate 8-OHdGTP opposite template A at 60% efficiency of normal T incorporation and 2-OHdATP opposite template T, G or C at substantial levels. Hence this enzyme is suggested to participate in oxidative mutagenesis through this ability.

Mutation in *POLH* and Its Effects

Maher et al[49] were the first to show that the frequency of UV induced mutation is 25-fold higher in XP-V cells than in normal human cells. Their subsequent studies revealed that caffeine enhanced the cytotoxic and mutagenic effects of UV in XP-V.[50] In XP-V patients, a large number of different mutations in this gene have been described,[14,51,52] these include single-base pair substitutions, small insertions and deletions resulting in frame-shifts and nonsense mutations (Fig. 1). In case of nonsense mutation, the prematurely terminated proteins, encoded by these mutant alleles, are unable to be transported to nuclei because they lack the nuclear localization signal, although they might have DNA polymerase activities. Other patients have missense mutations in the N-terminal region coding for the active site of the enzyme (approximately 350 amino acids long).[51] In an analysis of the *POLH* coding region, it was concluded that neither mutations nor polymorphism are required for the development of human squamous cell carcinoma.[53]

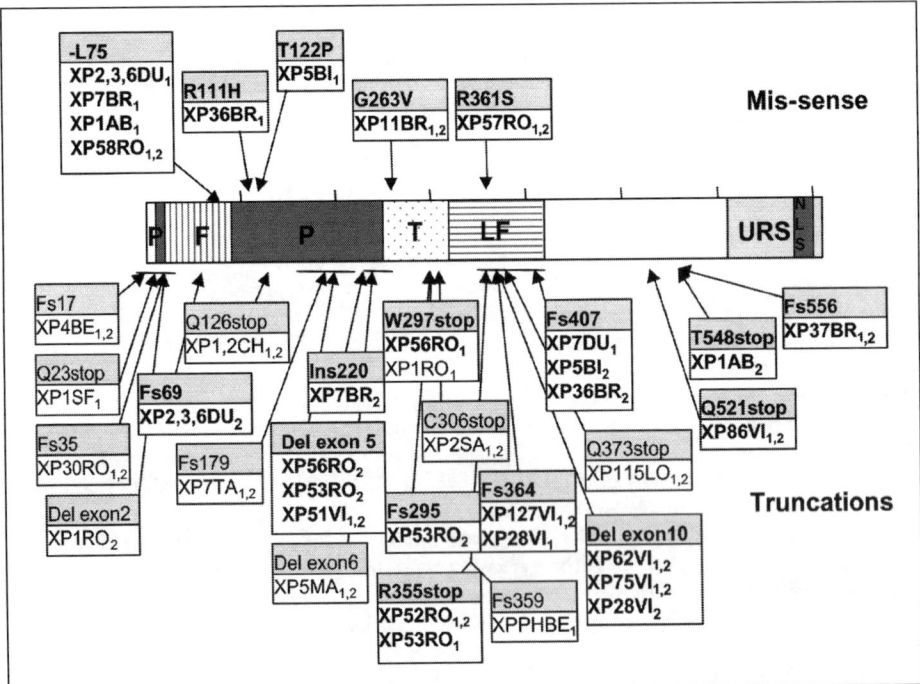

Figure 1. Sites of mutations identified in polη. Cell strain designations are indicated in boxes, with those used in the study.[79] Others are from references 14, 15 and 52. Subscripts denote different alleles. Horizontal lines indicate deletions. Single amino acid changes are shown in the upper part of the diagram and truncations are shown in the lower part. The central part of the diagram shows the different domains in the protein (see ref. 79). Reproduced with permission from: Broughton BC et al. Proc Natl Acad Sci USA 2002; 99:815-820;[80] ©2002 National Academy of Sciences, USA.

Mechanism of Mutagenesis in Pol η Mutant Strains— The Roles of Other Polymerases

In cells with nonmutant pol η when the replication polymerase, together with its replication complex, encounters damage, it stops further synthesis. This is assumed to send a signal to pol η and the enzyme bypasses the dimer accurately by inserting correct 2 As opposite the TT-CPD. Its subsequent step has not been confirmed but it is likely that the pol η complex, at this stage, is dissociated and then the DNA region is bound by another complex involving DNA polymerase δ and/or ε. According to the proposed model the DNA which is already attached with PCNA, also gets attached to this region forming a complex. In case of no other DNA damage, the replicative polymerase complex will continue synthesis without inserting any wrong nucleotides i.e., error proof synthesis will be achieved.

In pol η mutant cells DNA synthesis is carried out as usual by the replication polymerase complex and this complex again stops at the site of TT-CPD. Possibly the subsequent step is that the TT-CPD is attended by another set of enzymes, pol ι, pol κ, pol ζ and Rev1 for repair to take place. These enzymes possibly cooperate to bypass the TT-CPD and in doing so they introduce errors which ultimately lead to mutation. Hence repair of DNA in pol η mutants, carried out by other polymerases, is highly error-prone and mutations are introduced at a high frequency.[49] Additionally Rev1, by interacting with pol η, pol ι, pol κ, pol ζ may play important roles in mutagenic bypass.[19,54-56] Reduction in Rev1 expression results in reduced UV-induced mutations indicating that it participates in error-prone TLS past UV lesions.[57] In a recent study it has been shown that pol ι can insert G or T opposite the 3′ nucleotide of TT or TU-CPD.[58] It was concluded that pol ι causes UV-induced mutations in pol η mutants, although insertion of G against U is nonmutagenic because U is produced by C as a result of its deamination.[59] The introduction of error by pol ι is not only at the site of DNA damage but this polymerase can introduce spontaneous mutation too. The error frequency opposite G or C is approximately 10^{-2}.

Interaction of Pol η with Other Proteins

Pol η interacts with Y-family polymerases, pol ι and Rev1.[17,19,56] Pol ι and pol η form replication foci with PCNA which is largely dependent on pol η.[17] Likewise the foci formation by Rev1 is also largely pol η dependent (Masutani et al unpublished observation). Pol η forms foci with replication complexes in nuclei, which are enhanced by treating cells with DNA damaging agents such as UV. Pol η can interact with PCNA, the sliding clamp of DNA replication polymerase and this interaction may contribute to foci formation.[18] Responding to DNA replication arrest, PCNA is monoubiquitinated. Two ubiquitin binding domains, UBM and UBZ, which are conserved in all human or mammalian Y family polymerases have been identified.[21] These domains are essential for binding of pol η to ubiquitin. Subsequently polη accumulates in replication complexes and interacts with monoubiquitinated PCNA. It has further been shown that the UBZ domain of pol η is essential for efficient restoration of a normal response to UV irradiation in XP-V fibroblasts. In this situation the UBZ domain adopts a C(2)H(2) zinc finger structure.[60] Pol ι, pol κ and Rev1 also interact with ubiquitin.[61-63] Pol ι has been shown to interact noncovalently with free polyubiquitination chains and also with monoubiquitinated PCNA. The monoubiquitination of PCNA is catalyzed by the Rad6/Rad18 complex. It is lysine 164 of PCNA that is used to covalently bind ubiquitin for the process of ubiquitination.[64]

Nakajima et al[65] have found that human RAD18 accumulates very rapidly and remains for a long time at sites of different types of DNA damage including UVC- and X-ray-induced damage. Accumulation of RAD18 occurs at the site of damage even when replication is arrested and a small region containing a zinc finger motif, located in the middle of RAD18, is important and enough for replication independent damage accumulation. RAD18 also physically interacts with pol η.[20] The interaction of pol η with Rad6/Rad18 is increased on the chromatin if DNA replication is arrested.[66] These protein-protein interactions and post-translational modifications are likely to be important to guide pol η to the arrested replication machinery.

Mouse, Plant and Microbial Models for Pol η

A mouse homologue equivalent to human pol η (also known as pol η) has been identified and studied.[67] In addition, several other eukaryotes, *Aspergillus nidulans, Schizosaccharomyces pombe, Brugia malayi, Caenorhabditis elegans, Trypnosoma cruzi, Arabidopsis thaliana* and *Drosophila melanogaster* have been identified to have the homologue of pol η.[68]

Lin et al,[67] in their studies of XP-V deficient mice, showed that homozygous mice for null mutation in pol η are viable, fertile and do not show any apparent spontaneous physiological defect during the first year of life. Nevertheless fibroblasts from these mice showed enhanced sensitivity to UV and all mice having pol η deficiency developed skin tumors after UV irradiation. In contrast, the wild type littermate controls did not show such tumors. These results confirm that in mice, as in humans, pol η-dependent bypass of CPD suppresses UV induced skin cancer. It is interesting to note that 37.5% of heterozygous mice also developed skin cancer 5 months after a 5-months exposure to UV. The results may imply for heterozygous humans for mutation in pol η, an increased risk of skin cancer compared to homozygous normal humans.[69] In another study a double mutant mouse with deficiency in Pol η and Pol ι, had a slightly earlier onset of skin tumor formation compared to mice with a single gene mutation.[70] Furthermore, pol ι deficiency leads to the development of mesenchymal tumors such as sarcoma; these are not observed in *Pol η⁻/⁻* mice. A comparative study was carried out on UV induced mutagenesis in mice deficient in pol η or pol ι and mice having mutation in both the genes;[71] the *Hprt* site was used and it was shown that in pol η deficient mice UV induced mutation frequency had increased. In contrast, mutation frequency in cultured cells, derived from pol ι mutant mice, had strongly reduced, suggesting that pol ι is involved in bypassing incorrectly through the UV photoproducts.

Mice deficient in pol η exhibited an 80% reduction of A:T mutations, suggesting that pol η is the main DNA polymerase producing A:T mutations in somatic hypermutations.[72,73] Similar to pol η deficient mice, mice deficient in MSH2 or MSH6 also show a reduction of A:T mutations. The MSH2/MSH6 complex binds to mispaired bases including U:G pair, which could be generated by activation induced deaminase-mediated deamination of C. MSH2/MSH6 interacts with pol η, suggesting that these components cooperate in somatic hypermutation (SHM).[74] Pol θ has also been found to be involved in SHM in association with pol η.[55]

In *Saccharomyces cerevisiae* the equivalent of pol η is Rad30. The yeast Rad30 product has been purified and tested against different kinds of DNA damage.[75] In this organism, as in humans, the enzyme efficiently inserts two As opposite TT-CPD.[76] It also efficiently and accurately bypasses 8-oxoguanine, incorporating C opposite the lesion; although A and G are inserted with a reduced efficiency. Opposite acetyl aminofluorine-modified G, a C is inserted and a G is inserted opposite an apurinic/apyrimidinic (AP) site.[75] In the latter case, however, the enzyme is unable to extend the DNA synthesis any further and it lacks nuclease of proof-reading activity. Skoneczna et al[77] carried out a detailed analysis of TAP-tagged Rad30 from *S. cerevisiae* and showed that the level of this protein is post-translationally regulated via ubiquitination and proteosome mediated degradation. The half life of the enzyme is 20 min and it is increased when there are proteosomal defects. They also showed that UV irradiation causes transient stabilization of Rad30, which, in turn, leads to temporary accumulation of the enzyme in the cell; hence the proteolysis process plays an important role in regulating the cellular abundance of Rad30. Although generally pol η suppresses UV induced mutagenesis, in yeast this enzyme may introduce frameshift mutations by generating some classes of +1 frame shift.[78]

A gene, *AtPOLH*, in *Arabidopsis thaliana* has been identified which encodes a protein containing several sequence motifs characteristic of pol η homologues.[7] The gene contains 14 exons and 13 introns and is expressed in different plant tissues. A cloned *AtPOLH* cDNA in an expression vector was transformed into *S. cerevisiae;* when this strain was exposed to UV, its survival reached wild type level suggesting that *AtPOLH* can complement the deficiency of pol η in *S. cerevisiae.* Pol η in *C. elegance* has been shown to be important for UV damage tolerance during early embryogenesis but not so much for later stages.[79]

Conclusion

During the last three decades our understanding of molecular basis of XP has improved considerably and based on this knowledge, patients suffering from XP have been classified to belong to defect in one of two major routes for DNA repair: NER, in which mutation in any one of the seven XP genes (XP-A to XP-G) can lead to XP phenotype, primarily sensitivity of skin cells to UV light, leading to skin cancer. The other is defect in *POLH*, responsible to synthesize polymerase η. Studies of pol η, its purification and sub-cellular localization, protein/protein interactions and knockout model organisms revealed that this enzyme has the ability to accurately bypass the TLS across UV-induced CPD; thus tolerating DNA replication arrest at the lesions. The accuracy of TLS by pol η is important in that it can prevent mutagenesis and subsequent development of cancer. However, pol η is an error-prone enzyme intrinsically. It can induce somatic hypermutation in Ig variable genes.[27] Thus, an important biological role of pol η appears to promote variations in the immune system.

Yet we are some way away to unravel the complete picture of this enzyme; for example it is still unknown how replicative and TLS polymerases are taking their places in DNA before and after incorporating nucleotides opposite lesions, although recent findings suggest that post-translational modifications, specially ubiquitination and deubiquitination of PCNA and TLS polymerases may be playing important roles in the process. It is also important to learn about the switching over and selection mechanism between various polymerases and clarify what happens in mutant *POLH* cells when the replication complex arrives at CPD site and stops.

The next step therefore in this field is to extend our studies in further understanding of the exact DNA repair mechanism by this and other associated enzymes. Also it is important to study the mechanisms of regulation of these double-edged enzymes, specifically in the cells challenged by DNA damaging agents and without damage. These studies should be helpful in diagnosis, treatment and prevention of DNA damage related diseases which are too many to be listed here.

Acknowledgement

This work was supported by Grants-in-Aid for Science Research from the Ministry of Education, Culture, Sports, Science and Technology of Japan and by Solution Oriented Research for Science and Technology from the Japan Science and Technology. One of the authors (SIA) is indebted to the University of Osaka for providing a Senior Scientist/Visiting Professor's fellowship, during which this work was accomplished.

References

1. Hebra F, Kaposi M. On diseases of the skin, including the exanthemata. New Sydenham Soc 1874; 61:252-258.
2. Rothwell PJ, Waksman G. Structure and mechanisms of DNA polymerases. Adv Protein Chem 2005; 71:401-440.
3. Pavlov YI, Shcherbakova PV, Rogozin IB. Roles of DNA polymerases in replication, repair and recombination in eukaryotes. Int Rev Cytol 2006; 255:41-132.
4. Sweasy JB, Lauper JM, Eckert KA. DNA polymerases and human diseases. Radiat Res 2006; 166:693-714.
5. Yamada A, Masutani C, Iwai S et al. Complementation of defective translesion synthesis and UV light sensitivity in xeroderma pigmentosum variant cells by human and mouse DNA polymerase η. Nuc Acids Res 2000; 28:2473-2480.
6. Ishikawa T, Uematsu N, Mizukoshi T et al. Mutagenic and nonmutagenic bypass of DNA lesions by Drosophila DNA polymerases dpolη and dpolι. J Biol Chem 2001; 276:15155-15163.
7. Santiago MJ, Alejandre-Duran E, Ruiz-Rubio M. Analysis of UV-induced mutation spectra in Escherichia coli by DNA polymerase η from Arabidopsis thaliana. Mut Res 2006; 601:51-60.
8. Boudsocq F, Iwai S, Hanaoka F et al. Sulfolobus solfataricus P2 DNA polymerase IV (Dpo4): an archaeal DinB-like DNA polymerase with lesion-bypass properties akin to eukaryotic pol η. Nucleic Acids Res 2001; 29:4607-4616.
9. Madril AC, Johnson RE, Washington MT et al, Fidelity and damage bypass ability of Schizosaccharomyces pombe Eso1 protein, comprised of DNA polymerase eta and sister chromatid cohesion protein Ctf7. J Biol Chem 2001; 276:42857-42862.

10. Tanaka K, Yonekawa T, Kawasaki Y et al. Fission yeast Eso1p is required for establishing sister chromatid cohesion during S phase. Mol Cell Biol 2000; 20:3459-3469.

11. Ohmori H, Friedberg E, Fuchs RPP et al. The Y family of DNA polymerases. Molec Cell 2001; 8:7-8.

12. Goodman MF. Error-prone repair DNA polymerases in prokaryotes and eukaryotes. Ann Rev Biochem 2002; 71:17-50.

13. Masutani C, Araki M, Yamada A et al. Xeroderma pigmentosum variant (XP-V) correcting protein from HeLa cells has a thymine dimer bypass DNA polymerase activity, EMBO J 1999; 18:3491-3501.

14. Masutani C, Kusumoto R, Yamada A et al. The XPV (xeroderma pigmentosum variant) gene encodes human DNA polymerase η. Nature 1999; 399:700-704.

15. Yuasa M, Masutani C, Eki T et al. Genomic structure, chromosomal localization and identification of mutations in the xeroderma pigmentosum variant (XPV) gene. Oncogene 2000; 19:4721-4728.

16. Gratchev A, Strein P, Utikal J. Molecular genetics of Xeroderma pigmentosum variant. Expt Dermatol 2003; 12:529-536.

17. Kannouche P, Fernandez de Henestrosa AR, Coull B et al. Localization of DNA polymerases η and ι to the replication machinery is tightly co-ordinated in human cells. EMBO J 2003; 22:1223-1233.

18. Kannouche PL, Wing J, Lehmann AR. Interaction of human DNA polymerase η with monoubiqui-tinated PCNA: a possible mechanism for the polymerase switch in response to DNA damage. Mol Cell 2004; 14:491-500.

19. Ohashi E, Murakumo Y, Kanjo N et al. Interaction of hREV1 with three human Y-family DNA poly-merases. Genes Cells 2004; 9:523-531.

20. Watanabe K, Tateishi S, Kawasuji M et al. Rad18 guides pol η to replication stalling sites through physical interaction and PCNA monoubiquitination. EMBO J 2004; 23:3886-3896.

21. Bienko M, Green CM, Crosetto N et al. Ubiquitin-binding domains in Y-family polymerases regulate translesion synthesis. Science 2005; 310:1821-1824.

22. Trincao J, Johnson RE, Escalante CR et al. Structure of the catalytic core of S. cerevisiae DNA polymerase η: implications for translesion DNA synthesis. Mol Cell 2001; 8:417-426.

23. Matsuda T, Bebenek K, Masutani C et al. Low fidelity DNA synthesis by human DNA polymerase η. Nature 2000; 404:1011-1013.

24. Johnson RE, Washington MT, Prakash S et al. Fidelity of human DNA polymerase η. J Biol Chem 2000; 275:7447-7450.

25. Kusumoto R, Masutani C, Shimmyo S et al. DNA binding properties of human DNA polymerase η: implications for fidelity and polymerase switching of translesion synthesis. Genes Cells 2004; 9:1139-1150.

26. McCulloch SD, Kokoska RJ, Masutani C et al. Preferential cis-syn thymine dimer bypass by DNA polymerase η occurs with biased fidelity. Nature 2004; 428:97-100.

27. Zeng X, Winter DB, Kasmer C et al. DNA polymerase η is an A-T mutator in somatic hypermutation of immunoglobulin variable genes. Nat Immunol 2001; 2:537-541.

28. Rogozin IB, Pavlov YI, Bebenek K et al. Somatic mutation hotspots correlate with DNA polymerase η error spectrum. Nat Immunol 2001; 2:530-536.

29. McIlwraith MJ, Vaisman A, Liu Y et al. Human DNA polymerase η promotes DNA synthesis from strand invasion intermediates of homologous recombination. Mol Cell 2005; 20:783-792.

30. Kawamoto T, Araki K, Sonoda E et al. Dual roles for DNA polymerase η in homologous DNA recom-bination and translesion DNA synthesis. Mol Cell 2005; 20:793-799.

31. Haracska L, Prakash S, Prakash L. Replication past O(6)-methylguanine by yeast and human DNA polymerase η. Mol Cell Biol 2000; 20:8001-8007.

32. Kusumoto R, Masutani C, Iwai S et al. Translesion synthesis by human DNA polymerase η across thymine glycol lesions. Biochemistry 2002; 41:6090-6099.

33. Choi JY, Guengerich FP. Adduct size limits efficient and error-free bypass across bulky N2-guanine DNA lesions by human DNA polymerase η. J Mol Biol 2005; 352:72-90.

34. Choi JY, Chowdhury G, Zang H. Translesion synthesis across O6-alkylguanine DNA adducts by recombinant human DNA polymerases. J Biol Chem 2006; 281:38244-38256.

35. Choi JY, Zang H, Angel KC et al. Translesion synthesis across 1, N2-ethanoguanine by human DNA polymerases, Chem Res Toxicol 2006; 19:879-886.

36. Choi JY, Stover JS, Angel KC et al. Biochemical basis of genotoxicity of heterocyclic arylamine food mutagens: Human DNA polymerase eta selectively produces a two-base deletion in copying the N2-guanyl adduct of 2-amino-3-methylimidazo[4,5-f]quinoline but not the C8 adduct at the NarI G3 site. J Biol Chem 2006; 281:25297-25306.

37. Perrino FW, Harvey S, Blans P et al. Polymerization past the N2-isopropylguanine and the N6-isopro-pyladenine DNA lesions with the translesion synthesis DNA polymerases η and ι and the replicative DNA polymerase α. Chem Res Toxicol 2005; 18:1451-1461.

38. Masutani C, Kusumoto R, Iwai S et al. Mechanisms of accurate translesion synthesis by human DNA polymerase η. EMBO J 2000; 19:3100-3109.

39. Vaisman A, Takasawa K, Iwai S et al. DNA polymerase ι-dependent translesion replication of uracil containing cyclobutane pyrimidine dimers. DNA Repair (Amst) 2006; 5:210-218.

40. Albertella MR, Green CM, Lehmann AR et al. A role for polymerase η in the cellular tolerance to cisplatin-induced damage. Cancer Res 2005; 65:9799-9806.

41. Chen YW, Cleaver JE, Hanaoka F et al. A novel role of DNA polymerase η in modulating cellular sensitivity to chemotherapeutic agents. Mol Cancer Res 2006; 4:257-265.

42. Yasui M, Suzuki N, Laxmi YR et al. Translesion synthesis past tamoxifen-derived DNA adducts by human DNA polymerases η and κ. Biochemistry 2006; 45:12167-12174.

43. Suzuki N, Yasui M, Santosh Laxmi YR et al. Translesion synthesis past equine estrogen-derived 2'-deoxy-cytidine DNA adducts by human DNA polymerases η and κ. Biochemistry 2004; 43:11312-11320.

44. Yasui M, Laxmi YR, Ananthoju SR et al. Translesion synthesis past equine estrogen-derived 2'-deoxy-adenosine DNA adducts by human DNA polymerases η and κ. Biochemistry 2006; 45:6187-6194.

45. Vu B, Cannistraro VJ, Sun L et al. DNA synthesis past a 5-methylC-containing cis-syn-cyclobutane pyrimidine dimer by yeast pol η is highly nonmutagenic. Biochemistry 2006; 45:9327-9335.

46. Kokoska RJ, McCulloch SD, Kunkel TA. The efficiency and specificity of apurinic/apyrimidinic site bypass by human DNA polymerase η and Sulfolobus solfataricus Dpo4. J Biol Chem 2003; 278:50537-50545.

47. Barone F, McCulloch SD, Macpherson P et al. Replication of 2-hydroxyadenine-containing DNA and recognition by human MutSα. DNA Repair (Amst) 2007; 6:355-366.

48. Shimizu M, Gruz P, Kamiya H et al. Efficient and erroneous incorporation of oxidized DNA precursors by human DNA polymerase η. Biochemistry 2007; 46:5515-5522.

49. Maher VM, Ouellette LM, Curren RD et al. Frequency of ultraviolet light-induced mutations is higher in xeroderma pigmentosum variant cells than in normal human cells. Nature 1976; 261:593-595.

50. Maher VM, Ouellette LM, Curren RD et al. Caffeine enhancement of the cytotoxic and mutagenic effect of ultraviolet irradiation in a xeroderma pigmentosum variant strain of human cells. Biochem Biophys Res Commun 1976; 71:228-234.

51. Broughton BC, Cordonnier A, Kleijer WJ et al. Molecular analysis of mutations in DNA polymerase η in Xeroderma pigmentosum-variant patients. Proc Natl Acad Sci USA 2002; 99:815-820.

52. Johnson RE, Kondratick CM, Prakash S et al. hRAD30 mutations in the variant form of Xeroderma pigmentosum. Science 1999; 285:263-265.

53. Glick E, White LM, Elliott NA et al. Mutations in DNA polymerase η are not detected in squamous cell carcinoma of the skin. Int J Cancer 2006; 119:2225-2227.

54. Murakumo Y, Ogura Y, Ishii H et al. Interactions in the error-prone post-replication repair proteins hREV1, hREV3 and hREV7. J Biol Chem 2001; 276:35644-35651.

55. Masuda Y, Ohmae M, Masuda K et al. Structure and enzymatic properties of a stable complex of the human REV1 and REV7 proteins. J Biol Chem 2003; 278:12356-12360.

56. Tissier A, Kannouche P, Reck MP et al. Colocalization in replication foci and interaction of human Y-family members, DNA polymerase pol η and REV1 protein. DNA Repair (Amst) 2004; 3:1503-1514.

57. Clark DR, Zacharias W, Panaitescu L et al. Ribozyme-mediated REV1 inhibition reduces the frequency of UV-induced mutations in the human HPRT gene. Nucleic Acids Res 2003; 31:4981-4988.

58. Wang Y, Woodgate R, McManus TP et al. Evidence that in Xeroderma pigmentosum variant cells, which lack DNA polymerase η, DNA polymerase ι causes the very high frequency and unique spectrum of UV-induced mutations. Cancer Res 2007; 67:3018-3026.

59. Vaisman A, Masutani C, Hanaoka F et al. Efficient translesion replication past oxaliplatin and cisplatin GpG adducts by human DNA polymerase η. Biochemistry 2000; 39:4575-4580.

60. Bomar MG, Pai MT, Tzeng SR et al. Structure of the ubiquitin-binding zinc finger domain of human DNA Y-polymerase η. EMBO Rep 2007; 8:247-251.

61. Plosky BS, Vidal AE, Fernandez de Henestrosa AR et al. Controlling the subcellular localization of DNA polymerases ι and η via interactions with ubiquitin. EMBO J 2006; 25:2847-2855.

62. Bi X, Barkley LR, Slater DM et al. Rad18 regulates DNA polymerase κ and is required for recovery from S-phase checkpoint-mediated arrest. Mol Cell Biol 2006; 26:3527-3540.

63. Guo C, Tang TS, Bienko M et al. Ubiquitin-binding motifs in REV1 protein are required for its role in the tolerance of DNA damage. Mol Cell Biol 2006; 26:8892-8900.

64. Hoege C, Pfander B, Moldovan GL et al. RAD6-dependent DNA repair is linked to modification of PCNA by ubiquitin and SUMO. Nature 2002; 419:135-141.

65. Nakajima S, Lan L, Kanno S et al. Replication dependent and—independent responses of RAD18 to DNA damage in human cells. J Biol Chem 2006; 281:34687-34695.

66. Yuasa MS, Masutani C, Hirano A et al. A human DNA polymerase η complex containing Rad18, Rad6 and Rev1; proteomic analysis and targeting of the complex to the chromatin-bound fraction of cells undergoing replication fork arrest. Genes Cells 2006; 11:731-744.

67. Lin Q, Clark AB, McCulloch SD et al. Increased susceptibility to UV-induced skin carcinogenesis in polymerase η-deficient mice. Cancer Res 2006; 66:87-94.

68. McDonald JP, Rapic-Otrin V, Epstein JA et al. Novel human and mouse homologs of Saccharomyces cerevisiae DNA polymerase eta. Genomic 1999; 60:20-30.

69. Itoh T, Linn S, Kamide R et al. Xeroderma pigmentosum variant heterozygotes show reduced levels of recovery of replicative DNA synthesis in the presence of caffeine after ultraviolet irradiation. J Invest Dermatol 2000; 115:981-985.

70. Ohkumo T, Kondo Y, Yokoi M et al. UVB radiation induces epithelial tumors in mice lacking DNA polymerase η and mesenchymal tumors in mice deficient for DNA polymerase ι. Mol Cell Biol 2006; 26:7696-7706.

71. Dumstorf CA, Clark AB, Lin Q et al. Participation of mouse DNA polymerase ι in strand-biased mutagenic bypass of UV photoproducts and suppression of skin cancer. Proc Natl Acad Sci, USA 2006; 103:18083-18088.

72. Martomo SA, Yang WW, Wersto RP et al. Different mutation signatures in DNA polymerase η- and MSH6-deficient mice suggest separate roles in antibody diversification. Proc Natl Acad Sci, USA 2005; 102:8656-8661.

73. Delbos F, De Smet A, Faili A et al, Contribution of DNA polymerase eta to immunioglobulin gene hypermutation in the mouse. J Expt Med 2005; 201:1191-1196.

74. Wilson TM, Vaisman A, Martomo SA et al. MSH2-MSH6 stimulates DNA polymerase η, suggesting a role for A:T mutations in antibody genes. J Exp Med 2005; 201:637-645.

75. Yuan F, Zhang Y, Rajpal DK et al. Specificity of DNA lesion bypass by the yeast DNA polymerase η. J Biol Chem 2000; 275:8233-8239.

76. Washington MT, Johnson RE, Prakash S et al. Accuracy of thymine-thymine dimer bypass by Saccharomyces cerevisiae DNA polymerase η. Proc Natl Acad Sci, USA 2000; 97:3094-3099.

77. Skoneczna A, McIntyre J, Skoneczny M et al. Polymerase η is a short-lived, proteasomally degraded protein that is temporarily stabilized following UV irradiation in Saccharomyces cerevisiae. J Mol Biol 2007; 366:1074-1086.

78. Abdulovic AL, Jinks-Robertson S. The in vivo characterization of translesion synthesis across UV-induced lesions in Saccharomyces cerevisiae: insights into Pol ζ- and Pol η-dependent frameshift mutagenesis. Genetics 2006; 172:1487-1498.

79. Ohkumo T, Masutani C, Eki T et al. Deficiency of the Caenorhabditis elegans DNA polymerase η homologue increases sensitivity to UV radiation during germ-line development. Cell Struct Funct 2006; 31:29-37.

80. Broughton BC, Cordonier A, Kleijer et al. Molecular analysis of mutations in DNA polymerase in Xeroderma pigmentosum-variant patients. Proc Natl Acad Sci USA 2002; 99:815-820.

CHAPTER 11

Other Proteins Interacting with XP Proteins

Steven M. Shell and Yue Zou*

Introduction

Genetic defects in Nucleotide excision repair (NER) lead to the clinical disorder xeroderma pigmentosum (XP) in humans which is characterized by dramatically increased sensitivity to UV light and a predisposition to development of skin cancers.[1,2] NER is a major mechanism of DNA repair in cells for the removal of a large variety of bulky DNA lesions induced by environmental genotoxic agents and chemicals. The molecular basis of XP has been attributed to mutations in any of the eight XP genes, XPA through G whose products are required for NER-mediated removal of DNA damage and XP-variant (XPV). The XP proteins involved in NER can be divided into three groups based on their activity in the NER process. XPA, XPC and XPE are required for sensing DNA damage and initiating the repair process. XPB and XPD, components of the basal transcription factor TFIIH, are helicases that create a DNA strand opening surrounding the adducted base(s) during NER. XPG and XPF are the endonucleases that perform the dual incisions to release the damaged strand and allow resynthesis using the nondamaged strand as a template.[3-5] Protein-protein interactions are integral for the correct assembly of the pre-incision complex and for the positioning of the nucleases prior to incision. However, these proteins have been found to form complexes with other proteins not directly involved in the NER mechanism. This chapter describes these proteins and their interactions and discusses their effects on the XP proteins, DNA repair, and genome stability.

Finding DNA Damage: XPA, XPC and XPE

Damage recognition, the first step of the NER process, is performed by three XP proteins: XPA, XPC and XPE. These proteins function to recognise and verify DNA damage, recruit and assemble other repair factors and initiate cell cycle control pathways. Deficiency in XPA results in the most severe XP phenotype.[4,5] The protein interaction, associated with this group of XP proteins is absolutely essential for the efficient repair of DNA damage and cellular DNA damage responses.

XPC, a 106 kDa protein, is the primary damage recognition factor required for the global genome NER (GG-NER) repair pathway, one of the two subpathways of NER (the other is transcription-coupled NER or TC-NER).[6,7] XPC forms a tight heterodimeric complex with HR23B, a 53 kDa protein and this interaction has been found to be indispensable for the stability of each protein as well as XPC's damage recognition function.[6] Due to the nature of the interaction between XPC and HR23B, this complex will be referred to as XPC for the remainder of this chapter. XPC has recently been shown to form a complex with centrin 2, a centrosomal and calcium binding protein involved in centrosome duplication;[8,9] this occurs through direct interaction

*Corresponding Author: Yue Zou—Department of Biochemistry and Molecular Biology, James H. Quillen College of Medicine, East Tennessee State University, Johnson City, Tennessee 37614, USA. Email: zouy@etsu.edu

Molecular Mechanisms of Xeroderma Pigmentosum, edited by Shamim I. Ahmad and Fumio Hanaoka. ©2008 Landes Bioscience and Springer Science+Business Media.

with XPC. Although this interaction is not required for NER in vitro, it forms a heterotrimeric complex with XPC and stimulates damage recognition, thus providing an extra sensitivity for DNA damage.[8] In GG-NER, XPC also interacts with the XPE, XPB and the basal transcription factor TFIIH.[5,7,10,11] In this case, XPE serves a similar role to that of centrin 2 to stimulate XPC's ability to recognize certain types of DNA damage. Unlike centrin 2, however, XPE is required for efficient GG-NER activity.[11] The involvement of XPC in GG-NER is regulated also by its sumoylation with the small ubiquitin-like SUMO-1 modifier; this modification is promoted by the recruitment of XPA to the damage site. UV-induced sumoylation of XPC is linked to the stabilization of the protein, as well as the dissociation of XPC from the damage site.[13] Once bound to damaged DNA, XPC recruits the basal transcription factor TFIIH through interaction with the p62 subunit of TFIIH, an interaction that serves to position TFIIH prior to opening the damage site.[7] The XPC-TFIIH interaction also serves a role in linking GG-NER to the cell cycle checkpoint kinase Ataxia-Telangiectasia Mutated (ATM).[14] This interaction requires the NER nuclease XPG to prevent apoptosis in the presence of certain DNA lesions.[14] XPC also has been linked to the base excision repair (BER) pathway via an interaction with thymine DNA glycosylase (TDG). Although XPC is not a required factor for BER, the XPC-TDG interaction is believed to stimulate recognition of the damaged base by TDG.[15]

XPE (also known as DDB-2) is a 48kDa protein and is part of the heterodimeric UV-damaged DNA binding complex (UV-DDB). XPE interacts with DDB-2, a 178kDa protein, to form the UV-DDB damage recognition complex that stimulates the damage recognition step in GG-NER.[16] XPE interacts with a variety of proteins to modulate cellular responses to DNA damage. Interactions between XPE and the CBP/p300[17,18] and STAGA[18,19] chromatin remodeling complexes also are critical for the efficient removal of damage by the GG-NER pathway. These complexes acetylate histones, thereby relaxing the local chromatin super structures to allow access to the DNA for processes such as repair or transcription. This is supported by the finding that XPE associates with monoubiquitinated histone H2A following UV irradiation. This ubiquitination is a modification that leads to relaxation of chromatin structure.[20] In addition, XPE interacts with transcriptional cofactor E2F1 to inhibit production of replication factors and arrest cell cycle progression.[18,21] Along with its role in transcriptional repression, XPE interacts with the Cullin4-ROC1-COP9 signalsome, which is part of the E3 ubiquitin ligase complex.[22,23] This pathway mediates 26S proteosome degradation of ubiquitinated proteins and the interaction with XPE may serve to modulate protein degradation in response to UV irradiation.[18,24] Recently, XPE was found to interact with c-Abl tyrosine kinase and this interaction has been linked to modulation of Cullin4 targeting of XPE for degradation by the 26S proteosome following UV irradiation.[22]

XPA, a 32 kDa zinc metalloprotein, is a recognition factor required for both GG-NER and TG-NER activities and is believed to play a role in verification of DNA damage.[25,26] To date, the only known function for the XPA protein is in mediating NER and its protein interaction partners are limited primarily to factors required for damage removal.[25-27] The primary XPA interaction occurs with the single-stranded DNA binding protein, replication protein A (RPA).[5,27-29] Following UV irradiation, both XPA and RPA are recruited to the damage site via interactions with XPC and TFIIH and, once assembled, form a tight complex that remains associated with the damage site throughout processing.[4,5] Rad14 the yeast homologue of XPA, also forms a tight complex with Rad1-Rad10 nuclease, the yeast counterpart of XPF-ERCC1.[30] A recent in vitro study has suggested that the high affinity of XPA for the damage site may not be dependent on the damaged base but rather on the pre-incision DNA structures generated by local unwinding of the DNA at and around the damage.[31] Thus, XPA may play a structural role in maintaining the pre-incision DNA bubble and positions the remaining NER factors, particularly XPF-ERCC1,[30] for final incisions while RPA may protect the undamaged strand (which will be used as the template for resynthesis following incisions) from nuclease attack.[32]

Other XPA interacting proteins include ATR (ATM and RAD3-Related),[33] a DNA damage checkpoint kinase of the phosphoinositide 3-kinase-like kinase (PIKK) family, XAB1 (XPA-binding protein 1)[34,35] and XAB2 (XPA-binding protein 2).[36,37] The interaction of XPA with ATR may be responsible for the rapid translocation of XPA from the cytoplasm to the nucleus in response to UV irradiation as this activity is dependent on ATR and can be abolished using either ATR inhibitors or siRNA-mediated knockdown of the kinase.[38] Although the mechanism of the translocation remains unclear, it is possible that the XPA binding protein XAB1, a cytoplasmic GTPase, may be involved. XAB1 binds to XPA via the nuclear localization signal located in the N-terminal region of XPA and is believed to help shuttle XPA, though the nuclear pore by virtue of GTP hydrolysis.[34,35] However, this mechanism remains speculative. XPA also has been shown to be phosphorylated by ATR in cells upon UV-irradiation, and silencing this phosphorylation, for example by mutation, makes the cells significantly more sensitive to UV. The mechanism, however, is still unknown.[33]

XPA interaction with XAB2 (XPA-binding protein 2) was identified by yeast-two hybrid screening.[36] XAB2 interacts with a variety of proteins including RNA Pol II and is active in mRNA splicing.[36,37] While its role in NER is not clear, it is believed that it promotes TC-NER through its dual interaction between RNA, Pol II and XPA. However, attempts to study this protein interaction in vivo, using transgenic mice, have proven difficult as deletion of XAB2 is embryonically lethal.[37]

Preparing the Site: XPB and XPD

Following the initial damage recognition step, opening of the DNA at and around the damage is accomplished by the basal transcription factor TFIIH.[4,5] The helicase activity of TFIIH resides in two XP proteins: XPB and XPD.[39] Although they play critical roles in NER, their activities are also critical for the transcriptional role of TFIIH in mRNA synthesis. Therefore each helicase forms various protein-protein interactions that will be described later in this section.

XPD, an 87 kDa protein, is a $5' \rightarrow 3'$ ATP-dependent helicase and serves as the prominent helicase in NER, although it plays only a minor role in transcription.[22] XPD is part of the Cdk Activating Complex (CAK) component of TFIIH, containing the additional subunits cdk7, cyclin H and MAT1. XPD is bound to the CAK complex via an interaction with the coiled-coil region of the MAT1 subunit.[40] The CAK complex interacts with the TFIIH core complex via interaction of XPD with the N-terminal domain of p44 which acts as a bridge between the two complexes.[41,42] Besides its helicase activity and the structural role in binding together the TFIIH supercomplex, XPD has been shown to interact with hMMS19, a transcription factor found to stimulate estrogen receptor-mediated activation of the ERα promoter via stimulation of AF-1 activity.[43,44] hMMS19 has been shown also to play a role in NER. However this function is still not completely understood.[43,44] XPD, as well as XPB, interacts with the p53, and this interaction has been demonstrated to inhibit the helicase activity of both XPD and XPB while reducing the rate of apoptosis. However, the mechanism of this protection is still unclear.[45-49]

XPB, a 90 kDa protein, is a $3' \rightarrow 5'$ ATP hydrolysis-driven helicase although it has much weaker activity compared to that of XPD;[22] however, mutations in XPB are among the most rare in XP indicating XPB plays an extremely critical role in both NER and transcription.[50] In accordance with this observation, XPB interacts with a variety of different repair and transcription factors and it is these interactions, not XPB's helicase activity, that are critical for TFIIH activity. XPB is part of the TFIIH core complex and is positioned via interactions with the p62, p52, p44 and p8 proteins.[51] Mutations in XPB are believed to disrupt these interactions, leading to the destabilization of TFIIH.[50,51] The interaction between XPB and p52 has been demonstrated as critical for XPB's role in promoter melting and any defect in this interaction have serious consequences for transcription.[52] Lin et al demonstrated that XPB helicase activity also is modulated by the transcription factor TFIIE β subunit to promote transcription initiation by TFIIH.[53]

Although the major NER helicase is XPD, modulation of XPB in NER is critical for the efficient removal of DNA lesions. Recently the XPB-p8 interaction has been shown to be important for NER.[22] p8 stimulates the helicase activity of both XPB (through direct interaction) and XPD (via interaction through p44) to promote DNA strand opening, a step necessary for assembly of pre-incision complexes.[22] The C-terminal of XPB can be phosphorylated by an as yet unidentified PP2A-related protein kinase and this phosphorylation is essential for the $5'$ incision by XPG.[54]

Table 1. **Protein-protein interactions of the XP complementation groups**

Protein	MW	Function	Interacting Proteins	References
XPA	32kDa	Damage Sensor	ATR: Cell-cycle Arrest Kinase	38, 33
			RPA: single-strand DNA binding protein	4, 5, 27, 28, 94
			TFIIH: Transcription/Repair complex	4, 5
			XAB1: GTPase	34, 35
			XAB2: mRNA splicing	36, 37
			XPF: Endonuclease	30
XPB	90kDa	3' → 5' DNA Helicase	PP2A-related kinase: XPB phosphorylation	54
			p210BCR/ABL: tyrosine kinase	58
			p53: Transcritption/apoptosis factor	47-49
			Rad52: Homologous recombination factor	57
			SUG1: Ubiquitin ligase	55, 56
			TFIIE: Transcription Complex	53
			XPC: DNA damage sensor	7
			XPG: Endonuclease	51
XPC	106kDa	Damage Sensor	ATM: Cell-cycle arrest kinase	14
			Centrin 2: Centrosome duplication	8, 9, 95
			SUMO-1: Sumoylation	13
			N-methylpurine DNA glycosylase: BER repair protein	15
			S5a: 26S proteosome component	12
			XPB: 3' → 5' DNA Helicase	7
			XPE: DNA damage sensor	10, 11
			XPG: Endonuclease	96
XPD	87kDa	5' → 3' DNA Helicase	hMMS19: Transcription Factor	43, 44
			p44: TFIIH component	42, 97
			p53: Transcription/apoptosis factor	45, 46
			MAT1: TFIIH component	40
XPE	48kDa	Damage Sensor	c-ABL: tyrosine kinase	22
			CBP/p300: Histone acetylase	17, 18
			Cullin4-COP9: E3 Ubiquitin ligase components	22, 23
			E2F1: Transcription Factor	18. 21
			STAGA: Histone acetylase	18, 19
			uH2A: monoubiquitinated Histone H2A	20
			XPC: UV-DNA damage binding protein	10, 11
XPF	103kDa	Nuclease	αSPIIΣ: Nuclear structural protein	76, 77
			FANC-A: Crosslink repair factor	76, 77
			Msh2: Mismatch recognition factor	75, 78
			Rad51: Homologous recombination factor	73
			Rad52: Homologous recombination factor	74
			RPA: Single-strand DNA binding protein	72
			TRF2: Telomere elongation factor	80, 81
			XPA: DNA damage recognition factor	30

continued on next page

Table 1. Continued

Protein	MW	Function	Interacting Proteins	References
XPG	133kDa	Nuclease	Nth1: DNA glycosylase	67-69
			PCNA: DNA replication elongation factor	61, 65
			RNA Pol II: mRNA polymerase	66
			RPA: Single-strand DNA binding protein	61, 72
			TFIIH: Transcription/Repair complex	59, 61-64
			XPA: DNA damage recognition factor	61
XPV	79kDa	Y-Family DNA Polymerase	PCNA: DNA replication factor	86-89
			Polι: Y-family DNA polymerase	85
			Rad18: Ubiquitin ligase	90
			Rad51: Homologous recombination factor	93
			Rev1: Deoxycytidyl transferase	92

Another critical interaction of XPB is with hSUG1, a component of the 26S proteosome.[55,56] Lommel et al found that in yeast the XPB-SUG1 interaction modulates the degradation of RAD4, an XPC homologue and yielded increased repair efficiency.[55]

XPB and XPD have also been shown to directly associate with the recombination factor Rad52 and it is believed that this interaction serves to couple transcription with the error-free homologous recombinational repair pathway.[57] Recently, XPB has been demonstrated to associate with the p210BCR/ABL tyrosine kinase and this interaction promoted DNA repair by modulating the association of TFIIH with PCNA.[58] Although the exact mechanism of this modulation is unclear, it appears that XPB may modulate the role of TFIIH in homologous recombinational repair as well as in NER.

Cutting It Out: XPG and XPF

In NER, removal of the damaged oligonucleotides from the genome requires dual incisions to be made flanking the damaged site. This process is accomplished by the remaining two XP proteins: XPG and XPF.[4,5] Following opening of the DNA strand in and around the damage by TFIIH and dissociation of XPC/HR23B (in the case of GG-NER) the XPG and XPF nucleases are recruited to the damage site and positioned via protein-protein interactions with XPA, RPA and components of the TFIIH complex.[5] However, these two nucleases have also been reported to interact with other proteins in repair and transcription.

XPG, a 133 kDa protein, is a structure-specific nuclease that cleaves the damaged strand 3' to the damaged site at the single-strand/double-strand DNA junction. XPG also serves a structural role in that helps in positioning the XPF/ERCC1 endonuclease for the 5' incision.[59,60] The XPA/RPA and TFIIH complexes are required for recruitment and positioning of the nuclease prior to incision.[61-63] The p62 subunit of TFIIH contains a pleckstrin/phosphotyrosine homology domain that has been shown to specifically interact with the XPG nuclease.[64] Also, as described above, XPG interacts with the XPB helicase and the phosphorylation state of XPB modulates XPG nuclease activity.[54] Interaction of XPG with RPA is required for completing the incision.[5,61] In addition, XPG shows a high degree of homology with the FEN-1 endonuclease, which is involved in processing of Okazaki fragments during DNA replication. Like FEN-1, XPG is able to associate with the replication elongation factor PCNA and this interaction is significant during resynthesis of the excised region in NER.[61,65] It has been found that XPG could recognize and interact with RNA Pol II at stalled transcription bubbles.[6] This finding may suggest a DNA damage recognition mechanism in which XPG initiates the recruitment of repair factors to remodel the stalled transcription bubble at lesions encountered by the polymerase II.[66] Like XPC, XPG also interacts with proteins from the BER pathway, in this case Nth1. Nth1 is a DNA glycosylase-AP lyase involved in repairing thymine glycol lesions and other types of BER lesions. However, its affinity for the damaged base is greatly increased in the presence of XPG protein.[67-69] In its interaction with Nth1, however, XPG does not incise the strand and serves only to stimulate Nth1.[67-69]

XPF, a 103 kDa protein, is the second endonuclease required for NER. Like XPG, it is a structure-specific nuclease that recognizes the single-strand/double strand DNA junction 5′ to the adducted base(s).[70] Together with its interaction partner, ERCC1, XPF cleaves the damaged strand in the region of the junction. XPF requires XPG and RPA to be bound to the substrate before it can be recruited and correctly positioned to make the cleavage.[71,72] XPF also interacts with the homologous recombinational repair proteins RAD51 and RAD52 and is involved in the repair of interstrand crosslinks.[73,74] Furthermore, XPF interacts with the FANC-A, MSH2 and nonerythroidαSPIIΣ protein, further demonstrating its role in the removal of interstrand crosslinks.[75-78] A recently identified interesting role for XPF is its interaction with the telomere repeat binding factor TRF2. This factor is responsible for the conversion of telomeric TTAGGG repeats tracts to T-loop structures that protect the telomeres from inadvertent double-strand break repair. Also, it has been speculated that TRF2 performs a strand-break sensing role in the nonhomologous end joining (NHEJ) repair pathway.[79] Although it is still not clear what role TRF2 plays in repair of strand breaks, it is believed that its interaction with XPF promotes trimming of the ends in preparation for end joining.[80,81]

Getting Past It: XPV

XPV (also known as pol η) is a 79 kDa protein and a member of the Y-Family DNA polymerases.[82] XPV has been identified as a member of the XP group of proteins although it does not participate in NER pathway.[83] Defects in XPV lead to development of the xeroderma pigmentosum variant disorder characterized by sensitivity to UV-irradiation despite an active NER repair pathway.[82,83] The XPV polymerase is a trans-lesion polymerase that allows DNA synthesis to bypass UV photoproducts in an error-free manner [84-86] and has been shown to physically interact with pol ι, another damage bypass polymerase.[85] XPV associates with the replication machinery during DNA replication and interacts specifically with mono-ubiquitinated PCNA through two interaction sites on the XPV.[85-89] This interaction is promoted by the Rad18 ubiquitin ligase that forms a complex with PCNA and XPV.[90] Rad18 monoubiquitinates PCNA following replication-fork stalling which promotes polymerase switching from pol δ to polymerase η allowing trans-lesion synthesis past the damage site.[86,90] Along with PCNA and Rad18, XPV also associates with Rev1 at stalled replication centers. Rev1, a deoxycytidyl transferase,[91] is thought to play a structural role in providing a scaffolding for polymerase η to bypass certain lesions and this function is independent of its enzymatic activity.[92] XPV also is involved in the recombinational double strand break repair via direct interaction with Rad51.[93] Rad51 recruits polymerase η to the D-loop recombination intermediate, a structure that XPV can use as a primer to initiate DNA synthesis.[93]

Conclusions

It is clear that XP protein-protein interactions mediate the progression of nucleotide excision repair by promoting recognition of damaged DNA, recruitment of repair factors and ultimately the removal of the damaged strand and resynthesis of a new oligonucleotides patch. However, the XP proteins, required for NER, are also responsible for many other DNA metabolic processes, including repair, gene transcription and cell cycle control (Table 1). Although much has been learned about how the XP proteins mediate the repair of damaged DNA, it is still unclear how repair affects and is affected by many cellular nonNER pathways. Continued research in how template DNA is handed off between competing pathways via protein-protein interactions in response to DNA damage is required to fully understand how genome integrity is maintained.

Acknowledgements
We thank Dr. Phillip Musich for his critical reading of this manuscript.

References

1. Kraemer KH, Lee MM, Andrews AD et al. The role of sunlight and DNA repair in melanoma and nonmelanoma skin cancer. The Xeroderma pigmentosum paradigm. Arch Dermatol 1994; 130:1018-1021.

2. Kraemer KH, Lee MM, Scotto J. Xeroderma pigmentosum. Cutaneous, ocular and neurologic abnormalities in 830 published cases. Arch Dermato 1987; 123:241-250.

3. Lehmann AR. DNA repair-deficient diseases, Xeroderma pigmentosum, Cockayne syndrome and trichothiodystrophy. Biochimie 2003; 85:1101-1111.

4. Park CJ and Choi BS. The protein shuffle. Sequential interactions among components of the human nucleotide excision repair pathway. FEBS J 2006; 273:1600-1608.

5. Riedl T, Hanaoka F and Egly JM. The comings and goings of nucleotide excision repair factors on damaged DNA. EMBO J 2003; 22:5293-5303.

6. Masutani C, Sugasawa K, Yanagisawa J et al. Purification and cloning of a nucleotide excision repair complex involving the Xeroderma pigmentosum group C protein and a human homologue of yeast RAD23. EMBO J 1994; 13:1831-1843.

7. Yokoi M, Masutani C, Maekawa T et al. The Xeroderma pigmentosum group C protein complex XPC-HR23B plays an important role in the recruitment of transcription factor IIH to damaged DNA. J Biol Chem 2000; 275:9870-9875.

8. Araki M, Masutani C, Takemura M et al. Centrosome protein centrin 2/caltractin 1 is part of the Xeroderma pigmentosum group C complex that initiates global genome nucleotide excision repair. J Biol Chem 2001; 276:18665-18672.

9. Popescu A, Miron S, Blouquit Y et al. Xeroderma pigmentosum group C protein possesses a high affinity binding site to human centrin 2 and calmodulin. J Biol Chem 2003; 278:40252-40261.

10. Wijnhoven SW, Hoogervorst EM, de Waard H et al. Tissue specific mutagenic and carcinogenic responses in NER defective mouse models. Mut Res 2006; 614:77-94.

11. Sugasawa K, Okuda Y, Saijo M et al. UV-induced ubiquitylation of XPC protein mediated by UV-DDB-ubiquitin ligase complex. Cell 2005; 121:387-400.

12. Hiyama H, Yokoi M, Masutani C et al. Interaction of hHR23 with S5a. The ubiquitin-like domain of hHR23 mediates interaction with S5a subunit of 26 S proteasome. J Biol Chem 1999; 274:28019-28025.

13. Wang QE, Zhu Q, Wani G et al. DNA repair factor XPC is modified by SUMO-1 and ubiquitin following UV irradiation. Nucleic Acids Research 2005; 33:4023-4034.

14. Colton SL, Xu XS, Wang YA et al. The involvement of ataxia-telangiectasia mutated protein activation in nucleotide excision repair-facilitated cell survival with cisplatin treatment. J Biol Chem 2006; 281:27117-27125.

15. Shimizu Y, Iwai S, Hanaoka F et al. Xeroderma pigmentosum group C protein interacts physically and functionally with thymine DNA glycosylase. EMBO J 2003; 22:164-173.

16. Patterson M, Chu G. Evidence that Xeroderma pigmentosum cells from complementation group E are deficient in a homolog of yeast photolyase. Mol Cell Biol 1989; 9:5105-5112.

17. Datta A, Bagchi S, Nag A et al. The p48 subunit of the damaged-DNA binding protein DDB associates with the CBP/p300 family of histone acetyltransferase. Mutat Res 2001; 486:89-97.

18. Kulaksiz G, Reardon JT, Sancar A. Xeroderma pigmentosum complementation group E protein (XPE/DDB2):purification of various complexes of XPE and analyses of their damaged DNA binding and putative DNA repair properties. Mol Cell Biol 2005; 25:9784-9792.

19. Martinez E, Palhan VB, Tjernberg A et al. Human STAGA complex is a chromatin-acetylating transcription coactivator that interacts with premRNA splicing and DNA damage-binding factors in vivo. Mol Cell Biol 2001; 21:6782-6795.

20. Kapetanaki MG, Guerrero-Santoro J, Bisi DC et al. The DDB1-CUL4ADDB2 ubiquitin ligase is deficient in Xeroderma pigmentosum group E and targets histone H2A at UV-damaged DNA sites. PNAS 2006; 103:2588-2593.

21. Hayes S, Shiyanov P, Chen X et al. DDB, a putative DNA repair protein, can function as a transcriptional partner of E2F1. Mol Cell Biol 1998; 18:240-249.

22. Chen X, Zhang J, Lee J et al. A kinase-independent function of c-Abl in promoting proteolytic destruction of damaged DNA binding proteins. Mol Cell 2006; 22:489-499.

23. Chen X, Zhang Y, Douglas L et al. (2001) UV-damaged DNA-binding proteins are targets of CUL-4A-mediated ubiquitination and degradation. J Biol Chem 2001; 276:48175-48182.

24. Groisman R, Polanowska J, Kuraoka I et al. The ubiquitin ligase activity in the DDB2 and CSA complexes is differentially regulated by the COP9 signalosome in response to DNA damage. Cell 2003; 113:357-367.

25. Batty DP, Wood RD. Damage recognition in nucleotide excision repair of DNA. Gene 2000; 41:193-204.

26. Cleaver JE, States JC. The DNA damage-recognition problem in human and other eukaryotic cells:the XPA damage binding protein. Biochem J 1997; 328:1-12.
27. Lee SH, Kim DK, Drissi R. Human Xeroderma pigmentosum group A protein interacts with human replication protein A and inhibits DNA replication. J Biol Chem 1995; 270:21800-21805.
28. Mer G, Bochkarev A, Gupta R et al. Structural basis for the recognition of DNA repair proteins UNG2, XPA and RAD52 by replication factor RPA. Cell 2000; 103:449-456.
29. Zou Y, Liu Y, Wu X et al. Functions of human replication protein A (RPA):from DNA replication to DNA damage and stress responses. J Cell Physiol 2006; 208:267-273.
30. Guzder SN, Sommers CH, Prakash L et al. Complex formation with damage recognition protein Rad14 is essential for Saccharomyces cerevisiae Rad1-Rad10 nuclease to perform its function in nucleotide excision repair in vivo. Mol Cell Biol 2006; 26:1135-1141.
31. Yang Z, Roginskaya M, Colis LC et al. Specific and Efficient Binding of Xeroderma Pigmentosum Complementation Group A to Double-Strand/Single-Strand DNA Junctions with 3'- and/or 5'-ssDNA Branches. Biochemistry 2006; 45:15921-15930.
32. Missura M, Buterin T, Hindges R et al. Double-check probing of DNA bending and unwinding by XPA-RPA: an architectural function in DNA repair. EMBO J 2001; 20:3554-3564.
33. Wu X, Shell SM, Yang Z et al. Phosphorylation of nucleotide excision repair factor Xeroderma pigmentosum group A by ataxia telangiectasia mutated and Rad3-related-dependent checkpoint pathway promotes cell survival in response to UV irradiation. Cancer Res 2006; 66:2997-3005.
34. Lembo F, Pero R, Angrisano T et al. MBDin, a novel MBD2-interacting protein, relieves MBD2 repression potential and reactivates transcription from methylated promoters. Mol Cell Biol 2003; 23:1656-1665.
35. Nitta M, Saijo M, Kodo N et al. A novel cytoplasmic GTPase XAB1 interacts with DNA repair protein XPA. Nucleic Acids Res 2000; 28:4212-4218.
36. Nakatsu Y, Asahina H, Citterio E et al. XAB2, a novel tetratricopeptide repeat protein involved in transcription-coupled DNA repair and transcription. J Biol Chem 2000; 275:34931-34937.
37. Yonemasu R, Minami M, Nakatsu Y et al. Disruption of mouse XAB2 gene involved in premRNA splicing, transcription and transcription-coupled DNA repair results in preimplantation lethality. DNA Repair 2005; 4:479-491.
38. Wu X, Shell SM, Liu Y et al. ATR-dependent checkpoint modulates XPA nuclear import in response to UV irradiation. Oncogene 2007; 26:757-764.
39. Dip R, Camenisch U, Naegeli H. Mechanisms of DNA damage recognition and strand discrimination in human nucleotide excision repair. DNA Repair 2004; 3:1409-1423.
40. Sandrock B, Egly JM. A yeast four-hybrid system identifies Cdk-activating kinase as a regulator of the XPD helicase, a subunit of transcription factor IIH. J Biol Chem 2001; 276:35328-35333.
41. Coin F, Marinoni JC, Rodolfo C et al. Mutations in the XPD helicase gene result in XP and TTD phenotypes, preventing interaction between XPD and the p44 subunit of TFIIH. Nat Genet 1998; 20:184-188.
42. Kellenberger E, Dominguez C, Fribourg S et al. Solution structure of the C-terminal domain of TFIIH P44 subunit reveals a novel type of C4C4 ring domain involved in protein-protein interactions. J Biol Chem 2005; 280:20785-20792.
43. Seroz T, Winkler GS, Auriol J et al. Cloning of a human homolog of the yeast nucleotide excision repair gene MMS19 and interaction with transcription repair factor TFIIH via the XPB and XPD helicases. Nucleic Acids Res 2000; 28:4506-4513.
44. Wu X, Li H, Chen JD. The human homologue of the yeast DNA repair and TFIIH regulator MMS19 is an AF-1-specific coactivator of estrogen receptor. J Biol Chem 2001; 276:23962-23968.
45. Robles AI, Harris CC. p53-mediated apoptosis and genomic instability diseases. Acta oncologica (Stockholm, Sweden) 2001; 40:696-701.
46. Spillare EA, Wang XW, von Kobbe C et al. Redundancy of DNA helicases in p53-mediated apoptosis. Oncogene 2006; 25:2119-2123.
47. Jaitovich-Groisman I, Benlimame N, Slagle BL et al. Transcriptional regulation of the TFIIH transcription repair components XPB and XPD by the hepatitis B virus x protein in liver cells and transgenic liver tissue. J Biol Chem 2001; 276:14124-14132.
48. Leveillard T, Andera L, Bissonnette N et al. Functional interactions between p53 and the TFIIH complex are affected by tumour-associated mutations. EMBO J 1996; 15:1615-1624.
49. Wang XW, Forrester K, Yeh H et al. Hepatitis B virus X protein inhibits p53 sequence-specific DNA binding, transcriptional activity and association with transcription factor ERCC3. PNAS 1994; 91:2230-2234.
50. Coin F, Proietti De Santis L, Nardo T et al. p8/TTD-A as a repair-specific TFIIH subunit. Mol Cell 2006; 21:215-226.

51. Hall H, Gursky J, Nicodemou A et al. Characterization of ERCC3 mutations in the Chinese hamster ovary 27-1, UV24 and MMC-2 cell lines. Mutat Res 2006; 593:177-186.

52. Jawhari A, Laine JP, Dubaele S et al. p52 Mediates XPB function within the transcription/repair factor TFIIH. J Biol Chem 2002; 277:31761-31767.

53. Lin YC, Gralla JD. Stimulation of the XPB ATP-dependent helicase by the beta subunit of TFIIE. Nucleic Acids Res 2005; 33:3072-3081.

54. Coin F, Auriol J, Tapias A et al. Phosphorylation of XPB helicase regulates TFIIH nucleotide excision repair activity. EMBO J 2004; 23:4835-4846.

55. Lommel L, Chen L, Madura K et al. The 26S proteasome negatively regulates the level of overall genomic nucleotide excision repair. Nucleic Acids Res 2000; 28:4839-4845.

56. Weeda G, Rossignol M, Fraser RA et al. The XPB subunit of repair/transcription factor TFIIH directly interacts with SUG1, a subunit of the 26S proteasome and putative transcription factor. Nucleic Acids Res 1997; 25:2274-2283.

57. Liu J, Meng X and Shen Z. Association of human RAD52 protein with transcription factors. Biochem Biophys Res Commun 2002; 297:1191-1196.

58. Canitrot Y, Falinski R, Louat T et al. p210 BCR/ABL kinase regulates nucleotide excision repair (NER) and resistance to UV radiation. Blood 2003; 102:2632-2637.

59. Dunand-Sauthier I, Hohl M, Thorel F et al. The spacer region of XPG mediates recruitment to nucleotide excision repair complexes and determines substrate specificity. J Biol Chem 2005; 280:7030-7037.

60. Lieber MR. The FEN-1 family of structure-specific nucleases in eukaryotic DNA replication, recombination and repair. Bioessays 1997; 19:233-240.

61. Thorel F, Constantinou A, Dunand-Sauthier I et al. Definition of a short region of XPG necessary for TFIIH interaction and stable recruitment to sites of UV damage. Mol Cell Biol 2004; 24:10670-10680.

62. Iyer N, Reagan MS, Wu KJ et al. Interactions involving the human RNA polymerase II transcription/nucleotide excision repair complex TFIIH, the nucleotide excision repair protein XPG and Cockayne syndrome group B (CSB) protein. Biochemistry 1996; 35:2157-2167.

63. Zotter A, Luijsterburg MS, Warmerdam DO et al. Recruitment of the nucleotide excision repair endonuclease XPG to sites of UV-induced dna damage depends on functional TFIIH. Mol Cell Biol 2006; 26:8868-8879.

64. Gervais V, Lamour V, Jawhari A et al. TFIIH contains a PH domain involved in DNA nucleotide excision repair. Nat Struct Mol Biol 2004; 11:616-622.

65. Gary R, Ludwig DL, Cornelius HL et al. The DNA repair endonuclease XPG binds to proliferating cell nuclear antigen (PCNA) and shares sequence elements with the PCNA-binding regions of FEN-1 and cyclin-dependent kinase inhibitor p21. J Biol Chem 1997; 272:24522-24529.

66. Sarker AH, Tsutakawa SE, Kostek S et al. Recognition of RNA polymerase II and transcription bubbles by XPG, CSB and TFIIH: insights for transcription-coupled repair and Cockayne Syndrome. Mol Cell 2005; 20:187-198.

67. Bessho T. Nucleotide excision repair 3' endonuclease XPG stimulates the activity of base excision repair enzyme thymine glycol DNA glycosylase. Nucleic Acids Res 1999; 27:979-983.

68. Klungland A, Hoss M, Gunz D et al. Base excision repair of oxidative DNA damage activated by XPG protein. Mol Cell 1999; 3:33-42.

69. Oyama M, Wakasugi M, Hama T et al. Human NTH1 physically interacts with p53 and proliferating cell nuclear antigen. Biochem Biophys Res Commun 2004; 321:183-191.

70. Park CH, Bessho T, Matsunaga T et al. Purification and characterization of the XPF-ERCC1 complex of human DNA repair excision nuclease. J Biol Chem 1995; 270:22657-22660.

71. Volker M, Mone MJ, Karmakar P et al. Sequential assembly of the nucleotide excision repair factors in vivo. Mol Cell 2001; 8:213-224.

72. de Laat WL, Appeldoorn E, Sugasawa K et al. DNA-binding polarity of human replication protein A positions nucleases in nucleotide excision repair. Genes Dev 1998; 12:2598-2609.

73. Cummings M, Higginbottom K, McGurk CJ et al. XPA versus ERCC1 as chemosensitising agents to cisplatin and mitomycin C in prostate cancer cells: role of ERCC1 in homologous recombination repair. Biochem Pharmacol 2006; 72:166-175.

74. Motycka TA, Bessho T, Post SM et al. Physical and functional interaction between the XPF/ERCC1 endonuclease and hRad,52. J Biol Chem 2004; 279:13634-13639.

75. Lan L, Hayashi T, Rabeya RM et al. Functional and physical interactions between ERCC1 and MSH2 complexes for resistance to cis-diamminedichloroplatinum(II) in mammalian cells. DNA Repair 2004; 3:135-143.

76. Sridharan D, Brown M, Lambert WC et al. Nonerythroid alphaII spectrin is required for recruitment of FANCA and XPF to nuclear foci induced by DNA interstrand cross-links. J Cell Sci 2003; 116:823-835.

77. Thompson LH, Hinz JM, Yamada, N. A et al. How Fanconi anemia proteins promote the four Rs: replication, recombination, repair and recovery. Environ Mol Mutagen 2005; 45:128-142.

78. Zhang N, Kaur R, Lu X et al. The Pso4 mRNA splicing and DNA repair complex interacts with WRN for processing of DNA interstrand cross-links. J Biol Chem 2005; 280:40559-40567.

79. Bradshaw PS, Stavropoulos DJ and Meyn MS. Human telomeric proetin TRF2 associates with genomic double-strand breaks as an early response to DNA damage. Nat Gen 2005; 37:193-197.

80. McDaniel LD, Schultz RA and Friedberg EC. TERF2-XPF: caught in the middle; beginnings from the end. DNA Repair 2006; 5:868-872.

81. Munoz P, Blanco R, Flores JM et al. XPF nuclease-dependent telomere loss and increased DNA damage in mice overexpressing TRF2 result in premature aging and cancer. Nat Genet 2005; 37:1063-1071.

82. Masutani C, Kusumoto R, Yamada A et al. The XPV (Xeroderma pigmentosum variant) gene encodes human DNA polymerase eta. Nature 1999; 399:700-704.

83. Lehmann AR, Kirk-Bell S, Arlett CF et al. Xeroderma pigmentosum cells with normal levels of excision repair have a defect in DNA synthesis after UV-irradiation. PNAS 1975; 72:219-223.

84. Johnson RE, Prakash S and Prakash L. Efficient Bypass of a Thymine-Thymine Dimer by Yeast DNA Polymerase, Poleta. Science 1999; 283:1001-1004.

85. Kannouche P, Fernandez de Henestrosa AR, Coull B et al. Localization of DNA polymerases eta and iota to the replication machinery is tightly co-ordinated in human cells. EMBO J 2002; 21:6246-6256.

86. Kannouche PL, Wing J and Lehmann AR. Interaction of human DNA polymerase eta with monoubiquitinated PCNA: a possible mechanism for the polymerase switch in response to DNA damage. Mol Cell 2004; 14:491-500.

87. Bienko M, Green CM, Crosetto N et al. Ubiquitin-binding domains in Y-family polymerases regulate translesion synthesis. Science 2005; 310:1821-1824.

88. Maga G and Hubscher U. Proliferating cell nuclear antigen(PCNA): a dancer with many partners. J Cell Sci 2003; 116:3051-3060.

89. Vidal AE, Kannouche P, Podust VN et al. Proliferating cell nuclear antigen-dependent coordination of the biological functions of human DNA polymerase iota. J Biol Chem 2004; 279:48360-48368.

90. Watanabe K, Tateishi S, Kawasuji M et al. Rad18 guides pol eta to replication stalling sites through physical interaction and PCNA monoubiquitination. EMBO J 2004; 23:3886-3896.

91. Lawrence CW, Gibbs PE, Murante RS et al. Roles of DNA polymerase zeta and Rev1 protein in eukaryotic mutagenesis and translesion replication. Cold Spring Harb Symp Quant Biol 2000; 65:61-69.

92. Tissier A, Kannouche P, Reck MP et al. colocalization in replication foci and interaction of human Y-family members, DNA polymerase pol eta and REV1 protein. DNA Repair 2004; 3:1503-1514.

93. McIlwraith MJ, Vaisman A, Liu Y et al. Human DNA polymerase eta promotes DNA synthesis from strand invasion intermediates of homologous recombination. Mol Cell 2005; 20:783-792.

94. Rademakers S, Volker M, Hoogstraten D et al. Xeroderma pigmentosum group A protein loads as a separate factor onto DNA lesions. Mol Cell Biol 2003; 23:57555767.

95. Nishi R, Okuda Y, Watanabe E et al. Centrin 2 stimulates nucleotide excision repair by interacting with Xeroderma pigmentosum group C protein. Mol Cell Biol 2005; 25:5664-5674.

96. Araujo SJ, Nigg EA, Wood RD. Strong functional interactions of TFIIH with XPC and XPG in human DNA nucleotide excision repair, without a preassembled repairosome. Mol Cell Biol 2001; 21:2281-2291.

97. Seroz T, Perez C, Bergmann E et al. p44/SSL1, the regulatory subunit of the XPD/RAD3 helicase, plays a crucial role in the transcriptional activity of TFIIH. J Biol Chem 2000; 275:33260-33266.

CHAPTER 12

The Nucleotide Excision Repair of DNA in Human Cells and Its Association with Xeroderma Pigmentosum

Alexei Gratchev*

Introduction

Throughout their lifespan all free-living organisms encounter diverse chemical and physical environmental and endogenous factors leading to DNA damage. Since DNA is a highly reactive macromolecule, these damages may affect both bases and the sugar-phosphate backbone and may lead to a severe dysfunction or death of organism. Already, at the early stages of the evolution, the mechanisms dedicated for repairing various DNA damages were developed. Cellular responses to DNA damage may be classified into two major groups: (i) tolerance and (ii) repair mechanisms.[1] Tolerance mechanism, represented for example by translesion DNA synthesis, gives the cell a possibility to function further as if no damage had occurred. For the damages that cannot be tolerated, the repair mechanisms developed restore the structure of DNA molecule as close to its natural state as prior to damage. To fulfil this task repair mechanisms demonstrate high level of "intelligence" composed of outstanding sensitivity and specificity. One of the most important repair mechanisms is nucleotide excision repair (NER) which is a complex system responsible for recognition and removal of a wide spectra of DNA lesions. In eukaryotes components of NER may also be involved in other repair pathways and in various aspects of DNA metabolism. The importance of NER is supported by the fact that defects in NER cause an extreme ultraviolet (UV) sensitivity and lead to inherited disease such as xeroderma pigmentosum in humans.[2,3]

The most important physical factors damaging DNA molecule is UV radiation. It has been estimated that under the strong sunlight an exposed cell in the human epidermis develops about 40000 damaged sites in one hour, primarily from absorption of UV radiation by DNA.[15] UV leads to the formation of the most frequent mutagenic DNA lesions: cyclobutane pyrimidine dimer(CPD) and pyrimidine 6-4 pyrimidone (6-4PPs).[4,5] Another physical factor, ionising radiation, causes the formation of oxidised or reduced bases as well as single- and double-stranded breaks. This effect is similar to the result of intracellular oxidative metabolism that leads to the generation of reactive oxygen species (ROS). Chemical agents that damage DNA comprise a group of structurally diverse compounds that may either bind DNA directly or form DNA reactive species after metabolic activation. Typical members of this group are carcinogens like benzo(a)pyrene or acetylaminofluorene and chemotherapeutic drugs like cisplatin or tamoxifen.[1,3]

NER is capable of eliminating a broad spectra of DNA damage by the removal of a fragment of damaged strand that contains the lesion. It is highly conserved in eukaryotes and its action can be

*Alexei Gratchev—Klinik für Dermatologie, Venerologie und Allergologie Klinikum Mannheim gGmbH—Universitätsklinikum, Ruprecht-Karls-Universität Heidelberg, Theodor-Kutzer-Ufer 1-3, 68167 Mannheim, Germany. Email: alexei.gratchev@haut.ma.uni-heidelberg.de

Molecular Mechanisms of Xeroderma Pigmentosum, edited by Shamim I. Ahmad and Fumio Hanaoka. ©2008 Landes Bioscience and Springer Science+Business Media.

subdivided into following four steps: (i) recognition of the damage and assembly of pre-incision complex; (ii) excision of an oligonucleotide of 24-32 residues containing the damage; (iii) filling in the gap by DNA polymerases and (iv) ligation of the nick by DNA ligase 1 (Fig. 1).[3,6,7] NER is a primary repair system for CPD, 6-4PPs, as well as for following DNA adducts: benzo(a) pyrene-guanine adduct, acetylaminofluorene-guanine (AAF-G) and cisplatin-d(GpG) adduct. Out of three steps of NER the most puzzling and sophisticated is the step of damage recognition. The main problem of any DNA repair system is the high number of possible DNA lesions that can form within the life span of the cell and have to be repaired. To perform this task NER developed as

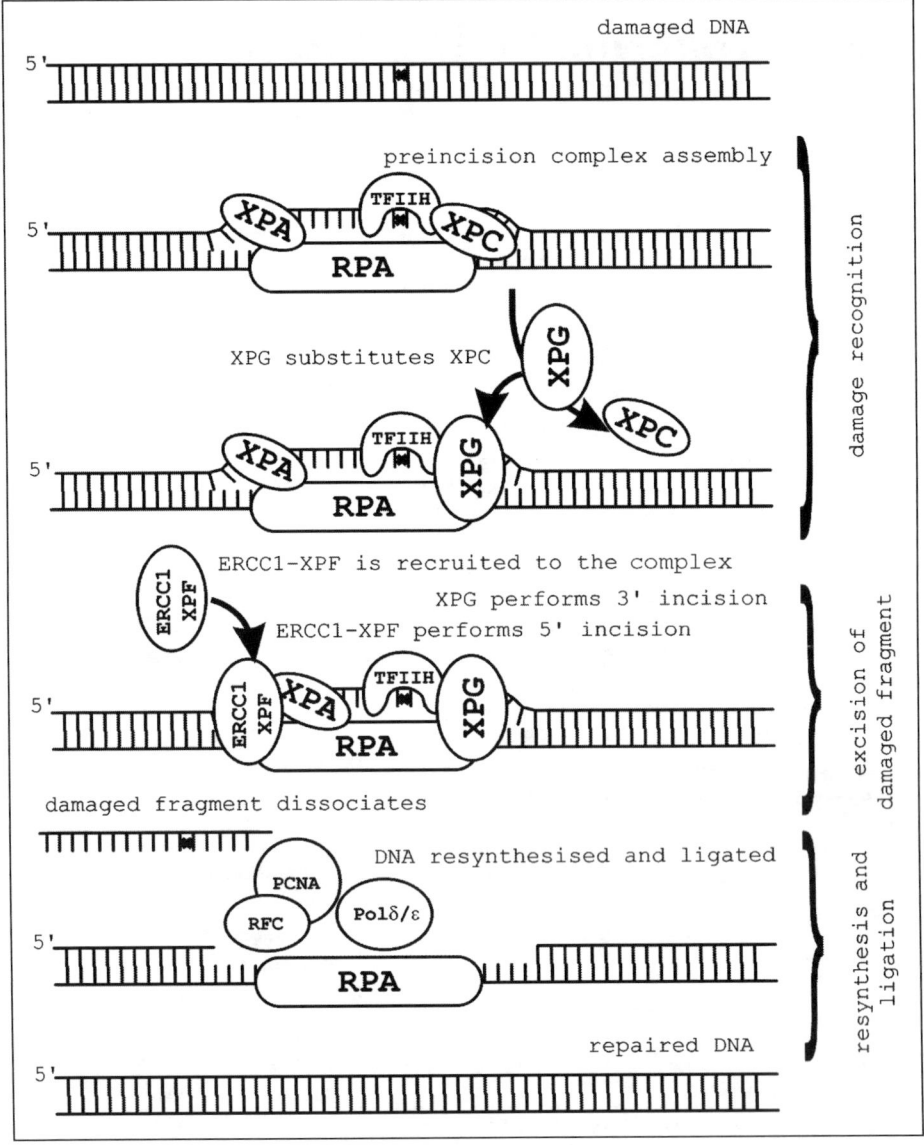

Figure 1. Model for NER in human cells.

a multiprotein system with a broad substrate range and is able to cope with almost infinite number of different lesions, avoiding normal bases.[8] To keep high efficiency, combined with acceptable specificity, NER makes use of cooperative binding and molecular matchmaking during the step of damage recognition.[1] Cooperative binding is the most common thermodynamic mechanism used by biological organisms for increasing specificity. In cooperative DNA-protein interactions (for detail see next paragraph), the binding of one protein to DNA facilitates the binding of either the second subunit of the same protein or of an unrelated protein. The binding sites of the monomers could be adjacent or overlapping and cooperative binding may include more than two proteins. After the binding of one protein to DNA, specific protein-protein interaction leads to a physical relocation of the second protein to the DNA or to the increase of the local concentration of the binding partner. Protein-protein interaction may also help to place the interacting proteins in the appropriate orientation with respect to the binding site.[1] The cooperativity of binding during pre-incision complex assembly is highly important, since the damage recognition is the critical step. All NER factors bind DNA and interact with at least one other repair factor, but there are differences in DNA substrate preferences and in the affinities of repair factor binding for each other. In human NER, cooperative binding to damaged DNA was shown for XPA + RPA (Replication protein A) binding to AAF-G adducts and 6-4PPs.[9] As well cooperative binding was suggested for XPC, which is usually in complex with Transcription factor IIH (TFIIH).[1,10] Another mechanism used by NER complex to increase the specificity and efficiency of damage recognition was termed molecular matchmaking.[11] Molecular matchmaking describes an ATP dependent process when a member of the complex uses the energy of ATP to recruit a protein that is unable to join the complex by itself.[11,12] Five criteria were suggested for the identification of a molecular matchmaker: (i) the affinity of matched protein in the absence of matchmaker must be insignificantly low; (ii) matchmaker must promote stable complex formation; (iii) one of the proteins involved must be able to hydrolyse ATP to obtain energy needed for the complex formation; (iv) matchmaker must form a complex with DNA and matched protein; (v) matchmaker must leave after complex stabilization.[1] In human cells XPC performs a matchmaker function delivering TFIIH and XPG to the pre-incision complex.[12]

It is well established that the core damage recognition and excision system in NER comprises 6 factors. These are XPC, TFIIH, XPA, RPA, XPG and ERCC1-XPF complex (Table 1). XPC, XPA and RPA recognize damage, TFIIH is involved in complex stabilization and unwinds the duplex and XPG and ERCC1-XPF make 3′ and 5′ incisions. After the dissociation of the 24-32 residues including damaged nucleotide(s), replication factors PCNA, RFC, DNA polymerase α or δ resynthesize and DNA ligase 1 ligates the new strand with the old one of DNA.[1,3,5]

XP Associated Genes and Their Roles in NER

XPC is a 106 kDa DNA binding protein that complexes with HR23B needed for its conformational activation and stabilization. XPC binds preferentially to a single stranded UV-damaged DNA[13,14] via its 140 amino acid internal DNA binding domain.[15] Via C-terminal domain XPC

Table 1. Core NER factors

NER Factor	Subunits	Function
XPC	XPC, HR23B	Damage recognition, molecular matchmaker
TFIIH	XPB, XPD, p34, p44, p62, p52, Mat1, Cdk7, Cyclin H	DNA unwinding, helicase, 3′-5′ helicase, 5′-3′ helicase
XPA	XPA	Damage recognition
RPA	RPA70, RPA32, RPA14	Damage recognition, DNA resynthesis
XPG	XPG	3′ incision
ERCC1-XPF	ERCC1, XPF	5′ incision

also interacts with TFIIH and XPB.[15] It was suggested that XPC-HR23B complex initiates NER by sensing and binding lesions and recruiting other factors of the system.[3]

TFIIH is a nine subunit complex that includes XPB (89 kDa) and XPD (80 kDa). It was first identified as a general transcription factor for RNA polymerase II.[16] TFIIH has multiple enzymatic activities. XPB and XPD exhibit DNA-dependent ATPase and helicase functions, XPB can unwind DNA in a 3'-5' direction and XPD in the opposite direction.[17,18] The enzymatic activities of other subunits have not been identified. It is likely that they function in the assembly and stabilization of TFIIH.[3]

XPA gene product has a crucial role at an early stage of NER.[19] Mutation in this gene leads to the most frequent and severe form of XP. XPA is a 31 kDa DNA-binding protein with a marked preference to damaged DNA.[20] Its C-terminal Zn^{2+}-finger containing minimal region is required for DNA binding and is essential for its function. XPA is essential for the assembly of the pre-incision complex, interacting with core repair factors ERCC1-XPF,[21,22] the p32 and p70 subunits of RPA[23] and TFIIH.[24]

Replication Protein A (RPA) is a heterotrimeric single stranded DNA binding protein essential for replication, recombination and repair. Human RPA is composed of three subunits 70, 32 and 14 kDa.[3] In NER the assembly of fully operating complex around the lesion requires RPA[25] which probably binds to the undamaged DNA strand. The size of opened repair intermediate is about 30 nucleotides, which corresponds to the size of the optimal DNA binding region of a single RPA heterotrimer. RPA binds to damaged DNA cooperatively with XPA and is also necessary for coordinated action of NER nucleases. It interacts with XPG, ERCC1-XPF, guiding them to their specific positions of action i.e., XPG to the 3' and ERCC1-XPF to 5' to the lesion.[26] Apart from this RPA provides ERCC1-XPF with strand specificity, by inhibiting the incision of undamaged strand.[26] Being involved also in replication, RPA is supposed to be involved in the resynthesis and ligation steps of NER. RPA was shown to stimulate the activity of polymerases δ and ε.[3]

XPG is a 135 kDa protein that belongs to the FEN-family of structure-specific nucleases. XPG makes an incision in the DNA backbone 3' to the lesion near the junction between single and double-stranded DNA.[27] However XPG is required not only for the 3' incision, but also for the full NER core assembly/stabilization and action. It was shown, that the presence of XPG is necessary for the generation of 5' incision by ERCC1-XPF,[28] as well interaction of XPG with TFIIH and RPA is needed for pre-incision complex stabilization.[3]

Excision Repair Cross-Complementing 1 is a complex of ERCC1 (33 kDa) and XPF (103 kDa) proteins form a stable complex via their C-terminal domains. The ERCC1-XPF is a structure specific endonuclease that incises a variety of DNA substrates. Incisions are always made in one strand of the duplex at the 5' side of the junction with ssDNA.[29] In NER, ERCC1-XPF complex makes the 5' incision consistent with its cleavage polarity. ERCC1-XPF has no structural function in pre-incision complex; therefore 5' incision can be made after the core assembly and 3' incision. ERCC1-XPF interacts with XPA and RPA that are responsible for its correct positioning and functioning.[3]

NER Pathway

A model proposed for NER is summarized (Fig. 1). The damage recognition factors XPC, RPA, XPA and TFIIH assemble at the damage site in a cooperative manner and form an unstable complex. Two subunits of TFIIH, XPB and XPD hydrolyze ATP and unwind the duplex at the damage site to form a repair bubble of about 20 nucleotides.[25,30] This unwinding leads to conformational changes in all the components of the complex and to its stabilization. XPC then uses the ATP hydrolytic activity of TFIIH to recruit XPG to the pre-incision complex. XPG then makes an incision at the 6th ± 3 phosphodiester bond 3' to the damage. ERCC1-XPF recruited to the complex makes a second incision at the 20th ± 3 phosphodiester bond 5' to the damage. This results in 24-32 nucleotides oligomer to be removed and most of the repair factors dissociated from the complex. Concomitantly, RPA binds to DNA and promotes the synthesis of DNA by repair synthesis proteins RFC, PCNA and polymerases δ and ε. Synthesized fragment is then sealed by DNA ligase 1.[1]

XP was reported initially in 1874 (Hebra and Kaposi) and the term "xeroderma pigmentosum" was introduced by Kaposi in 1882 (see Chapter 1). Xeroderma pigmentosum is a rare autosomal recessive disease that occurs in the United States at a frequency of approximately 1:250,000,[31] but with higher frequency in Japan (1:40,000).[32] This disorder is characterized by sensitivity to sunlight, photophobia, early onset of freckling and subsequent neoplastic changes on sun-exposed skin. In XP, the median age of onset of the cutaneous symptoms in sun exposed areas in general is 1-2 years and the median age of onset of nonmelanoma skin cancer is 8 years compared with 60 years in the general population.[33] The incidence of primary cutaneous neoplasms including melanoma is approximately 2000-fold higher in patients with XP than in normal individuals.[34] Neurological symptoms reported in 15% of all patients with XP and range from mild (e.g., isolated hyporeflexia) to severe (e.g., progressive mental retardation).[33,35,36]

In XP, the skin abnormalities result from exposure to sunlight and are caused by an inability of the cell to respond adequately to UV-induced DNA damage. Since NER is a primary system, responsible for elimination of UV-induced CPDs and 6-4 PPs, the mutations of core NER factors are most frequently associated with XP. XP is genetically heterogeneous and has been classified into seven complementation groups XPA-XPG and XP-variant or XPV[37] (Table 2). The most frequent complementation group is A, followed by XPV and XPC. These three groups account for about 90% of all XP cases.[38] Different complementation groups demonstrate different capacity to repair UV induced damage that is equivalent of NER efficiency. Experimentally the efficiency of NER is usually assessed using the technique of unscheduled DNA synthesis (UDS) that provides information about the magnitude of DNA synthesis, happening without normal DNA replication. Technically UDS is measured quantitatively by the incorporation of ^{3}H thymidine by cells that are not in the S phase of the cell cycle.[37] Measurement of UDS demonstrated, that fibroblast from patients with XP, except in XPE and XPV, show significantly reduced UDS (Table 2). Differences in repair capacity of cells from patients from different complementation groups results from the role of corresponding NER factor in the repair process. The best example is the lowest UDS values observed in case of XPA fibroblasts. The differences in repair capacity of the cells from the same complementation group results from different mutation that may cause an incomplete inactivation of the factor. In contrast to cells of XP complementation groups, associated with NER core factors, the cells from XPE and XPV patients demonstrate comparatively high levels of UDS (Table 2), which indicates that these factors are not directly related to NER. Indeed in vitro experiments with free DNA as a substrate, it has been shown that addition of XPE does not improve NER efficiency.

Table 2. Xeroderma pigmentosum complementation groups and related genes

Complementation Group	UDS	Official Gene Symbol (Aliases)	Component of NER
XPA	<10%	XPA (XP1, XPAC)	yes
XPB	3-7%	ERCC3 (XPB; BTF2; GTF2H; RAD25; TFIIH)	yes
XPC	10-20%	XPC (XP3, XPCC)	yes
XPD	25-50%	ERCC2 (EM9; TTD; XPD)	yes
XPE	40-50%	DDB1 (XPE; DDBA; XAP1; XPCE; XPE-BF; UV-DDB1)	no
XPF	10-20%	ERCC4 (XPF; RAD1)	yes
XPG	5-25%	ERCC5 (XPG; UVDR; XPGC; ERCM2)	yes
XPV	75-100%	POLH (XPV; XP-V; RAD30A)	no

Two potential functions of XPF are: (i) that XPE is involved in repair of UV induced chromatin damage,[3,5] and (ii) XPE mediates cellular response to DNA damage regulating cell cycle and apoptosis.[39] The XPV was initially termed "pigmented xerodermoid" since its clinical parameters differ significantly from that of classical XP[40] and the UDS demonstrated by XPV fibroblasts was within the normal range. Later, however, pigmented xerodermoid was reclassified as a variant of XP. The clarity came with the identification of the XPV associated gene—polymerase η (POLH). POLH is a translesion DNA polymerase that is capable of inserting correct nucleotides opposite the damaged template.[40] This property of POLH makes it a part of DNA damage tolerance, rather than a repair system.

References

1. Reardon JT, Sancar A. Nucleotide excision repair. Prog Nucleic Acid Res Mol Biol 2005; 79:183-235.
2. Cleaver JE. Defective repair replication of DNA in xeroderma pigmentosum. Nature 1968; 218(5142):652-656.
3. de Laat WL, Jaspers NG, Hoeijmakers JH. Molecular mechanism of nucleotide excision repair. Genes Dev 1999; 13(7):768-785.
4. Kim JK, Choi BS. The solution structure of DNA duplex-decamer containing the (6-4) photoproduct of thymidylyl(3'—>5')thymidine by NMR and relaxation matrix refinement. Eur J Biochem 1995; 228(3):849-854.
5. Ura K, Hayes JJ. Nucleotide excision repair and chromatin remodeling. Eur J Biochem 2002; 269(9):2288-2293.
6. Sancar A. DNA excision repair. Annu Rev Biochem 1996; 65:43-81.
7. Batty DP, Wood RD. Damage recognition in nucleotide excision repair of DNA. Gene 2000; 241(2):193-204.
8. Sancar A, Lindsey-Boltz LA, Unsal-Kacmaz K et al. Molecular mechanisms of mammalian DNA repair and the DNA damage checkpoints. Annu Rev Biochem 2004; 73:39-85.
9. Wakasugi M, Sancar A. Order of assembly of human DNA repair excision nuclease. J Biol Chem 1999; 274(26):18759-18768.
10. Reardon JT, Sancar A. Thermodynamic cooperativity and kinetic proofreading in DNA damage recognition and repair. Cell Cycle 2004; 3(2):141-144.
11. Sancar A, Hearst JE. Molecular matchmakers. Science 1993; 259(5100):1415-1420.
12. Wakasugi M, Sancar A. Assembly, subunit composition and footprint of human DNA repair excision nuclease. Proc Natl Acad Sci USA 1998; 95(12):6669-6674.
13. Sugasawa K, Shimizu Y, Iwai S et al. A molecular mechanism for DNA damage recognition by the xeroderma pigmentosum group C protein complex. DNA Repair (Amst) 2002; 1(1):95-107.
14. Masutani C, Sugasawa K, Yanagisawa J et al. Purification and cloning of a nucleotide excision repair complex involving the xeroderma pigmentosum group C protein and a human homologue of yeast RAD23. EMBO J 1994; 13(8):1831-1843.
15. Uchida A, Sugasawa K, Masutani C et al. The carboxy-terminal domain of the XPC protein plays a crucial role in nucleotide excision repair through interactions with transcription factor IIH. DNA Repair (Amst) 2002; 1(6):449-461.
16. Egly JM. The 14th Datta Lecture. TFIIH: from transcription to clinic. FEBS Lett 2001; 498(2-3):124-128.
17. Schaeffer L, Roy R, Humbert S et al. DNA repair helicase: a component of BTF2 (TFIIH) basic transcription factor. Science 1993; 260(5104):58-63.
18. Schaeffer L, Moncollin V, Roy R et al. The ERCC2/DNA repair protein is associated with the class II BTF2/TFIIH transcription factor. EMBO J 1994; 13(10):2388-2392.
19. Tanaka K, Miura N, Satokata I et al. Analysis of a human DNA excision repair gene involved in group A xeroderma pigmentosum and containing a zinc-finger domain. Nature 1990; 348(6296):73-76.
20. Jones CJ, Wood RD. Preferential binding of the xeroderma pigmentosum group A complementing protein to damaged DNA. Biochemistry 1993; 32(45):12096-12104.
21. Li L, Elledge SJ, Peterson CA et al. Specific association between the human DNA repair proteins XPA and ERCC1. Proc Natl Acad Sci USA 1994; 91(11):5012-5016.
22. Li L, Peterson CA, Lu X et al. Mutations in XPA that prevent association with ERCC1 are defective in nucleotide excision repair. Mol Cell Biol 1995; 15(4):1993-1998.
23. Li L, Lu X, Peterson CA et al. An interaction between the DNA repair factor XPA and replication protein A appears essential for nucleotide excision repair. Mol Cell Biol 1995; 15(10):5396-5402.

24. Park CH, Mu D, Reardon JT et al. The general transcription-repair factor TFIIH is recruited to the excision repair complex by the XPA protein independent of the TFIIE transcription factor. J Biol Chem 1995; 270(9):4896-4902.

25. Evans E, Moggs JG, Hwang JR et al. Mechanism of open complex and dual incision formation by human nucleotide excision repair factors. EMBO J 1997; 16(21):6559-6573.

26. de Laat WL, Appeldoorn E, Sugasawa K et al. DNA-binding polarity of human replication protein A positions nucleases in nucleotide excision repair. Genes Dev 1998; 12(16):2598-2609.

27. Matsunaga T, Mu D, Park CH et al. Human DNA repair excision nuclease. Analysis of the roles of the subunits involved in dual incisions by using anti-XPG and anti-ERCC1 antibodies. J Biol Chem 1995; 270(35):20862-20869.

28. Mu D, Hsu DS, Sancar A. Reaction mechanism of human DNA repair excision nuclease. J Biol Chem 1996; 271(14):8285-8294.

29. Sijbers AM, van der Spek PJ, Odijk H et al. Mutational analysis of the human nucleotide excision repair gene ERCC1. Nucleic Acids Res 1996; 24(17):3370-3380.

30. Mu D, Wakasugi M, Hsu DS et al. Characterization of reaction intermediates of human excision repair nuclease. J Biol Chem 1997; 272(46):28971-28979.

31. Robbins JH, Kraemer KH, Lutzner MA et al. Xeroderma pigmentosum. An inherited diseases with sun sensitivity, multiple cutaneous neoplasms and abnormal DNA repair. Ann Intern Med 1974; 80(2):221-248.

32. Takebe H, Nishigori C, Satoh Y. Genetics and skin cancer of xeroderma pigmentosum in Japan. Jpn J Cancer Res 1987; 78(11):1135-1143.

33. Kraemer KH, Lee MM, Scotto J. Xeroderma pigmentosum. Cutaneous, ocular and neurologic abnormalities in 830 published cases. Arch Dermatol 1987; 123(2):241-250.

34. Bootsma D, Hoeijmakers JH. The genetic basis of xeroderma pigmentosum. Ann Genet 1991; 34(3-4):143-150.

35. Jung EG. Xeroderma pigmentosum: heterogeneous syndrome and model for UV carcinogenesis. Bull Cancer 1978; 65(3):315-321.

36. Mimaki T, Itoh N, Abe J et al. Neurological manifestations in xeroderma pigmentosum. Ann Neurol 1986; 20(1):70-75.

37. Moriwaki S, Kraemer KH. Xeroderma pigmentosum—bridging a gap between clinic and laboratory. Photodermatol Photoimmunol Photomed 2001; 17(2):47-54.

38. Norgauer J, Idzko M, Panther E et al. Xeroderma pigmentosum. Eur J Dermatol 2003; 13(1):4-9.

39. Kulaksiz G, Reardon JT, Sancar A. Xeroderma pigmentosum complementation group E protein (XPE/DDB2): purification of various complexes of XPE and analyses of their damaged DNA binding and putative DNA repair properties. Mol Cell Biol 2005; 25(22):9784-9792.

40. Gratchev A, Strein P, Utikal J et al. Molecular genetics of Xeroderma pigmentosum variant. Exp Dermatol 2003; 12(5):529-536.

Roles of Oxidative Stress in Xeroderma Pigmentosum

Masaharu Hayashi*

Abstract

Tissue damage caused by oxidative stress has been implicated in aging, carcinogenesis, atherosclerosis and neurodegeneration. In xeroderma pigmentosum (XP) and Cockayne syndrome (CS), oxidative stress is associated with promoted occurrence of skin cancers and progressive neurodegeneration, because decreased DNA repair and persistent DNA damage can result in augmented oxidative nucleotide damage. Oxidative nucleotide damage has been investigated mainly in isolated human skin and blood cells or their cell lines, in which CS cells may be more sensitive to oxidative DNA lesions than XP cells. However, cells from patients with XP group A (XPA) show defective repair of 8, 5'-(S)-cyclo-2'-deoxyadenosine, a free radical-induced endogenous DNA lesion and antioxidant system seems to be disturbed variously in cells from XP patients. We have neuropathologically investigated the involvement of oxidative stress in the brains of XPA and CS autopsy cases and clarified the enhanced lipid peroxidation and protein glycation in the pallidal and cerebellar degeneration. Also, oxidative nucleotide damage with reduced expression of superoxide dismutases has been identified in the basal ganglia lesions, lending further weight involvement of oxidative stress in neurodegeneration in XPA patients. Additionally, we are developing ELISA analysis of oxidative stress markers in the urine and cerebrospinal fluid from XP patients, which will aid with further data on oxidative stress in pathogenesis of XP.

Reactive Oxygen Species and Oxidative Stress

Molecules or molecular fragments with one or more unpaired electrons are called free radicals. Tissue damage caused by oxidative stress, is involved in various pathological processes including aging, carcinogenesis, atherosclerosis, diabetes mellitus, inflammatory diseases and neurodegeneration.[1] Oxidative stress originates from an imbalance between the production of reactive oxygen species (ROS) such as superoxide anion (O_2^-) and hydrogen peroxide (H_2O_2), in addition to reactive nitrogen species (RNS) such as nitrite ion (NO_2^-) and peroxynitrite anion ($ONOO^-$) and antioxidant capacities of cells and organs. A wide range of antioxidants are: several vitamins (A, C and E) and endogenous antioxidative scavengers, including catalase, superoxide dismutase (SOD) and glutathione peroxidase (GPX). SOD converts O_2^- into H_2O_2, which is rapidly reduced by catalase and glutathione peroxidase.[2] When the production of ROS exceeds its breakdown or detoxification, the balance shifts towards accumulation of ROS, leading to oxidative stress. Oxidative stress to lipids, proteins and nucleosides results in accumulation of substrate-specific substances known as oxidative stress markers.[3] The most important product of ROS is lipid peroxidation, which causes disruption of cell membrane thereby leading to their destruction.

*Masaharu Hayashi—Tokyo Metropolitan Institute for Neuroscience, Department of Clinical Neuropathology, 2-6 Musashi-dai, Fuchu-shi, Tokyo 183-8526, Japan.
Email: mahayasi@tmin.ac.jp

Molecular Mechanisms of Xeroderma Pigmentosum, edited by Shamim I. Ahmad and Fumio Hanaoka. ©2008 Landes Bioscience and Springer Science+Business Media.

One of the major products of lipid peroxidation in cell membrane is 4-hydroxy nonenal protein (4-HNE), which is well known to be a toxic product in oxidative damage.[4] Oxidative damages to DNA and RNA produce 8-hydroxy-2'-deoxyguanosine (8-OHdG) and 8-hydroxyguanosine (8-OHG), respectively, which are known as markers for evaluation of oxidative DNA damage.[5] Also advanced glycation end products (AGE) is considered as a marker of protein damage by gly-coxidation and generation of AGE has been delineated in aging, atherosclerosis and progression of diabetes mellitus.[6]

Oxidative Damage at the Cellular Level in Xeroderma Pigmentosum

Xeroderma pigmentosum (XP), Cockayne syndrome (CS) and trichothiodystrophy (TTD) are associated with disturbance in nucleotide excision repair (NER) system,[7] but the latter two disorders will not be discussed in this chapter. Complementation studies by using cell hybridization have revealed the existence of eight genes in XP (groups A-G and a variant) and two in CS (A and B). NER includes global genome repair and transcription-coupled repair (TCR), which involves two CS genes (*CSA* and *CSB*) and several XP genes (especially *XP-A* to *XP-G*). It is likely that decreased DNA repair and persistent DNA damage can result in augmented oxidative nucleotide damage in XP, CS and TTD. Since oxidative stress can be associated with initiation, promotion and progression processes during carcinogenesis, analysis of oxidative DNA stress is important for patients with XP, which are predisposed to skin cancers, or less frequently cancers in the internal organs.[8]

Oxidative nucleotide damage has been investigated mainly in isolated skin and blood cells or their cell lines.[9-14] Oxidative DNA lesions, 8-oxoguanine and thymidine glycol, were shown to be removed by the XP protein-dependent excision system in the cell lines, although XP cell lines are not hypersensitive to X-ray killing.[9] Also it is reported that the removal of 8-oxoguanine by TCR in human cells requires the XPB and XPD proteins as well as XPG and CSB.[10] Proliferating cell nuclear antigen (PCNA) is an important component of NER and the efficient formation of PCNA, following oxidative damage induced by hydrogen peroxide, was reduced in CSB cells but not XPA cells, suggesting that the PCNA-dependent repair may not require XPA protein.[11] Also primary fibroblasts from CS patients, but not XPA cell lines, are defective in cellular repair of 8-hydroxyguanine and 8-hydroxyadenine resulting from oxidative nucleotide damage caused by ionizing radiation.[12] Taking these findings together, CS cells seem to be more sensitive to oxidative DNA lesions than XP cells. Nevertheless, the real-time quantification and identification of free radicals in fibroblasts by microelectrodes demonstrated a significant increase of ROS in the XPA and XPD cells.[13] Also, 8, 5'-(S)-cyclo-2'-deoxyadenosine, a free radical-induced endogenous DNA lesion, is repaired by NER and SV40-transfomed human cell lines from patients with XPA shows defective repair of 8, 5'-(S)-cyclo-2'-deoxyadenosine, but not of the *cis, syn*-cyclobutane thymidine dimer, suggesting that 8, 5'-(S)-cyclo-2'-deoxyadenosine has the properties that could contribute to XP neurodegeneration.[14] Further studies on other XP complementation groups and/or the accumulation of cyclo-deoxyadenosine in XP neurons themselves remain to be investigated. Generally in the experiments using human cells, the results tend to be heterogeneous and variable due to the differences in target cells, types of oxidative stress and measurement methods of oxidative products. Embryonic fibroblasts from CS gene-deficient mice (CSB$^{-/-}$) are hypersensitive to oxidative lesions by gamma-rays-irradiaiton, while embryonic stem cells from XPA gene-deficient mice (XPA$^{-/-}$ mice) had a moderate gamma-ray-sensitivity.[15]

Antioxidant system has also been examined in isolated fibroblasts and cell lines from mice. Epidermal activities of SOD and catalase but not GPX showed a significant decrease in two XP families, one of XP-C and another of XP-D.[16] Vuillaume et al showed that UV induced five and three times more H_2O_2 production in XP cell lines compared with TTD or controls, respectively, although catalase transcription had no differences between normal, XP and TTD cell lines.[17] Subsequently, the same research group reported that decreased levels of NADPH in fibroblasts from XP patients are responsible for the low catalase activity and addition of exogenous NADPH can restore the enzyme activity.[18]

Oxidative Stress in Neurodegenerative Disorders

Free radicals are abundantly produced in the central nervous system, because neurons consume great amounts of molecular oxygen, neuronal mitochondria generate a large amount of O_2^-,[1] and the brain per se stores a readily bioavailable source of iron.[19] Therefore, oxidative stress has been suggested to be one of predisposing factors for neurodegeneration in several adult-onset neurological diseases such as Alzheimer's, Parkinson's and amyotrophic lateral sclerosis.[1,20] Increased lipid peroxidation, protein oxidation and increase of oxidized nucleosides has been found in cell culture, postmortem brains and cerebrospinal fluids from patients suffering from Alzheimer's disease and trangenic animal models.[21] Oxidative stress has been implicated in loss of dopaminergic neurons and disease progression in Parkinson disease, irrespective of differences in genetic background.[22] In addition, mutations in the gene encoding copper-zinc SOD (Cu/ZnSOD) is responsible for about 20% of patients with familial amyotrophic lateral sclerosis and multiple disease-causing mutants are demonstrated to be recruited to mitochondria, but only in affected tissues including the spinal motor neurons.[23] Similarly, oxidative stress has been reported to be involved in various child onset brain disorders, including Down syndrome and periventricular leukomalacia.[24,25]

We have neuropathologically examined the involvement of oxidative stress in neurodegeneration in child onset neurodegenerative disorders.[26-30] Subacute sclerosing panencephalitis is caused by the persistent measles virus infection in the central nervous system and characterized by slowly progressive and fatal clinical course, severe brain atrophy and extended demyelination. Nuclei immunoreactive for 8-OHdG was found in the cortical neurons and subsequently occurred in the glial cells in the demyelinated white matters, which demonstrated augmented lipid peroxidation. These indicate contribution of oxidative stress in neurodegeneration in subacute sclerosing panencephalitis.[26] In spinal muscular atrophy, a childhood counterpart of amyotrophic lateral sclerosis, oxidative nucleoside damage was observed in the motor cortex, lateral thalamus and cerebellum,[27] in addition to the atrophic motor neurons in the disease type having more protracted clinical course.[28] Neuronal ceroid-lipofuscinosis is a group of fatal neurodegenerative and lysosomal storage disorders and is clinically characterized with progressive developmental deterioration, visual loss and/or uncontrolled myoclonic epilepsy. Autopsy cases of late infantile form showed increased lipid peroxidation and oxidative DNA damage in the cerebral cortex, while they had increased protein glycation in the cerebellar cortex.[29] One case of rare protracted juvenile form showed only deposition of advance glycation end products (AGE) in the cerebellar cortex, indicating the possibility of difference in the degree of involvement of oxidative neuronal damage even in the same disease.[30]

Neuropathological Analysis on Oxidative Stress in the Brains of Autopsy Cases of Xeroderma Pigmentosum and Cockayne Syndrome

DNA cannot be adequately repaired in the brains in patients with XP or CS, which can lead to progressive neurological disturbances such as severe developmental delay, peripheral neuropathy, neuronal deafness, cerebellar ataxia, spasticity and extrapyramidal tract signs.[31] Protection from ultraviolet light can prevent development of skin lesions, whereas neurodegeneration cannot be stopped. We have neuropathologically investigated autopsy cases of XP group A (XPA) and CS,[32] and found the occurrence of apoptotic neuronal death in the hippocampal and cerebellar lesions.[33]

In order to investigate the involvement of oxidative stress in neurodegeneration of XPA and CS, we immunohistochemically examined the depositions of oxidative products in proteins and lipids in the brains of five autopsy subjects each of XPA, CS and controls.[34] More nitrotyrosine, AGE and 4-HNE were remarkably deposited in the globus pallidus in CS than XPA (Fig. 1A). In CS and XPA, both were frequently recognized in the pseudocalcified foci, neuropil free minerals and more rarely foamy spheroids. In addition, the deposition of HNE was observed in the cerebellar dentate neurons in both disorders, suggesting that oxidative stress can be involved in the pallidal

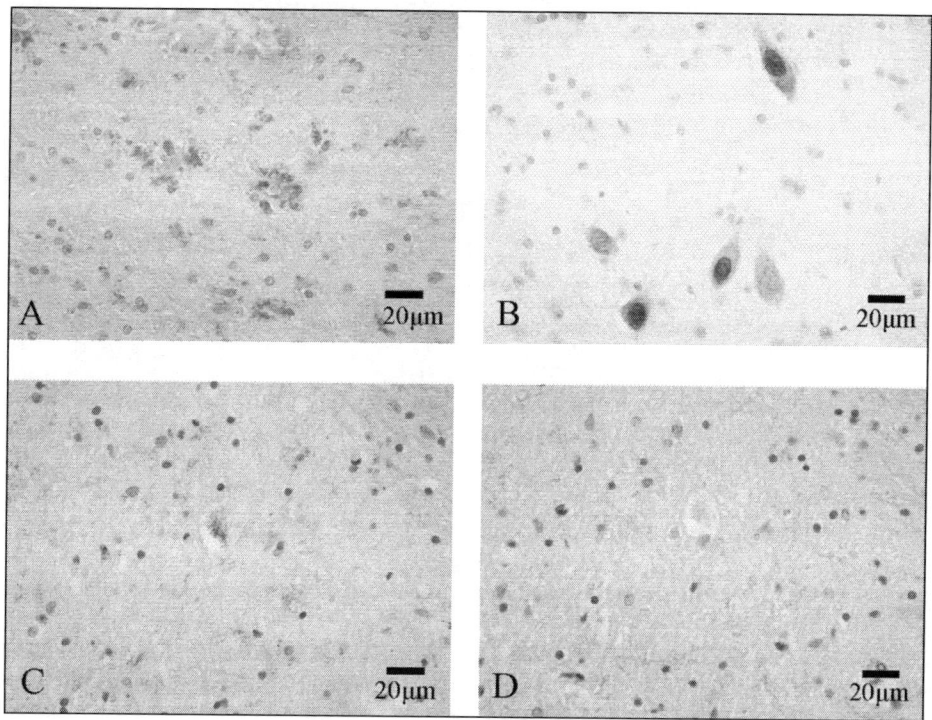

Figure 1. Deposition of oxidative products in the globus pallidus. A) The minerals were immunoreactive for 4-hydroxy nonenal in the external segment of globus pallidus in the 24-year-old female XPA case. B) Some of the pallidal neurons had nuclei immunoreactive for 8-hydroxy-2′-deoxyguanine in the external segment of globus pallidus in the 19-year-old male XPA case. C,D) The neuron had cytoplasmic immunoreactivity for 8-hydroxyguanosine (C) in the external segment of globus pallidus in the 26-year-old female XPA case (D), which vanished completely after the pretreatment with RNase.

and cerebellar degeneration in XPA and CS. Subsequently we examined the deposition of oxidative products in nucleotides and expression of two types of SOD in the XPA and CS subjects.[35] Cases of XPA but not CS demonstrated nuclear deposition of 8-OHdG and cytoplasmic deposition of 8-OHG, in DNA and RNA, respectively, in the globus pallidus (Fig. 1B-D). Also most XPA cases exhibited reduced cytoplasmic immunoreactivity for Cu/ZnSOD in the neurons of the cerebral and cerebellar corteces in addition to the basal ganglia and a few XPA cases also showed reduced immunoreactivity for MnSOD in the mitochondria of the neurons (Fig. 2). In contrast, five CS cases demonstrated comparatively preserved immunoreactivity for both SODs, suggesting that oxidative damage to nucleotides with disturbed SOD expression can be involved in neurodegeneration in XPA but not CS. Simultaneously, we examined the expression of neurotransmitters and neuropeptides in the basal ganglia and thalamus in XPA autopsy cases, because oxidative neuronal damage was most predominant in the basal ganglia.[36] The large cell neurons in the putamen were preferentially reduced, the immunoreactivity for tyrosine hydroxylase in the dopaminergic afferent and efferent pathways were severely affected, whereas reflecting the dopaminergic afferent and efferent pathways were severely affected, whereas the expression of substance P and methionine-enkephalin in the globus pallidus and substantia nigra, which are involved in the efferent pathways in the basal ganglia were comparatively spared. The selective damage to dopamine system in the basal ganglia seemed to be related to clinical abnormalities such as rigidity and laryngeal dystonia in patients with XPA.

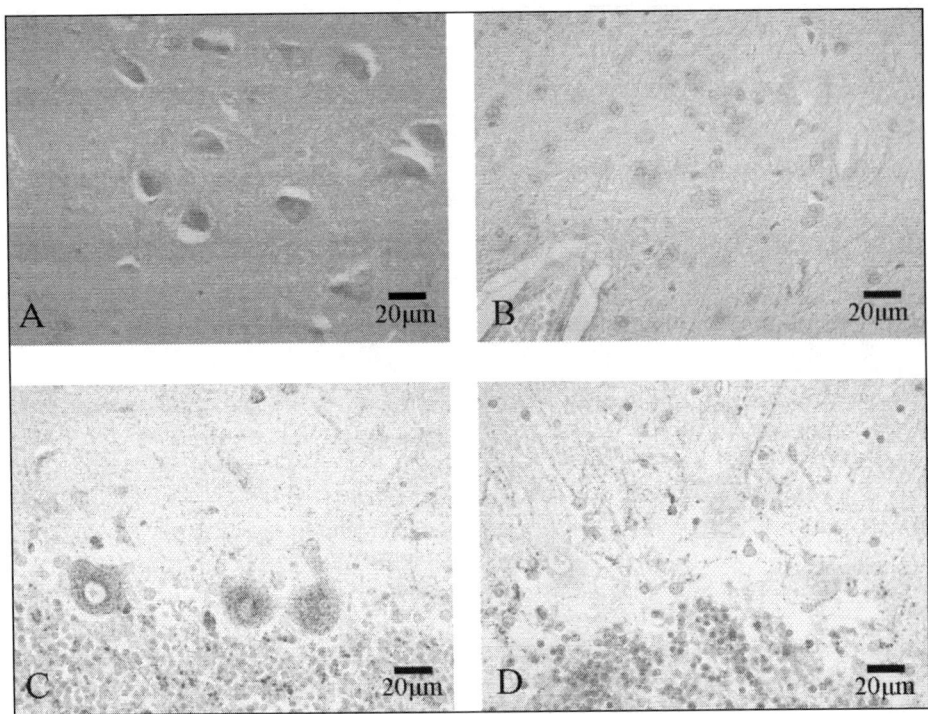

Figure 2. Expressions of superoxide dismutase (SOD). A) Neurons were immunoreactive for Cu/ZnSOD in the CA4 of the hippocampus in the 21-year-old male control. B) The hippocampal neurons immunoreactive for Cu/ZnSOD, were decreased in reactivity in the 24-year-old female XPA case. C) Purkinje cells were immunoreactive for MnSOD in the cerebellar cortex in the 29-year-old female control. D) Immunoreactivity for MnSOD was reduced in the Purkinje cells in the cerebellar cortex in the 26-year-old XPA case.

Preliminary ELISA Analysis of Oxidative Stress Markers in Urinary Samples from Patients with Xeroderma Pigmentosum

Oxidative stress markers are available to evaluate the levels of oxidative DNA damage and lipid peroxidation in the samples of urine, serum and cerebrospinal fluid from very low birth weight infants and children with neurological disorders.[37-40] We preliminary examined the early and late-stage markers for lipid peroxidation, hexanoyl-lysine adduct and acrolein-lysine adduct, respectively, in addition to 8-OHdG in urine samples from four XPA patients, one XPD patient and 17 healthy controls aged 3-81 years (Table 1). Samples were obtained from each subject in the morning and immediately stored at −80°C until analysis. The patients and/or their parents understood the purpose of this study and approved the offer of urine samples. Levels of urinary oxidative stress biomarkers, 8-OHdG, HEL and ACR were determined using an enzyme-linked immunosorbent assay (ELISA) kit. We also measured urinary creatinine (mg/dl) by a standard automated colorimetric assay. Each value was expressed relative to urinary creatinine to adjust for muscle mass. XPA patient aged 29 years with long disease duration, suffering from diabetes mellitus, showed a remarkable increase over the mean plus 2SD of data in 17 controls in both urinary 8-OHdG and hexanoyl-lysine adduct (Table 1, upward arrows), probably due to severe systemic complications. It is intriguing that young XPD patient aged 9 years also demonstrated a significant increase of urinary 8-OHdG (Table 1, upward arrow), although he only shows mild

Table 1. Preliminary ELISA analysis of urinary oxidative stress markers in xeroderma pigmentosum

Subjects	Age(yr)/Sex	Motor Ability	Mental Ability	Neurological Deafness	8-Hydroxy-2'-Deoxyguanosine (ng/mg Creatinine)	Hexanoyl-Lysine Adduct (pmol/mg Creatinine)	Acrolein-Lysine Adduct (nmol/mg Creatinine)
XPA	29/Male	Bedridden	Respond to pain	Severe	23.4↑	254.8↑	190.3
XPA	26/Male	Bedridden	Respond to orders	Severe	15.3	149.0	76.0
XPA	16/Female	Walk with support	Can converse	Moderate	9.8	139.1	160.4
XPA	7/Male	Independent walk	Mild retardation	Mild	13.6	117.6	189.2
XPD	9/Male	Independent walk	Mild retardation	Mild	28.9↑	166.1	187.2
Controls	(n = 17)				10.5 ± 2.9	69.2 ± 37.7	144.9 ± 62.0

mental retardation and neurological deafness. ELISA analysis of oxidative stress markers in the cerebrospinal fluid samples from XPA patients is now in progress in our laboratory, which will provide useful data to clarify the involvements of oxidative stress in neurodegeneration in XPA.

References

1. Halliwell B. Oxidative stress and neurodegeneration: where are we know? J Neurochem 2006; 97(6):1634-1658.
2. Fridovich I. Superoxide radical and superoxide dismutases. Annu Rev Biochem 1995; 64:97-112.
3. Dalle-Donne I, Rossi R, Colombo R et al. Biomarkers of oxidative damage in human disease. Clin Chem 2006; 52(4):601-623.
4. Toyokuni S. Reactive oxygen species-induced molecular damage and its application in pathology. Pathol Int 1999; 49(2):91-102.
5. Toyokuni S, Tanaka T, Hattori Y et al. Quantitative immunohistochemical determination of 8-hydroxy-2'-deoxyguanosine by a monoclonal antibody N45.1: its application to ferric nitrilotriacetate-induced renal carcinogenesis model. Lab Invest 1997; 76(3):365-374.
6. Viassara H, Bucala R, Striker L. Pathogenic effects of advanced glycosylation: biochemical, biologic and clinical implications for diabetes and aging. Lab Invest 1994; 70(2):138-151.
7. Laposa RR, Cleaver JE. DNA repair on the brain. Proc Natl Acad Sci USA 2001; 98(23):12860-12862.
8. Nishigori C, Hattori Y, Toyokuni S. Role of reactive species in skin carcinogenesis. Antioxid Redox Signal 2004; 6(3):561-570.
9. Reardon JT, Bessho T, Kung HC et al. In vitro repair of oxidative DNA damage by human nuclotide excision repair system: possible explanation for neurodegeneration in xeroderma pigmentosum patients. Proc Natl Acad Sci USA 1997; 94(17):9463-9468.
10. Le Page F, Kwoh EE, Avrutskaya A et al. Transcription-coupled repair of 8-oxoguanine: requirement for XPG, TFIIH and CSB and implication for Cockayne syndrome. Cell 2000; 101(2):159-171.
11. Balajee AS, Dianova I, Bohr VA. Oxidative damage-induced PCNA complex formation is efficient in xeroderma pigmentosum group A but reduced in Cockayne syndrome group B-cells. Nucleic Acids Res 1999; 27(22):4476-4482.
12. Tuo J, Jaruga P, Rodrigue H et al. Primary fibroblast of Cockayne syndrome patients are defective in cellular repair of 8-hydroxyguanine and 8-hydroxyadenine resulting from oxidative stress. FASEB J 2003; 17(6):668-674.
13. Arbault S, Sojic N, Bruce D et al. Oxidative stress in cancer prone xeroderma pigmentosum fibroblasts. Real-time and single monitoring of superoxide and nitric oxide production with microelectrodes. Carcinogenesis 2004; 25(4):509-515.
14. Brooks PJ, Wise DS, Berry DA et al. The oxidative DNA lesion 8, 5'-(S)-cyclo-2'-deoxyadenosine is repaired by the nucleotide excision repair pathway and blocks gene expression in mammalian cells. J Biol Chem 2000; 275(29):22355-22362.
15. de Waard H, de Wit J, Gorgels TG et al. Cell type-specific hypersensitivity to oxidative damage in CSB and XPA mice. DNA repair 2003; 2(1):13-25.
16. Schallreuter KU, Pittelkow MR, Wood JM. Defects in antioxidant defense and calcium transport in the epidermis of xeroderma pigmentosum patients. Arch Dermatol Res 1991; 283(7):449-455.
17. Vuillaume M, Daya-Grosjean L, Vincens P et al. Striking differences in cellular catalase activity between two DNA repair-deficient diseases: xeroderma pigmentosum and trichothiodystrophy. Carcinogenesis 1992; 13(3):321-328.
18. Hoffschir F, Daya-Grosjean L, Petit PX et al. Low catalase activity in xeroderma pigmentosum fibroblasts and SV40-transformed human cell lines is directly related to decreased intracellular levels of the cofactor, NADPH. Free Radic Biol Med 1998; 24(5):809-816.
19. Zecca l, Youdim MB, Riederer P et al. Iron, brain ageing and neurodegenerative disorders. Nat Review Neurosci 2004; 5(11):863-873.
20. Simonian NA, Coyle JT. Oxidative stress in neurodegenerative diseases. Annu Rev Pharmacol Toxicol 1996; 36:83-116.
21. Nunomura A, Castellani RJ, Zhu X et al. Involvement of oxidative stress in Alzheimer disease. J Neuropathol Exp Neurol 2006; 65(7):631-641.
22. Jenner P. Oxidative stress in Parkinson's disease. Ann Neurol 2003; 53(suppl 3):S26-S38.
23. Liu J, Lillo C, Jonsson PA et al. Toxicity of familial ALS-linked SOD1 mutants from selective recruitment of spinal mitochondria. Neuron 2004; 43(1):5-17.
24. Nunomura A, Perry G, Pappolla MA et al. Neuronal oxidative stress precedes amyloid-β deposition in Down syndrome. J Neuropathol Exp Neurol 2000; 59(11):1011-1017.
25. Haynes RL, Baud O, Li J et al. Oxidative and nitrative injury in periventricular leukomalacia: a review. Brain Pathol 2005; 15(3):225-233.

26. Hayashi M, Arai N, Satoh J et al. Neurodegenerative mechanisms in subacute sclerosing panencephalitis. J Child Neurol 2002; 17(10):725-730.

27. Hayashi M, Araki S, Arai N et al. Oxidative stress and disturbed glutamate transport in spinal muscular atrophy. Brain Dev 2002; 24(8):770-775.

28. Araki S, Hayashi M, Tamagawa K et al. Neuropathological analysis in spinal muscular atrophy type II. Acta Neuropathol 2003; 106(5):441-448.

29. Hachiya Y, Hayashi M, Kumada S et al. Mechanisms of neurodegeneration in neuronal ceroid-lipofuscinosis. Acta Neuropathol 2006; 111(2):168-177.

30. Anzai Y, Hayashi M, Fueki N et al. Protracted juvenile neuronal ceroid lipofuscinosis—an autopsy report and immunohistochemical analysis. Brain Dev 2006; 28(7):462-465.

31. Bootsma D, Kraemer KH, Cleaver JE et al. Nucleotide excision repair syndromes: xeroderma pigmentosum, Cockayne syndrome and trichothiodystrophy. In: Scriber CR, Beaudet AL, Sly WS, Valle D, eds. The metabolic and molecular bases of inherited disease, 8th ed. Vol. I, New York: McGraw-Hill, Inc., 2001:677-703.

32. Itoh M, Hayashi M, Shioda K et al. Neurodegeneration in hereditary nucleotide repair disorders. Brain Dev 1999; 21(5):326-333.

33. Kohji T, Hayashi M, Shioda K et al. Cerebellar neurodegeneration in human hereditary DNA repair disorders. Neurosci Lett 1998; 243(1-3):133-136.

34. Hayashi M, Itoh M, Araki S et al. Oxidative stress and glutamate transport in hereditary nucleotide repair disorders. J Neuropathol Exp Neurol 2001; 60(4):350-356.

35. Hayashi M, Araki S, Kohyama J et al. Oxidative nucleotide damage and superoxide dismutase expression in the brains of xeroderma pigmentosum group A and Cockayne syndrome. Brain Dev 2005; 27(1):34-38.

36. Hayashi M, Araki S, Kohyama J et al. Brainstem and basal ganglia lesions in xeroderma pigmentosum group A. J Neuropathol Exp Neurol 2004; 63(10):1048-1057.

37. Matsubasa T, Uchino T, Karashima S et al. Oxidative stress in very low birth weight infants as measured by urinary 8-OHdG. Free Radic Res 2002; 36(2):189-193.

38. Tsukahara H, Haruta T, Ono N et al. Oxidative stress in childhood meningitis: measurement of 8-hydroxy-2'-deoxyguanosine concentration in cerebrospinal fluid. Redox Rep 2000; 5(5):295-298.

39. Tsukahara H, Haruta T, Todoroki Y et al. Oxidant and antioxidant activities in childhood meningitis. Life Sci 2002; 71(23):2797-2806.

40. Schulpis KH, Lazaropoulou C, Regoutas S et al. Valproic acid monotherapy induces DNA oxidative damage. Toxicology 2006; 217(2-3):228-232.

Xeroderma Pigmentosum:
Its Overlap with Trichothiodystrophy, Cockayne Syndrome and Other Progeroid Syndromes

W. Clark Lambert,* Claude E. Gagna and Muriel W. Lambert

Introduction

Although this volume is devoted to xeroderma pigmentosum (XP), there are, in fact, at least three disorders, XP, trichothiodystrophy, (TTD) and Cockayne syndrome (CS), the etiopathogeneses of which are involved with the same biochemical pathways and, in a number of cases, with the same gene(s).[1-5] In some instances, patients have unequivocal evidence of having more than one of these diseases, which are clinically quite heterogeneous. This chapter will review those unusual cases, known as XP/TTD or XP/CS overlap syndromes or XP/TTD or XP/CS overlap complexes.

In theory, there are two ways an individual could be affected by these very different diseases (i.e., XP and TTD or XP and CS). The first is that he or she could simply be extraordinarily unlucky and be struck by two genetic bullets. This idea implies that the individual is homozygous or hemizygous (in the case that an allele on a sex chromosome is involved) for defective alleles at two different loci. This occurrence, which the two of us (WCL and MWL)[6,7] have previously labeled "corecessive inheritance", would imply extremely high carrier frequencies for at least some of these defective alleles. However, there is no direct evidence that corecessive inheritance is responsible for these overlap syndromes, although there are a number of examples in which two individuals with identical mutations have very different clinical phenotypes, implying involvement of more than one gene.

The second way one could be affected by more than one of these disorders is that one is homozygous or hemizygous for defective alleles at a single locus, but the gene product associated with that locus is involved in more than one biochemical pathway within the cell. There is considerable evidence that this is the etiopathogenesis of at least some of the overlap syndromes, which, although quite rare, are much more common than would be expected, based on the frequencies of these syndromes (XP, TTD and CS) considered independently.

In particular, we will be concerned with two pathways, nucleotide excision repair (NER), which is defective in all complementation groups of XP except the variant group (XPV) and initiation and to some extent continuation of transcription especially transcription by RNA polymerase II, which transcribes mRNA for active genes within the cell. As is reviewed extensively elsewhere in this volume, the large RNA transcription subunit, TFIIH, is intimately involved in both pathways, so much so that it could just as easily be considered a DNA repair protein complex as an RNA transcription protein complex.

*Corresponding Author: W. Clark Lambert—Department of Pathology, UMDNJ-New Jersey Medical School, Newark, NJ 07103, USA. Email: lamberwc@umdnj.edu

Molecular Mechanisms of Xeroderma Pigmentosum, edited by Shamim I. Ahmad and Fumio Hanaoka. ©2008 Landes Bioscience and Springer Science+Business Media.

TFIIH is a complex containing at least 10 proteins, notably including XPB, XPD and XPG, mutations in which appears to account for almost all of these overlap syndromes. The consensus, if there may be said to be such a thing, is that mutations in these genes that cause XP are those that affect the NER mechanism (particularly global genomic NER (GG-NER)). On the other hand, those mutations that primarily affect transcription-coupled NER (TC-NER), as well as transcription and possibly other systems cause TTD or CS. Presumably the overlap syndromes are due to mutations that affect more than one pathway and we shall regard this as a working hypothesis.

It is highly likely, however, that for several reasons the above hypothesis is an oversimplification; first, there are many other biochemical pathways that are either known to include or are likely to be found to involve these same gene products. These include both short patch and long patch base excision repair (BER), mismatch repair (MMR), homologous recombination (HR) and nonhomologous end-joining (NHEJ) of DNA double strand breaks (DSBs). Other processes that interact with these include telomere maintenance and cell cycle control. Second, many nuclear proteins are, at least to some extent, interdependent for maintenance of stability and other functions, so that a protein may affect one or more of these pathways without actually being involved in it directly. Third, nuclear proteins/gene products may affect rates of transcription or translation of other genes or activation of their gene products. Fourth, the function of some of these proteins, or even the entire pathway in which they are functioning, remains unclear. The list goes on and on. We have shown, for example, that a major nuclear structural protein, α-spectrin II, is active in the Fanconi anemia pathway, the function of which is still poorly understood.[8]

There is yet another area of uncertainty that must be addressed before we attempt to review these overlap syndromes: the diagnosis of any of these is dependent on clinical assessment, on who is looking and how hard they are doing it. Is one adult patient with early freckling but no skin cancers comparable to another patient with severe dyspigmentation and dozens of cancers arising in childhood? In these classifications both are likely to be considered to have XP, with additional signs/symptoms, which are similarly variable, causing them to be classified as also having CS or TTD. One must review these reports, many of which are authored by biochemists and molecular biologists who are wonderfully competent in their fields but who are not necessarily similarly skilled clinicians, or who may not even have direct access to the relevant clinical information, with caution. There is also heterogeneity in competencies among physicians reporting these cases. For example, a neurologist is less likely to note freckling than a dermatologist who, in turn, is less likely to detect mild spasticity.

Although there are reports of XP/TTD and XP/CS overlap syndromes, we are not aware of a report of a "TTD/CS overlap syndrome" and "XP/TTD/CS overlap syndrome". However, this may have been overlooked because the authors of the reports were mostly focusing on XP/TTD overlap or XP/CS overlap rather than other possibilities and may therefore represent a pedagogic solution to a scientific problem. Clinically, there is considerable overlap between TTD and CS and both are presumably (see "working hypothesis" above), related to transcriptional deficiencies.

In the following we shall review not only XP/TTD and XP/CS overlap syndromes due to XPB, XPD and XPG mutations, but certain other overlaps as well. In particular, we shall examine overlaps of XP with progeroid syndromes. CS and TTD may themselves be considered progeroid syndromes and these syndromes, although imperfectly so, are nevertheless intriguing models for ageing. We shall also review the ultraviolet sensitivity syndrome (UVSS), which, in some cases, may be related to a lack of the CSB protein.

The Ultraviolet-Sensitive Syndrome (UVSS)

Fujiwara et al (1981) described a patient with clinical manifestations similar to XP, including acute sunburn following minimal sun exposure, freckling, xerosis (dryness) telangiectasias and dyspigmentation of sun-exposed skin but without development of tumors or neurological abnormalies.[9] In 1994, Itoh et al described additional patients.[10] A delayed recovery of RNA synthesis, following UV exposure, was noted in the patients' cells—a feature of CS. These authors proposed the new term "ultraviolet-sensitive syndrome (UVSS)" to denote these patients.[10,11]

Horibata et al (2004) found that cells from the original patient described by Fujiwara et al[9] were homozygous for a null mutation in *ERCC6/CSB* (the excision repair, complementation group 6/Cockayne syndrome, complementation group B gene).[12] A cell line from a different patient with the UVSS had no mutations in *ERCC6/CSB* and normal amounts of the ERCC6 protein, however, indicating genetic heterogeneity within the UVSS. The UVSS has not yet been reported to have any overlapping signs or symptoms of TTD or CS.

XP/Trichothiodystrophy (XP/TTD) Overlap Syndromes

The term, Trichothiodystrophy (TTD) was introduced by Price et al (1980) to describe a syndrome or syndromes characterized by the presence of brittle, sulfur-deficient hair.[13] The hair and nails of the patients suffering from this disease are brittle because they contain reduced levels of cysteine-rich matrix proteins. Affected individuals also have ichthyosis, which is usually apparent at birth and much less apparent afterwards and retarded mental and physical development. TTD is part of a more broadly defined entity known as the IBIDS (ichthyosis, brittle hair, impaired intelligence, decreased fertility and short stature) syndrome,[14] which itself is somewhat poorly defined due to clinical variation between patients.[15] The original case was described by Tay (1971) and the syndrome is also known as Tay syndrome.[16] The hairs in patients with TTD have a distinctive, diagnostically useful appearance on polarized light microscopy known as the "tiger-tail" anomaly; the disease may also be diagnosed by amino acid analysis of the patients' hair shafts, which show decreased sulphur and cysteine content.[13]

About half of patients with TTD have also photosensitivity (usually clinically manifested as sun-sensitivity) and an associated cellular defect in the transition-coupled repair (TCR) subtype of the nucleotide excision repair (NER) type of DNA repair.[17] However, like CS, to date no case of TTD has been reported to develop skin cancer.[15,17] The photosensitive form of TTD is also known as "trichothiodystrophy-photosensitive (TTDP)".[18]

The TTD phenotype can be caused by mutations in at least three different genes. The overwhelming majority of patients have mutations in *ERCC2/XPD* (the Excision Repair Cross-Complementing 2/Xeroderma Pigmentosum Complementation Group D gene). A single family has had, instead, biallelic mutations in the *ERCC3/XPB* (the Excision Repair Cross-Complementing 3/Xeroderma Pigmentosum Complementation Group B gene) (see below). These two genes encode the two helicase subunits of the transcription/DNA repair factor, TFIIH (Transcription Factor for RNA polymerase II, subunit H, which itself is composed of at least 10 subunit proteins.). A different TTD complementation group was identified by Stefanini et al (1993), when they showed that cells from patients in this new group could complement all XP complementation groups and that this new complementation was not intragenic.[18] Cells from this complementation group were originally identified as "TTD1BR". It has been suggested that it may be identified as "TTD2", with the other known TTD cases at the time (all due at that time to mutations in *ERCC3/XPD*) identified as in complementation group "TTD1". Following the revision in nomenclature proposed by Lehmann, et al (1994), this new group is now known as TTD-A.[19] However, these alternative nomenclatures may be seen in the literature. This new complementation group of TTD (TTT-A) has been found to be caused by a mutation in the gene (*GTF2H5/TTD-A*) encoding a newly identified, tenth subunit of TFIIH, TFB5.[20] The overall levels of TFIIH in TTDA cells are also significantly depressed, presumably because TFB5, in addition to any other functions it may have, acts as a stabilization factor for TFIIH.[21] It is not known whether the depressed level of TFIIH or the defect in TFB5, or a combination of these, is responsible for the disease.

Depending on the mutation and possibly other components of the genetic background in individual cases, mutations in the *XP-B* or *XP-D* genes may cause XP only, TTD only, or, for *XP-D*, an overlap syndrome incorporating elements of both diseases, known as the xeroderma pigmentosum/trichothiodystrophy overlap syndrome (XP/TTD syndrome). Mutations in these genes and also in the XP-G gene (*XP-G*, see below) may also cause CS (notably with the absence of certain features in many cases) or a crossover syndrome between XP and CS (see below).

XP-D/TTD Overlap Syndromes

XP-D is a subunit of TFIIH, which has functions in NER, basal transcription by RNA polymerases I and II and transcription activation.[22] Mutations in XPD that affect only NER are associated with XP; if transcription is also affected, they may be associated with TTD or the XP-D/TTD overlap syndrome.[23]

Broughton et al identified two patients with features of both XP and TTD in 2001.[24] One patient was a three year old female with sun sensitivity and mental and physical developmental delay. Cells cultured from this patient had almost total loss of NER. She had a compound heterozygous mutation in *ERCC2/XP-D*. The second patient was a 28 year old female with sun sensitivity, dyspigmentation and skin cancers in areas exposed to sunlight. She had a much higher level of NER. She had an arg112-to-his mutation in one allele of *ERCC2/XP-D*, previously seen in TTD and a leu485-to-pro mutation in the other allele.

XP/Cockayne Syndrome (XP/CS) Overlap Syndromes

Cockayne syndrome (CS) is a rare (occurring in approximately one individual per 100,000 live births) autosomal recessive disorder characterized in part by progressive neurodegeneration, a progerioid (premature aged, wizened) appearance and photosensitivity without development of skin cancer.[25,26] The neurodegeneration is associated with dysmyelination, rather than with neural cell degeneration, as is seen in cases of XP associated with neurological disease.[27] Three forms of CS have been delineated: CS Type I, also known as "classic CS", CS Type II, also known as "cerebro-oculo-facial-skeletal syndrome (COFS syndrome)", the "Pena-Shokeir syndrome, Type II" and "connatal CS," and CS Type III. Some authors distinguish between CS Type II and COFS; we shall consider them synonymous here. CS Type II is the most severe, with abnormalities manifested in utero and gross deformities present at birth. In CS Type I, abnormalities are not apparent at birth but appear in early childhood. In CS type III the disease is milder with later onset, in some cases as late as the third decade of life. The XP/CS overlap syndrome may be said to constitute a fourth form, but individual cases may fall within one of the other forms, particularly the COFS syndrome. When it occurs without sun-sensitivity, CS has a very variable phenotype.[28]

CS Type I is characterized by normal prenatal growth but with growth and developmental abnormalities arising in the first two years following birth. Following onset, growth retardation becomes severe, so that at the end of the course of the disease, height, weight and head circumference are well below the fifth percentile. A very characteristic finding is recession of the eyes into their sockets, due to lack of post-orbital fat. This, along with lack of fat in facial tissues, contributes to a "wizened" faces forming part of a progeroid phenotype. Dental caries, due to decreased salivation and caused by innervation of the salivary glands, is a frequent feature. Patients may have a characteristic "horse rider" gait. Neurodegeneration is progressive and death usually occurs by seven years of age.

COFS (CS Type II) is even more severe than CS Type I, with gross abnormalities becoming apparent in utero. These abnormalities are clearly manifested at birth and progress rapidly afterwards. CS Type III is typically much less severe with signs and symptoms occurring much later in life.

CS of any type tends to be associated with spasticity, increased deep tendon reflexes and joint disease. X-ray examination reveals calcifications of basal ganglia of the brain. Neuropathologic changes are associated with dysmyelination, rather than the neuronal degeneration characteristic of the neurodegenerative disease seen in some cases of XP. Approximately 75 percent of cases of CS have been found to be due to a biallelic mutation in the *ERCC6/CSB* gene. Most of the remaining 25 percent of cases are due to a mutation in the *ERCC8/CSA* gene. Mutations in *ERCC8/CSA* or *ERCC6/CSB* cause defective TCR of UV damage.[29] There may also be subtle transcription defects.[30-32]

Very few cases of CS are due to a mutation in the *XPB, XPD* or *XPG* genes; some of these also have features of XP and are thus examples of the XP/CS overlap syndrome. A tiny subset of cases is due to a gene(s) that has not yet been identified.

Patients with CS, not in complementation groups A or B, were formerly identified as in complementation group C of CS (i.e., as CS-C).[33] This group has been abandoned.[19] Some of these patients had biallelic mutations in *XPD*.

Patients with the XP/CS overlap syndrome have very variable phenotypes and it is perhaps unwise to attempt to generalize regarding them. However, their XP features at a minimum usually include early freckling, perhaps accompanied by dryness of the skin in sun-exposed areas and perhaps skin cancers. Their CS features tend to include mental retardation, spasticity, short stature and hypogonadism, but not skeletal dysplasia, although there are exceptions.

XP-B/CS Overlap Syndrome(s)

To date, only seven mutations in *ERCC3/XPB* have been identified in six families.[34] Each family has one or more individuals with either XP (one family), TTD (one family), or the XP/CS overlap syndrome (four families). The very limited range of these mutations may be related to the fact that the XPB protein is a helicase, active in both NER and transcription and essential for both of them. In contrast, the XPD protein, although a helicase active in both NER and transcription, is essential only for the former.[34,35] Thus a mutation in *XPB* is much more likely to be lethal than one in *XPD*.

The six families with XPB mutations have been extensively studied by Oh et al (2006).[34] The first XPB patient identified (Oh et al,[34] Family III),[36,37] was found to have a severe form of the XP/CS overlap syndrome.[38] XP signs/symptoms included extreme sensitivity to sunlight, with blistering noted in infancy, dyspigmentation and multiple skin cancers with onset at age 15 years. CS signs/symptoms included severe dwarfism, microcephaly, microphthalmia, sensorineural deafness, severe mental retardation (IQ of 40 but with good social skills), corneal scarring, cataracts, pigmentary retinal degeneration, optic atrophy, abnormal (flat) retinogram, hyperreflexia, ataxia, decreased nerve conduction velocity, enlarged cerebral ventricles (brain), basal ganglia calcifications and immature sexual development. The patient died at 33 years of age of cardiovascular disease.

A second family (Oh et al,[34] Family II) had two brothers with mild XP/CS overlap syndrome. They were originally reported by Scott et al[39] with the XPB classification subsequently carried out by Vermeulen et al (1994).[40] The XP features in these patients consisted of severe sunburn at approximately six weeks of age and early freckling-like dyspigmentation. Neither had developed skin cancer when studied at ages 38 and 41, respectively. However, CS features were somewhat more severe, consisting of short stature, sensorineural hearing loss which began at the age of four years and late onset hyperreflexia, other clinical neurological deficiencies, decreased nerve conduction velocity, enlarged central ventricles, dysmyelinating neuropathy, retinopathology and immature sexual development.

A third family (Oh et al,[34] Family IV) consisted of a 27 year old man with a severe form of XP/CS overlap syndrome from Slovenia.[41] XP features included sun hypersensitivity, dyspigmentation and atrophy of sun-exposed skin, telangiectasias and recurrent eyelid squamous cell carcinoma. CS features included growth and mental retardation, sensorineural deafness, optic atrophy, pigmentary retinopathy, ataxia, decreased nerve conduction velocity, enlargement of central ventricles, cerebellar atrophy and calcifications of the basal ganglia. He died of end-stage renal failure at 31 years of age.

A fourth family (Family V), consisted of a 10 year old girl with severe XP/CS overlap syndrome from Germany.[34] XP signs/symptoms consisted of severe sunburn at age two weeks and freckling on sun-exposed skin since early childhood. She had not developed any skin cancers when evaluated at 10 years of age. CS signs/symptoms consisted of increasing growth retardation, bird-like faces, progressive hearing loss beginning at age 7 years and a progressive loss of mental development. Although she could walk at age 20 months, she subsequently developed ataxia, abnormal balance, central coordination disability and intermittent tremor. She had hyperopia but not optic nerve atrophy or cataracts. Patellar and Achilles tendon reflexes were absent; the Babrinski sign was negative.

A fifth (Family I) and sixth (Family VI) families reported by Oh et al[34] had XP, only and TTD, only, respectively.

All of these XPB families' cells were hypersensitive to UV irradiation and showed very low unscheduled DNA synthesis (UDS) levels (reduced to 4-10 % of normal levels) following UV irradiation. Recovery of RNA synthesis (RRS) was also low following UV irradiation. Oh et al also showed that more severely affected patients with XP-B mutations had less residual XP-B activity than more severely affected ones.[34]

XP-D/CS Overlap Syndrome(s)

There are three reports of the XP/CS overlap syndrome falling within the D complementation group of XP.[42-44]

Cells derived from individuals who have clinical features of CS and also skin changes, including skin cancers, typical of XP, have been found to have biochemical features typical of XP and to fall within the D complementation group of XP.[45] Some patients with the COFS phenotype have cells with cellular functional assays indicative of complementation group D of XP (increased sensitivity to UV radiation, decreased UDS following UV radiation and complementation of these features by the XP-D gene or its gene product).[46]

In cell survival assays, cells from XP-D/CS patients show markedly increased hypersensitivity to UV radiation, compared to XP cells within the same complementation group.[42,47] This is associated with uncontrolled DNA breakage, unrelated to sites of UV radiation induced pyrimidine dimers or pyrimidine-pyrimdone 6-4 photo-products, found specifically in cells derived from XP/CS overlap syndrome patients. As a rule, the sites of mutations in the *XPD* gene are specific for one or other of the clinical phenotypes.

Whereas two patients with the XP/CS overlap syndrome have had mutations affecting the C-terminal third of the XPD protein,[43,48] Fujimoto et al (2005) described two additional patients with identical mutations affecting the ATPase domain, near the N terminus, on the opposite end of the XPD protein.[42] They also showed the breakdown of DNA in these patients' cells that they had previously shown in the other two XP-D/CS overlap syndrome cases. Despite their molecular and cellular characteristics, however, these patients differed substantially in their clinical characteristics.

XP-G/CS Overlap Syndrome(s)

Cells derived from individuals, who have not only clinical features of CS but also skin changes, including skin cancers, typical of XP have been found to have biochemical features typical of XP and to fall within the G complementation group of XP.[49] Nouspikel et al (1997), reported that cells from XP-G patients, with severe early onset CS, produce terminated and unstable XPG proteins and cells from a pair of mildly affected XP-G siblings, without signs or symptoms of CS, are able to synthesize a full length product from one allele with a missence mutation.[50] (Portions of this paper were subsequently withdrawn by the editor, but those portions putting forth the above data and conclusions were defended as correct by the original authors, less the author of the withdrawn part of the paper.[51]) These results were subsequently confirmed by Lalle et al (2002)[52] and by Emmert et al (2002) who reported severe XP-G/CS overlap syndrome in two patients with *XPG* mutations expected to lead to severely truncated proteins in both alleles and only much milder XP-G in a third patient with mutations expected to retain residual activity in one allele.[53] Nouspikel et al (1997) concluded from these data that the XP-G protein must have a second function in addition to its role as a structure-specific nuclease in NER.[50,54] XP-G has also been proposed as functioning in transcription coupled base excision repair (TC-BER).

XP-H/CS Overlap Syndrome(s)

The proposed XP-H complementation group consisted of a single patient with severe XP/CS overlap syndrome.[55] Following some interchange of publications,[56,57] it was withdrawn and assigned to XP complementation group D.[58] (Xeroderma pigmentosum complementation group I has also been withdrawn.)

XP/CSB Overlap Syndrome(s)

Cells derived from persons with an XP clinical phenotype with neurodegenerative changes have been found to have a cellular phenotype typical of CS and a mutation in the *ERCC6/CSB* gene.[59,60]

XP/Progeroid Overlap Syndromes

TTD and CS are both examples of "progeroid syndromes", diseases that resemble, but do not precisely equate with accelerated ageing. They are also called "segmental" progeroid syndromes, indicating that they are incomplete mimics of full blown progeroid syndromes. Many of these diseases are associated with DNA repair defects.[61,62]

XP-F/Progeroid Overlap Syndrome

Niedernhofer et al recently described a boy with signs and symptoms of both xeroderma pigmentosum and a segmental progeroid syndrome.[63] He had a null mutation for *XP-F*. The gene product, protein XP-F, pairs with another protein, ERCC1, in a heteroduplex that endonucleolytically cleaves DNA containing single stranded lesions, such as UV radiation induced pyrimidine dimers and pyrimidine-pyrimidone 6-4 photoproducts, or DNA interstrand cross-links 5' to the lesion. He developed sunburns following minimal sun exposure. Signs of premature ageing appeared in late in childhood, before puberty and he died at 15 years of age. This syndrome is proposed to involve suppression of the somatotroph axis in a complex interplay. The authors show data obtained from a mouse model in which the *ERCC1* gene is null. They show convincing evidence that the phenotype of this mouse model is similar to this boy's disease. However, this is a difficult paper to interpret, in part because of this complexity. Even more recently, Jaspers et al (2007) reported a patient with a biallelic mutation in *ERCC1* who had signs and symptoms of the COFS syndrome.[64] There was a severe clinical phenotype despite a mild cellular phenotype, suggesting an additional role for ERCC1 beyond that in NER. The *ERCC1* deficient mouse model was also described as having a phenotype resembling COFS syndrome. These papers, published virtually simultaneously, one proposing the *ERCC1-/-* mouse as a model for a progeroid syndrome,[63] and the other proposing the *ERCC1-/-* mouse as a model for the COFS syndrome,[64] have several authors in common, all of whom are excellent investigators. This underscores the difficulties involved in drawing meaningful conclusions from these overlap syndromes and animal and cellular (including yeast) models that are used to study them. Perhaps one explanation for these problems involves involvement of additional genes.[6,7]

References
1. Wattendorf DJ, Kraemer KH. Xeroderma pigmentosum. Gene Reviews (accessible online at www.genereview.org).
2. Lehmann AR. DNA repair-deficient diseases, Xeroderma pigmentosum, Cockayne syndrome and trichothirodystrophy. Biochemie 2003; 85:1101-1111.
3. Kraemer KH, Patrones NJ, Schiffmann R et al. Xeroderma pigmentosum, trichothdystrophy and Cockayne syndrome: A complex genotype-phenotype relationship. Neuroscience, 2007 (Epub ahead of print).
4. Wood RD. Seven genes for three diseases. Nature, 199; 350:190.
5. Cleaver JE, Thompson LH, Richardson AS et al. A summary of mutations in the UV-sensitive disorders: xeroderma pigmentosum, Cockayne syndrome and trichothiodystrophy. Hum Mut 1999; 14:9-22.
6. Lambert WC, Lambert MW. Corecessive inheritance: A model for DNA repair, xeroderma pigmentosum and mutagenesis. Mutat Res 1985; 145:227-234.

7. Lambert WC, Lambert MW. Corecessive inheritance: A model for DNA repair and other surveillance genes in higher eukaryotes. Mutat Res 1992; 273:179-192.

8. Lambert MW, Lambert WC. Defects in chromatin-associated DNA repair mechanisms in human genetic disease. Progr Nucl Acids Res Molec Biol 1999; 64:257-310.

9. Fujiwara Y, Ichihashi M, Kano Y et al. A new human photosensitive subject with a defect in the recovery of DNA synthesis after ultraviolet light. J Invest Dermatol 1981; 77:256-263.

10. Itoh T, Ono T, Yamaizumi M. A new UV-sensitive syndrome not belonging to any complementation groups of xeroderma pigmentosum or Cockayne syndrome: Siblings showing biochemical characteristics of Cockayne syndrome without typical clinical manifestations. Mutat Res 1994; 314:233-248.

11. Itoh T, Fujiwara Y, Ono T et al. UV(S) syndrome: A new general category of photosensitive disorders with defective DNA repair, is distinct from xeroderma pigmentosum variant and rodent complementation group I. Am J Hum Genet 1995; 56:1267-1276.

12. Horibata K, Iwamoto Y, Kuraoka I et al. Complete absence of Cockayne syndrome group B gene product gives rise to UV-sensitive syndrome but not Cockayne syndrome. Proc Nat Acad Sci USA 2004; 101:15410-15415.

13. Price VH, Odom RB, Ward WH et al. Trichothiodystrophy: Sulfur-deficient brittle hair as a marker for a neuroectodermal symptom complex. Arch Dermatol 1980; 116:1375-1384.

14. Jorizzo JL, Atherton DJ, Crounse RG et al. Ichthyosis, brittle hair, impaired intelligence, decreased fertility and short stature (IBIDS syndrome). Brit J Dermatol 1982; 106:705-710.

15. Itin PH, Pittelkos MR. Trichothiodystrophy: Review of sulfur-deficient brittle-hair syndromes and association with the ectodermal dysplasias. J Am Acad Dermatol 1990; 22:705-717.

16. Tay CH. Ichyosiform erythroderma, hair shaft abnormalities and mental and growth retardation: A new recessive disorder. Arch Dermatol 1971; 104:4-13.

17. Stary A, Sarasin A. The genetic basis of xeroderma pigmentosum and trichothiodystrophy syndromes. Cancer Surv 1996; 26:155-171.

18. Stefanini M, Vermeulen W, Weeda G et al. A new nucleotide-excision-repair gene associated with the disorder, trichothiodystrophy. Am J Hum Genet 1993; 53:817-821.

19. Lehmann AR, Bootsma D, Clarkson SG et al. Nomenclature of the DNA repair genes. Mutat Res 1994; 315:41-42.

20. Giglia-Mari G, Coin F, Ranish JA et al. A new, tenth subunit of TFIIH is responsible for the DNA repair syndrome, trichothiodystrophy, group A. Nature Genet 2004; 36:714-719.

21. Vermeulen W, Bergmann E, Auriol J et al. Sublimiting concentration of TFIIH transcription/DNA repair factor causes TTD-A trichothiodystrophy. Mature Genet 2000; 26:307-313.

22. Keriel A, Stary A, Sarasin A et al. XPD mutations prevent TFIIH-dependent transactivation by nuclear receptors and phosphorylation of RAR alpha. Cell 2002; 109:123-135.

23. Lehmann AR: The xeroderma pigmentosum group D (XPD) gene: One gene, two functions, three diseases. Genes Dev 2001; 15:15-23.

24. Broughton BC, Berneburg M, Fawcett H et al. Two individuals with features of both xeroderma pigmentosum and trichothiodystrophy highlight the complexity of the clinical outcomes of mutations in the XPD gene. Hum Mol Genet 2001; 10:2539-2547.

25. Neilan EG. Cockayne syndrome. Gene Reviews, 2006, Available online at www.genetests. org

26. Nance MA, Berry SA. Cockayne syndrome: Review of 140 cases. Am J Med Genet 1992; 42:68-84.

27. Rapin I, Lindenbaum Y, Dickson DW et al. Cockayne syndrome and xeroderma pigmentosum. Neurology 2000; 55:1442-1449.

28. Mallery DL, Tanganelli B, Colella S et al. Molecular analysis of mutations in the CSB (ERCC6) gene in patients with Cockayne syndrome. Am J Hum Genet 1999; 62:77-85 [Erratum, Am J Hum Genet 1999; 64:1491].

29. Van Hoffen A, Natarajan AT, Mayne LN et al. Deficient repair of the transcribed strand of active genes in Cockayne syndrome cells. Nucleic Acids Res 1993; 21:5890-5895.

30. Tantin D. RNA polymerase II elongation complexes containing the Cockayne syndrome group B protein interact with a molecular complex containing the transcription factor II H components xeroderma pigmentosum B and p62 J Biol Chem 1998; 273:27794-27799.

31. Balajee AS, May A, Dianov GL et al. Reduced RNA polymerase II transcription in intact and permealised Cockayne syndrome group B-cells. Proc Nat Acad Sci USA 1997; 94:4306-4311.

32. Van Gool AJ, van der Horst GT, Citterio E et al. Cockayne syndrome: Defective repair of transcription. EMBO J 1997; 16:4155-4162.

33. Lehmann AR. Three complementation groups in Cockayne syndrome. Mutat Res 1982; 106:356.

34. Oh K-S, Khan SG, Jaspers NGJ et al. Phenotypic heterogeneity in the XPB DNA helicase gene (ERCC3): Xeroderma pigmentosum without and with Cockayne syndrome. Hum Mutat 2006; 27(11):1092-1103.

35. Winkler GS, Araujo SJ, Fiedler U et al. TFII H with inactive XPD helicace functions in transcription initation but is defective in DNA repair. J Biol Chem 2000; 275:4258-4266.
36. Noojin RO. Xeroderma pigmentosum treated with oral methoxalin. Arch Dermatol 1965; 92:422-423.
37. Robbins JH, Kraemer KH, Lutznem MA et al. Xeroderma pigmentosum: An inherited disease with sun sensitivity, multiple cutaneous neoplasms and abnormal DNA repair. Ann Intern Med 1974; 80:221-228.
38. Brumback RA, Yoder FW, Andrews AD et al. Normal pressure hydrocephalus: Recognition and relationships to neurological abnormalities in Cockayne's syndrome. Arch Neurol 1978; 35:337-345.
39. Scott RJ, Itin P, Kleijem WJ et al. Xeroderma pigmentosum-Cockayne syndrome complex in two patients; Absence of skin tumors despite severe deficiency of DNA excision repair. J Am Acad Dermatol 1993; 29:883-889.
40. Vermeulen W, Scott RJ, Rodgers S et al. Clinical heterogeneity within xeroderma pigmentosum associated with mutations in the DNA repair and transcription gene ERCC3. Am J Hum Genet 1994; 54:191-200.
41. Bartenjev I, Butina MR, Potocnick M. Rare case of Cockayne syndrome with Xeroderma pigmentosum. Acta Derm Venereol 2000; 80:213-214.
42. Fujimoto M, Leech SN, Theron T et al. Two new XPD patients compound heterozygous for the same mutation demonstrate diverse clinical features. J Invest Dermatol 2005; 125:86-92.
43. Broughton BC, Thompson AF, Harcourt SA et al. Molecular and cellular analysis of the DNA repair defect in a patient in xeroderma pigmentosum complementation group D who has the clinical features of xeroderma pigmentosum and Cockayne syndrome. Am J Hum Genet 1995; 56:167-174.
44. Lafforet D, Dupuy JA. Xeroderma pigmentosum and Cockayne syndrome. Pediatrics 1978; 61:675-676.
45. Van Hoffen A, Kalle WH, de Jong-Versteeg A et al. Cells from XP-D and XP-D-CS patients exhibit equally inefficient repair of UV-induced damage in transcribed genes but different capacity to recover UV-inhibited transcription. Nucleic Acids Res 1999; 27:2898-2904.
46. Graham JR Jr, Greenberg CR, Anyane-Yeboa K et al. Cranio-oculo-facial-skeletal syndrome caused by defective nucleotide excision repair. Am J Hum Genet 1998; 63: A33.
47. Berneburg M, Lowe JE, Nardo T et al. Ultraviolet damage causes uncontrolled DNA breaks in cells from patients with combined features of XP-D and Cockayne syndrome. EMBO J 2000; 19:1157-1166.
48. Takayma K, Salazar EP, Lehmann A et al. Defects in the DNA and transcription gene ERCC2 in the cancer-prone disorder xeroderma pigmentosum group D. Cancer Res 1995; 55:5656-5663.
49. Okinaka RT, Perez-Castro AV, Sena A et al. Heritable genetic alterations in a xeroderma pigmentosum group G/Cockayne syndrome pedigree. Mutat Res 1997; 385:107-114.
50. Nouspikel T, Lalle P, Leadon SA et al. A common mutational pattern in Cockayne syndrome patients from xeroderma pigmentosum group G: Implications for a second XPG function. Proc Nat Acad Sci USA 1997; 94:3116-3121.
51. Snyder SH [Editor]. Retraction for Nouspikel T, Lalle P, Leadon SA et al. A common mutational pattern in Cockayne syndrome patients from xeroderma pigmentosum group G: Implications for a second XPG function. Proc Nat Acad Sci USA 1997; 94:3116-3121. Proc Nat Acad Sci USA 2006; 103:19606.
52. Lalle P, Nouspikel T, Constantinou A et al. The founding members of xeroderma pigmentosum group G produce XPG protein with severely impaired endonuclease activity. J Invest Dermatol 2002; 118:344-351.
53. Emmert S, Slor H, Busch DB et al. Relationship of neurologic degeneration to genotype in three xeroderma pigmentosum group G patients. J invest Dermatol 2002; 9720982.
54. Clarkson SG. The XPG story. Biochemie, 2003; 85:1113-1121.
55. Moshell AN, Ganges MB, Lutzner MA et al. A new patient with both xeroderma pigmentosum and Cockayne syndrome establishes the new xeroderma pigmentosum complementation group H. In Friedberg EC, Bridges BA, Eds. Cellular Responses to DNA Damage. New York, Alan R. Liss (Publisher) 1983. pp. 209-213.
56. Johnson RT, Elliott GC, Squires S et al. Lack of complementation between xeroderma pigmentosum complementation groups D and H. Hum Genet 1989; 81:203-210.
57. Robbins JH. No lack of complementation for unscheduled DNA synthesis between xeroderma pigmentosum complementation groups D and H. [Letter] Hum Genet 1989; 84:99-100.
58. Robbins JH. Xeroderma pigmentosum complementation group H is withdrawn and assigned to group D. Hum Genet 1991; 88:242.
59. Greenshaw GA, Hebert A, Duke-Woodside ME et al. Xeroderma pigmentosum and Cockayne syndrome: Overlapping clinical and biochemical phenotype. Am J Hum Genet 1992; 50:677-689.
60. Colella S, Nardo T, Botta E et al. Identical mutations in the CSB gene associated with either Cockayne syndrome or the DeSanctis-Cacchione syndrome. Hum Mol Genet 2000; 9:1171-1175.

61. Andressoo J-O, Hoijmakers JHJ, Mitchell JR et al. Nucleotide excision repair disorders and the balance between cancer and ageing. Cell cycle 2006; 5:2886-2888.

62. Kirkwood T. Ageing: Too fast by mistake. Nature, 2006; 444:1015-1017.

63. Niedernhofer LJ, Garinis GA, Raams A et al. A new progerioid syndrome reveals that genotoxic stress suppresses the somatotroph axis. Nature 2006; 444:1038-1043.

64. Jaspers NGJ, Raams A, Silengo MC et al. First reported patient with human ERCC1 deficiency has cerebro-oculo-facial-skeletal syndrome with a mild defect in nucleotide excision repair and severe developmental failure. Am J Hum Genet 2007; 80:457-466.

CHAPTER 15

Population Distribution of Xeroderma Pigmentosum

Abdul Manan Bhutto* and Sandra H. Kirk

Introduction

Xeroderma pigmentosum (XP) is a rare autosomal recessive genetic disease caused by defects in the normal repair of DNA of various cutaneous and ocular cell types damaged by exposure to sunlight.[1-3] Hebra and Kaposi reported the disease initially in 1874.[4] It generally shows early onset of symptoms, hence mostly affecting children and is characterized by cutaneous and ocular pigmentary changes such as freckles, photophobia, conjunctivitis, corneal keratitis and ulcers. If the disease is not controlled at this level, there is a risk of developing malignancy in the future, the major cause of death amongst patients.

Population Distribution

The incidence of XP seems to vary across the globe. This could appear to be the case due to differences in reporting of primary mutation frequencies, or to the lack of accuracy of searching, investigating and diagnosing cases in different parts of the world. The result is that apparent figures for incidence have changed from time to time. In early studies the frequency of XP in Europe and USA was reported as 1:250000;[5] whereas the incidence reported in Japan and other countries is much higher, at 1:40000.[6,7] In other reports the frequency of XP was reported in the USA and Europe to be 1:1 million,[8] with even higher frequencies recorded in Netherlands, Isreal and Japan (1:100000).[9,10] Reports from other regions or countries including Egypt,[11] Germany,[12] Europe,[13] Israel,[14] Korea,[15,16] and China[17] show a considerable number of XP patients in those countries. These reports relate to the total of all complementation groups of the disorder and XP has been reported in all races including whites, Asians, blacks and Native Americans, albeit with slightly differing frequencies, as skin type plays some role in the induction of symptoms of this disease. The nonwhite population, with black or brown skin due to high amounts of epidermal melanin, has a lower incidence of developing sun-induced cutaneous complications such as sunburn, keratosis, freckles, wrinkles and rarely skin cancer. On the other hand, the fair-skinned (white) population, with a low amount of melanin in the epidermis, is at high risk of developing these acute and chronic problems associated with frequent exposure to sunlight. In general, the susceptibility of human skin to sun-induced acute and chronic damage is directly related to the intensity of UV radiation, reflecting the duration and frequency of sun exposure.

XP cases are frequently reported in the various states of India.[18] We (AMB and colleagues) presented a series of 36 cases of XP patients from Pakistan, who visited our department from 1995 to 2001. The frequency of XP patients in the general population in the southern region of Pakistan was about 1:100000.[19] This ratio is higher than in the USA, but lower than in Japan. Forty classic XP patients were reported from Algeria, showing high frequency of ocular and

*Corresponding Author: Abdul M. Bhutto—Department of Dermatology, Chandka Medical College, Larkana, Pakistan. Email: bhutto_manan1@yahoo.com

Molecular Mechanisms of Xeroderma Pigmentosum, edited by Shamim I. Ahmad and Fumio Hanaoka. ©2008 Landes Bioscience and Springer Science+Business Media.

neurological manifestations,[20] the latter being linked in particular to the XP-A complementation group. Twenty-four cases were reported from Libya in ten years, from 1981~1990,[21] with a high percentage of patients showing early onset of severe ophthalmic lesions; the majority of patients arising as offspring from consanguineous marriage. When a thorough epidemiological study was conducted in the department of Dermatology and Venereology of Ibn Rochd University Hospital of Casablanca, Morocco, a total of 120 patients were registered from 1990 to 2000.[22] Of these, 80% had developed cutaneous tumors such as basal cell carcinoma (BCC) and squamous cell carcinoma (SCC). Twelve cases were reported in the black South Africans population and it was concluded that XP in South African blacks is characterized by the very early development of cutaneous, ocular and tongue SCC and usually has a rapid and devastating course.[23] A retrospective study conducted from 1973 to 1998 involved a total of 216 XP patients, seen in Charles Nicolle Hospital Tunisia. Twelve out of 216 had developed melanoma and four of them showed metastasis.[24]

Genetics and Population

Incidence is determined by genetics and is exacerbated by consanguinity. The skin-related symptoms are reported to be affected by skin type, weather and latitude. In a comparative study between the two cities of Kwangju[15] and Pusan in Korea and Osaka, Japan, which are believed to have almost equal altitude and with similar weather conditions, a marked difference was observed in the frequency of patients with neurological abnormalities. A major discrepancy was that no consanguineous marriage in parents of patients was noted in Korea, whereas 35% of the Osaka families were consanguineous. This is probably due to the legal and traditional prohibition of consanguineous marriage in Korea. The differences observed likely reflect different founder effects in the two populations and a different balance of complementation group involvement, as not all groups show symptoms of neurological disorder. In the above comparative study the data revealed that the incidence of neurological abnormalities was not related to consanguinity and authors concluded that the reason for the absence of patients with neurological abnormalities could be due to different genetic complementation groups in Korea and Japan.[9] Consanguinity of the Japanese patients' parents was reported in 31% of cases, an elevated frequency often seen in recessive disorders. Parental consanguinity in Indian cases is reported to be as high as 40%. Nearly 20% of patients, including a high proportion of Japanese patients, had neurological abnormalities.[25] The pathology of peripheral nervous system (PNS) in 2 autopsied Japanese cases was presented in detail and it was suggested that the underlying pathogenetic mechanism is that of a neuropathy.[26]

Defective XP gene (of all types) frequency has been estimated at between 1 in 100 and 1 in 1000. Men and women are found to be affected with the same frequency as the disorder is autosomal. Inheritance is recessive and parents of affected individuals, who are thus obligate heterozygotes, are usually clinically normal, although a higher incidence of skin cancer is reported among XP heterozygotes.[27]

XP is classified into eight genetic complementation groups: XP-A to XP-G and a variant group XP-V.[28] Each group has a different causative gene. Seven (XP-A to XP-G) have defects in nucleotide excision repair and one, XP-V, is mutated in polymerase eta (η), affecting the accurate repair of cyclobutane pyrimidine dimers (CBD) and other kinds of damage by UV (see Chapter 10).

Complementation group-A comprises patients with the most severe neurological and somatic abnormalities (the De Sanctis-Cacchione syndrome) as well as patients with minimal or no neurologic abnormalities.[25,30] This is the most common group of XP in Japan and usually corresponds to the most severe clinical form of XP, which in most patients gives rise to involvement of the central and peripheral nervous system in addition to cutaneous lesions.[9] A study was conducted in a group of 33 Japanese XP-A patients to determine the frequency of convulsive episodes in this complementation group. It was found that Japanese XP-A patients had a low incidence of febrile seizures (3%), while exhibiting a higher incidence of epilepsy (15%) and it was found that deep convulsions do not occur until the brain is more developed.[31] XP-A patients display extreme sun sensitivity, showing characteristic skin alterations in very early childhood, with tumor formation commencing thereafter. This form is also present in the USA, Europe and the Middle East.

Approximately 90% of Japanese XP-A patients have the same single base substitution mutation in the causal gene.[32] It has been suggested that, as with other genetic disorders, different mutations in the XP group-A complementing group might induce different clinical features. This has served as the basis for development of a rapid diagnostic assay for Japanese XP-A patients using PCR analysis of a small sample of DNA.[33] A rare, single Japanese XPA patient, XP79KO, is reported as a compound heterozygote with a newly identified T to G transversion at the splice donor site in intron 1 in one allele and with the AlwNI mutation in another allele of the XP-A gene.[34] Most XP-A patients show multiple SCC, with BCC also found but at a lower frequency.

Complementation group B is composed of three patients in two kindreds who had cutaneous abnormalities and malignancies in conjunction with neurological and ocular abnormalities typical of Cockayne`s syndrome (CS).[5,35] A single case of XPB with trichothiodystrophy (TTD) is reported.[36]

Complementation group C patients exhibit ocular involvement but without neurological abnormalities.[5,30,37] This is the most common form in the USA, Europe and the Middle East (Egypt), but is rare in Japan.

Complementation group D has been described with four different clinical phenotypes. The patients of this group along with the cutaneous lesions may or may not develop late onset neurological abnormalities in their second decade of life.[5,30,37,38] Two XP-D patients have been reported with clinical symptoms of both XP and CS.[39,40] A hospital based study in a North Eastern Chinese population was conducted to determine the effect of the DNA repair gene XPD polymorphism on the risk of lung cancer. In the results no association of XPD Arg156Arg was found with the risk of lung cancer.[41] A thorough study was conducted to elucidate the association between the XPD Lys751Gln polymorphism and risk of a second primary cancer in individuals with NMSC. It was concluded that individuals with non-melanoma skin cancer (NMSC) who have the variant XPD Gln allele are at increased risk of developing a second primary cancer.[42]

Complementation group E was found in one patient in Europe and several others in Japan. These patients showed comparatively mild cutaneous abnormalities without neurological involvement.[30,37,38]

Complementation group F patients have been reported mainly from Japan.[38] These patients had mild clinical symptoms without neurological involvement or other cutaneous and ocular malignancies. A study of group F Japanese patients revealed very mild skin symptoms with no ocular or neurological abnormalities.[43] A first case of human inherited ERCC1 deficiency has been reported from Holland. This patient has cerebro-oculo-facio-skeletal syndrome. The heterodimer ERCC1-XPF is one of the two endonucleases required for NER. This finding represents a novel complementation group of patients with defective NER.[44]

Complementation group G patients are reported frequently from Europe and Japan. Most patients in this group suffer from neurological abnormalities without development of cancer. Fibroblasts from one of the patients were found to be hypersensitive to killing by ionizing radiation (X-rays as well as by UV radiation).[45] A couple of other cases belonging to group G have been reported from Japan and Europe and they exhibited the clinical features of XP and neurological abnormalities seen in CS.[46,47]

XP variant patients have also been reported in the USA, Europe and Japan.[9,48] The cutaneous and ocular abnormalities were severe in some patients but mild in others. Very few are also reported as having neurological abnormalities. An XP-V family with four affected siblings was identified in Germany. The affected individuals had undergone extensive sun exposure and showed late onset of cutaneous cancer.[49] XP-E and the variants regularly show significant affliction with BCC. This is not at all exclusive and one may also find SCC and malignant melanoma (MM) in some of the variants but with a low frequency.[10]

Population and Malignancy Type

It is reported that sunlight exposure is responsible for the induction of malignant melanoma as well as NMSC (BCC and SCC) in patients with XP.[50] Cutaneous melanomas are uncommon in the young and are seldom diagnosed in the early stages, perhaps due to a reluctance to accept this diagnosis in the early age group patients. de Sa et al[51] reported that 12 out of 32 patients (37.5%) with cutaneous melanomas were aged 12 years or younger. They suggested that prevention and early stage diagnosis depend upon the recognition that this disease is also present in the young age group.[51] People with white skin are the population more at risk from sun-induced MM and NMSC than people with pigmented skin. These patients are more at risk of developing cutaneous malignancy when they are severely sun exposed in early childhood. XP patients younger than 20 years of age have a 1000-fold increased risk of developing cutaneous malignancy over the normal population.[50,52] Six out of 36 of our Pakistani XP patients (16.7%) developed SCC between the ages of 3.5 and 14 years (mean 9.8 years).[19] Kraemer et al reported malignant skin neoplasms in 45% of XP patients from 20 different countries including the USA, Japan, Egypt, Turkey etc.[52] In other studies, Goyal et al observed these malignant changes in 60% of their patients.[18] In Tunisia, skin carcinomas are reported at exceptionally high frequency and are associated with the unusually high prevalence of XP.[53] Although malignant skin tumors are rare in childhood, the prognosis is relatively better compared to adults. MMs are the most frequently reported tumors in Turkey. In a study, it was observed that among 21 cases of malignant melanoma 43% patients had an underlying defect in their immune barriers which was thought to be responsible for the development of their malignancies.[54] Lip cancer is reported in two Zimbabwean XP patients.[55] Ocular neoplasms like SCC, BCC and malignant melanoma are also frequently reported from the conjunctiva, cornea and eyelids. A single rare case of MM was reported as developing from the iris.[56] The incidence of internal neoplasms is said to be 10-20 times higher in XP patients than in the normal population. Certain nucleotide excision repair (NER) genotypes appear to be associated with an altered risk of endometrial cancer. It is proposed that there are interactions between NER genotypes and DNA damage-causing exposures in the etiology of endometrial cancer.[57] Mortality occurs in XP patients due to either neurological complications[58] or actinic induced tumors. A study was carried out in British families whose patients suffered from Ataxia telangiectasia (AT) (67 in number) or xeroderma pigmentosum (48), to determine the mortality versus cancer incidence. Parents of all these patients were obligate heterozygotes and grandparents had a probability of heterozygosity of 0.5. While 14 AT patients had died over the same period, only 5 XP patients had died in the period covered in the study. The death of only one AT patient was from non-Hodgkin's lymphoma, and no XP patients died from internal malignancy.[59]

References

1. Cleaver JE. Defective repair replication of DNA in xeroderma pigmentosum. Nature 1968; 218:652-656.
2. Cleaver JE. DNA damage and repair in light-sensitive human skin disease. J Invest Dermatol 1970; 54:181-195.
3. Cleaver JE, Carter DM. Xeroderma pigmentosum variants: influence of temperature on DNA repair. J Invest Dermatol 1973; 60:29-32.
4. Hebra F, Kaposi M. On diseases of the skin including the exanthemata. Vol 3 Tay W, trans. London. The New Sydenham Society 1874; 61:252-258.
5. Robbins JH, Kraemer KH, Lutzner MA et al. Xeroderma Pigmentosum: an inherited disease with sun sensitivity, multiple cutaneous neoplasms and abnormal DNA repair. Ann Inter Med 1974; 80:221-248.
6. Neel JV, Kod ai M, Brewer R et al. The incidence of consanguineous matings in Japan: with remarks on the estimation of comparative gene frequencies and the expected rate of appearance of induced recessive mutations. Am J Hum Genet 1949; 1:156-178.
7. Thielmann HW, Edler L, Popanda O et al. Xeroderma pigmentosum patients from the federal republic of Germany: decrease in post-UV colony-forming ability in 30 xeroderma pigmentosum fibroblast strains is quantitatively correlated with a decrease in DNA-incising capacity. J Cancer Res Clin Oncol 1985; 109:227-240.
8. Kraemer KH. Heritable diseases with increased sensitivity to cellular injury. In: Fatzpatrick TB, Eisen AZ, Wolff K, Freedberg IM, Austen KF, eds. Dermatology in General Medicine, 4th ed. New York: McGraw-Hill, 1993:1974.
9. Takebe H, Nishigori C and Satoh Y. Genetics and skin cancer of xeroderma pigmentosum in Japan. Jpn J Cancer Res (Gann) 1987; 78:1135-1143.

10. Jung EG. Xeroderma Pigmentosum. Int J Dermatol 1986; 25:629-633.
11. Hashem N, Bootsma D, Keijzer W et al. Clinical characteristics, DNA repair and complementation groups in xeroderma pigmentosa patients from Egypt. Cancer Res 1980; 40:13-18.
12. Fischer E, Thielmann HW, Neundorfer B at al. Xeroderma pigmentosum patients from Germany: Clinical symptoms and DNA repair characteristics. Arch Dermatol Res 1982; 274:229-247.
13. Pawsey SA, Magnus IA, Ramsay CA. Clinical, genetic and DNA repair studies on a consecutive series of patients with xeroderma pigmentosum. Qurat J Med 1979; 190:179-210.
14. Kraemer KH, Slor H. Xeroderma pigmentosum. Clin Dermatol 1985; 3:33-69.
15. Hwang SW, Yoo YE, Kim YP. The genetics and clinical studies of xeroderma pigmentosum. Korean J Dermatol 1982; 20:879-884.
16. Park SD, Chung HY. Characterization of a Korean xeroderma pigmentosum cell strain, XP1SE, by somatic cell hybridization and complementation studies. Korean J Genetic 1982; 4:69-78.
17. Jiang Z, Hu Y, Chen Q et al. Study of DNA repair enzyme system. 1. Ultraviolet induced H-TdR unscheduled incorporation in xeroderma pigmentosum lymphocytes. Acta Genet. Sin 1981; 8:310-315.
18. Goyal JL, Rao VA, Srinivasan R et al. Oculocutaneous manifestations in xeroderma pigmentosa. Br J Ophthalmol 1994; 78:295-297.
19. Bhutto AM, Shaikh A, Nonaka S. Incidence of xeroderma pigmentosum in Larkana, Pakistan. Br J Dermatol 2005; 152:545-551.
20. Bouadjar B, Ait-Belkacem F, Daya-Grosjean L et al. Xeroderma pigmentosum. A study in 40 Algerian patients. (Article in French) Ann Dermatol Venereol 1996; 123:303-306.
21. Khatri ML, Shafi M, Mashina A. Xeroderma pigmentosum. A clinical study of 24 Libyan cases. J Am Acad Dermatol 1992; 26:75-78.
22. Moussaid L, Benchikhi H, Boukind EH et al. Cutaneous tumors during xeroderma pigmentosum in Morocco: study of 120 patients. Ann Dermatol Venereol 2004; 131:29-33.
23. Jacyk WK. Xeroderma pigmentosum in black Sounth Africans. Int J Dermatol 1999; 38:511-514.
24. Fazaa B, Zghal M, Bailly C et al. Melanoma in xeroderma pigmentosum: 12 cases. Ann Dermatol Venereol 2001; 128(4):503-506.
25. Kato T, Akiba H, Seiji M et al. Clinical and biological studies of 26 cases of xeroderma pigmentosum in northeast district of Japan. Arch Dermatol Res 1985; 277:1-7.
26. Kanda T, Oda M, Yonezawa M et al. H. Peripheral neuropathy in xeroderma pigmentosum. Brain 1990; 113:1025-1044.
27. Swift M, Chase C. Cancer in families with xeroderma pigmentosum. J Natl Cancer Inst 1979; 218:652-656.
28. Thompson LH. Nucleotide excision repair: Its relation to human disease. In: Nickoloff JA, Hoekstra M, eds. DNA Repair in Higher Eukaryotes. Totowa: Humana Press, 1998; Vol 2; pp 335-393.
29. Svoboda DL, Briley LP, Vos J-HM. Defective bypass replication of a leading strand cyclobutane thymine dimer in xeroderma pigmentosum variant cell extracts. Cancer Research 1998; 58:2445-2448.
30. Cleaver JE, Kraemer KH. Xeroderma Pigmentosum and Cockayne Syndrome. In: The metabolic and molecular bases of inherited disease, 7th ed, edited by Scriver CR, Beaudet AL, Sly WS, Valle D. New York: McGraw-Hill, 1995; Vol III, pp 4393-4419.
31. Kohyama J, Furushima W, Sugawara Y et al. Convulsive episodes in patients with group A xeroderma pigmentosum. Acta Neurol Scand 2005; 112(4):265-269.
32. Nishigori C, Moriwaki S, Takebe H et al. Gene alterations and clinical characteristics of xeroderma pigmentosum group A patients in Japan. Arch Dermatol 1994; 130:191-197.
33. Kore-eda S, Tanaka T, Moriwaki S et al. A case of xeroderma pigmentosum group A diagnosed with a polymerase chain reaction (PCR) technique. Arch Dermatol 1992; 128:971-974.
34. Tanioka M, Budiyant A, Ueda T et al. A novel XPA gene mutation and its functional analysis in a Japanese patient with xeroderma pigmentosum group A. J Invest Dermatol 2005; 125(2):244-2446.
35. Scott RJ, Itin P, Kleijer WJ et al. Xeroderma pigmentosum-Cockayne sundrome complex in two patients: Absence of skin tumors despite severe deficiency of DNA excision repair. J Am Acad Dermatol 1993; 29:883.
36. Weeda G, Evano E, Donker I et al. A mutation in the XPB/ERCC3 DNA repair transcripyion gene, associated with trichothiodystrophy. Am J Hum Genet 1997; 60:320.
37. Thielmann HW, Popanda O, Edler L et al. Clinical symptoms and DNA repair characteristics of xeroderma pigmentosum patients from Germany. Cancer Res 1991; 51:3456.
38. Kondo S et al. Late onset of skin cancers in 2 xeroderma pigmentosum group F siblings and a review of 30 Japanese xeroderma pigmentosum patients in groups D, E and F. Photodermatology 1989; 6:89.
39. Vermeulen W, Stefanini M, Giliani S et al. Xeroderma pigmentosum complementation group H falls into complementation group D. Mutat Res 1991; 255:201.

40. Broughton BC, Thompson AF, Harcourt SA et al. Molecular and cellular analysis of the DNA repair defect in a patient in xeroderma pigmentosum complementation group D who has the clinical features of xeroderma pigmentosum and cockayne syndrome. Am J Hum Genet 1995; 56:167-174.
41. Yin J, Li J, Ma Y et al. The DNA repair gene ERCC2/XPD polymorphism Arg 156Arg (A22541C) and risk of lung cancer in a Chinese population. Cancer Lett 2005; 223(2):219-226.
42. Brewster AM, Alberg AJ, Strickland PT et al. XPD polymorphism and risk of subsequent cancer in individuals with nonmelanoma skin cancer. Cancer Epidemiol Biomarkers Prev 2004; 13(8):1271-1275.
43. Yamamura K, Ichihashi M, Hiramoto T et al. Clinical and photobiological characteristics of xeroderma pigmentosum complementation group F: a review of cases from Japan. Br J Dermatol 1989; 121:471-480.
44. Jaspers NG, Raams A, Silengo MC et al. First reported patient with human ERCC1 deficiency has cerebro-oculo-facio-skeletal syndrome with a mild defect in nucleotide excision repair and severe developmental failure. Am J Hum Genet 2007; 80(3):457-466.
45. Arlett CF, Harcourt SA, Lehmann AR et al. Studies of a new case of xeroderma pigmentosum (XP3BR) from complementation group G with cellular sensitivity to ionizing radiation. Carcinogenesis 1980; 1:745.
46. Jaeken J et al. Clinical and biochemical studies in three patients with severe early infantile Cockayne syndrome. Hum Genet. 1989; 83:339.
47. Moriwaki S, Stefanini M, Lehmann AR et al. DNA repair and ultraviolet mutagenesis in cells from a new patient with xeroderma pigmentosum group G and Cockayne syndrome resemble xeroderma pigmentosum cells. J Dermatol Invest 1996; 107:647.
48. Mamada A, Miura K, Tsunoda K et al. Xeroderma pigmentosum variant associated with multiple skin cancers and a lung cancer. Dermatology 1992; 184:177-181.
49. Somos S, Schneider I, Rasko I. Xeroderma pigmentosum variant or pigmented xerodermoid. Anticancer Res 1997; 17:753-756.
50. Kraemer KH, Lee MM andrews AD et al. The role of sunlight and DNA repair in melanoma and nonmelanoma skin cancer. The xeroderma pigmentosum paradigm. Arch Dermatol 1994; 130:1018-1021.
51. de Sa BC, Rezze GG, Scramim AP et al. Cutaneous melanoma in childhood and adolescence: retrospective study of 32 patients. Melanoma Res 2004; 14(6):487-492.
52. Kraemer KH, Lee MM, Scotto J. Xeroderma pigmentosum. Cutaneous, ocular and neurologic abnormalities in 830 published cases. Arch Dermatol 1987; 123:241-250.
53. Stiller CA. International variations in the incidence of childhood carcinomas. Cancer Epidemiol Biomarkers Prev 1994; 3(4):305-310.
54. Varan A, Gokoz A, Akyuz C et al. Primary malignant skin tumors in children: etiology, treatment and prognosis. Pediatr Int 2005; 47(6):653-657.
55. Chidzonga MM. Lip cancer in Zimbabwe. Int J Oral Maxillofac Surg 2005; 34(2):149-151.
56. Johnson MW, Skuta GL, Kincaid MC et al. Malignant Melanoma of the iris in xeroderma pigmentosum. Arch Ophthalmol 1989; 107:402-407.
57. Weiss JM, Weiss NS, Ulrich CM et al. Nucleotide excision repair genotype and the incidence of endometrial cancer: effect of other risk factors on the association. Gynecol Oncol 2006; 103(3):891-896.
58. English JSC, Swerdlow AJ. The risk of malignant melanoma, internal malignancy and mortality in xeroderma pigmentosum patients. Br J Dermatol 1987; 117:457-461.
59. Pippard EC, Hall AJ, Barker DJ et al. Cancer in homozygotes and heterozygotes of ataxia-telangiectasia and xeroderma pigmentosum in Britain. Cancer Res 1988; 48(10):2929-2932.

Progress and Prospects of Xeroderma Pigmentosum Therapy

Alain Sarasin*

Introduction

Further to a full description of clinical features of xeroderma pigmentosum (XP) in Chapter 2, this disease is characterized by dry skin, hypo and hyper-pigmentation, actinic keratosis and skin cancers.[1,2] The genetic defect in nucleotide excision repair (NER) and polymerase η (XP-variant) renders the XP patients sensitive to sun leading to development of carcinoma and melanoma with more than one thousand times higher frequency than in the general population.[1] Among the eight complementation groups of XP (XPA to XPG and XP-V), only XPC, XPE, XP-V and 50% of XPG patients do not develop neurological disorders. The other groups, particularly the XPA, XPB, XPD and XPG/CS, develop severe form of neurological disorders in about 30% of patients.[3] The first signs of XP neurological disorders are the loss of deep tendon reflexes and hearing loss that can lead to death in the first two decades of their life.

Besides neurological disorders, for which no therapy has been envisaged yet, the major therapy proposed for XP is the protection towards UV exposure. Also early detection and treatment of skin abnormalities, basically pretumours and malignant tumours, are proposed. Attempts to correct the genetic repair deficiency have been carried out using either exogenous enzyme(s) able to repair DNA damage or gene therapy protocols.

Protection Towards UV Exposure

Because the XP cells are unable to repair DNA lesions induced by solar and artificial ultraviolet (UV), the most efficient control of XP is to fully protect the body from all kind of UV exposure. In theory it is easy but in practice fairly difficult to achieve throughout the life. One way to achieve this is either not to go out at all once sun is present (that is almost half of the life) or to be fully protected by using dresses that cover the entire body from top to toe including using sunglasses (Fig. 1A). Development of efficient protection of face, at the same time allowing for the child to be recognisable, is required to avoid psychological complexity including the loss of personality (Fig. 1B). The use of commercially produced sunscreens can give limited protection and can be fairly expensive.

To limit psychological problems, XP children may be allowed to get out, with a light protection, for a maximum of one hour after sunrise and one hour before sunset. During this period, the amount of harmful UVB and UVA is much reduced as compared to the middle part of the day. Even then the UV exposure should be carefully monitored by employing a good quality dosimeter, as the UV flux can depend on the location of the place on the earth. Equatorial regions of the earth receive more UV throughout the year than for example the Polar Regions.

*Alain Sarasin—Laboratory of Genomes and Cancers, FRE2939 CNRS, Institute Gustave Roussy, 39, rue Camille Desmoulins, 94805 Villejuif, France. Email: sarasin@igr.fr

Molecular Mechanisms of Xeroderma Pigmentosum, edited by Shamim I. Ahmad and Fumio Hanaoka. ©2008 Landes Bioscience and Springer Science+Business Media.

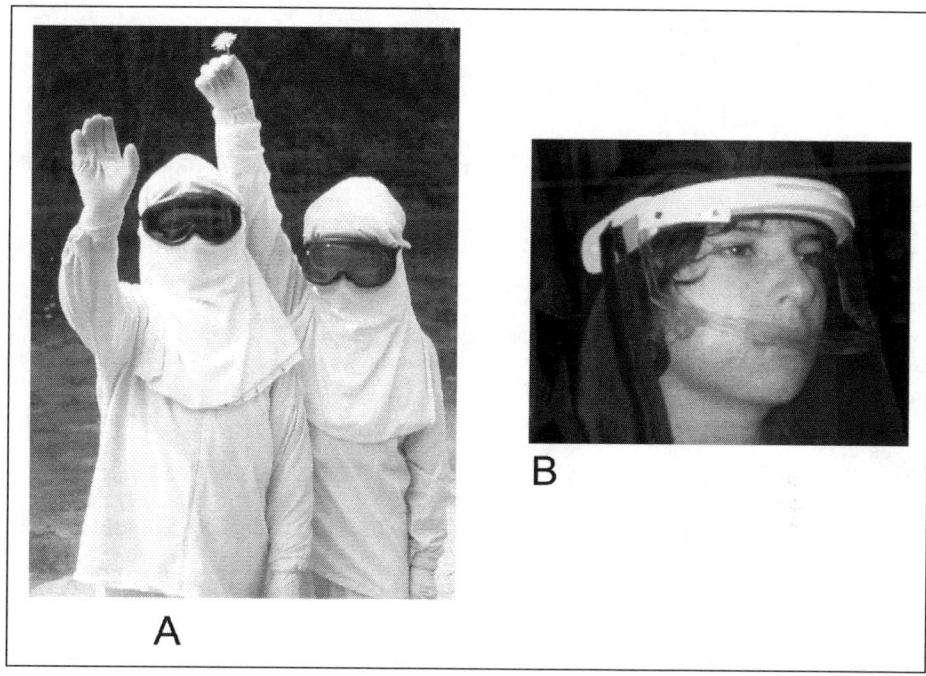

Figure 1. Protection of XP children from UV-exposure. A) Full protection of the whole body with special garment, including gloves and sunglasses. B) New translucent protections of face are developed in order to be able to recognize the child. Thanks to the French XP Association "Les enfants de la lune" for the photographs.

Classical XP Therapy

Usually several months to several years are required after the birth of an XP child for the full development of the disease. During this childhood period, the putative patients are un-intentionally exposed to sunlight, which leads to DNA damage and mutations, that are almost impossible to be repaired or rectified.

The subsequent appearance of the first symptoms of XP occurs typically between one and three years of age. About 50% of children present severe sunburns on minimal exposure to sunlight even in a cloudy winter day. The others may present numerous freckles and hyper-pigmented spots on sun-exposed parts of the body. All of them display some kind of photophobia toward visible light. This un-intentional exposure to UV, whatever the following treatment, leads to freckles, skin aging and skin cancers. Hence early detection of XP in children is very important. Parents of the children at XP risk, therefore should be trained for early XP detection.[4] At least surveillance for very early symptoms, such as photophobia, should be carried out during the first weeks of life of the newly born child in the families at risk before a biological testing be done.

The major life threat to XP patients is the development of skin cancer and its fast growth. XP leads to premalignant actinic keratosis (AK), malignant squamous cell carcinoma (SCC), basal cell carcinoma (BCC) and cutaneous melanoma (MM). Although these tumours are easy to detect and are relatively less prone to metastasis (except for MM), their accumulation on XP skin can lead to death of the patient. Therefore, XP patients should be seen regularly by trained dermatologists and pictures of their exposed skin should be taken every 3-6 months to clearly distinguish newly-formed abnormal skin growth.

Chemotherapy and Surgery

In case of early appearance of AK, BCC or SCC, the tumours can be locally treated by using chemotherapeutic drugs such as 5-fluorouracil (FU).[5] Combined action of the topical and systemic chemotherapy of FU has been shown to be partially effective in XP patients. FU can also be given in association with cisplatin. These treatments can minimise the risk of scarring on the body and face following multiple surgeries. The most efficient treatment is, however, surgical removal of tumours followed by pathological analysis of the sample to eventually adapt the therapy. In case of a very large number of tumours present on the exposed skin, full resurfacing of the damaged areas may be followed by autologous skin grafts.[6]

Tumours that cannot be removed surgically have been treated by ionizing radiations.[7] Because XP cells are very sensitive to certain chemotherapeutic drugs such as cis-platin that damage DNA as well as to ionizing radiations,[8] the use of these combined therapies should be carefully envisaged and doses should be minimised in compare to healthy normal subjects. Indeed, XP subjects are more prone than the general population to develop internal tumours such as brain and thyroid tumours and lymphomas.[1] In such cases efficient cocktails of antitumor drugs should be used, but it is absolutely vital that right dosage must be adapted taking in consideration of the DNA repair deficiency of these patients. A decrease of doses between 30 to 50% should sometimes be envisaged. Untimely rapid death of XP patients, due to high dosage of chemotherapeutic agents, has been reported (our unpublished data). Cooperation between dermatologists, oncologists and biologists is necessary to adapt the doses of radiations and the antitumor drugs according to the sensitivity of the patient's cells that must be quantified in vitro before their employment in vivo.

Chemoprevention

Chemoprevention by systemic use of retinoids has been successfully applied in XP patients to decrease the risk of skin cancer development.[9,10] Although retinoids have been shown to be fairly efficient in blocking skin cancer development, these drugs are highly toxic, particularly after prolonged use, leading to side effects such as damage to mucocutaneous layers, hair loss, hepatotoxicity, hyperlipidemia, bony changes, growth and developmental retardations. Moreover, following treatment interruption, a rapid induction of skin cancers has been observed almost immediately rendering the previous protection completely inefficient.[10] Hence retinoid treatment has been slowly abandoned by dermatologists for XP prevention.

In some cases Imiquimod 5% cream has been used as a topical immune response modifier to prevent the appearance of tumours.[11] A normal course of treatment ranges from 4 to 12 weeks. Once the inflammation has settled, there is generally a good response and an excellent cosmetic result. Combination of local application of antitumor cream and oral retinoids, for a limited period of time, has been proposed as an alternative to major surgery.[12]

Enzyme Therapy

XP cells are deficient in NER because one of the major proteins (XPA to XPG) does not function properly in the pathway able to faithfully remove bulky DNA lesions (Fig. 2). One possible therapy is to replace the defective repair protein(s) by supplying relevant protein from exogenous source.

Human enzymes are expensive to produce and probably difficult to use because they aggregate with each other forming complexes that scrutinise genomic DNA. Therefore, enzymes catalysing on their own, produced by genetically engineered lower organisms have been preferred. Many bacteria synthesise CPD-photolyase that is able to bind specifically to cyclobutane pyrimidine dimers (CPD), the major UV-induced DNA lesions. When illuminated with visible light this enzyme specifically monomerises the CPD and brings the DNA to its normal state without introducing any mistake. The phage enzyme, T4-endonuclease V, has been shown in vitro to be able to complement the UV-sensitivity of XP cells.[13]

The T4-endonuclease V (T4-endo) is able to cut the DNA backbone between the two pyrimidines at the site of UV-induced CPD. The T4-endonuclease exerts DNA glycosylase as well as

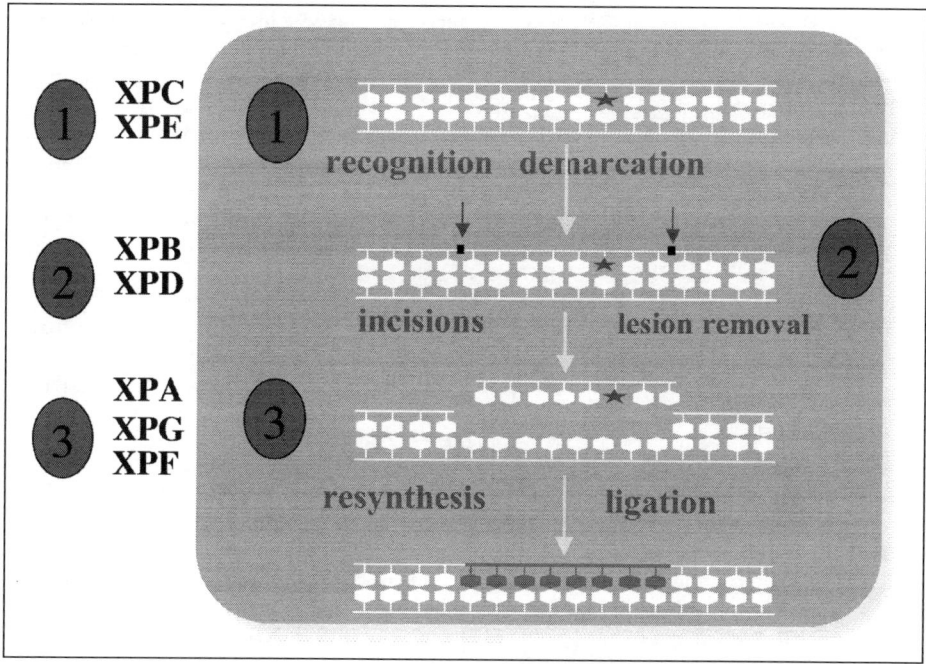

Figure 2. Schematic representation of the nucleotide excision repair pathway. The XP proteins are deficient in one of the complementation group XPA to XPG. They are all necessary for a full efficient repair pathway.

apurinic/apyrimidinic lyase activities.[14] Although this type of break is not a substrate for normal mammalian cell repair, following the action of T4-endonuclease, XP cells, in vitro, are able to process DNA repair.

Clinical trial, using liposome-encapsulated T4-endonuclease or CPD-photolyase, have been shown to prevent UVB-induced immunosuppression. A T4-endo (T4N5) liposome lotion has been used on skin of twenty XP patients for one year.[15] These patients showed a 68% decrease of AK and 30% decrease of BCC as compared to ten untreated XP controls. Although this result is very exciting for XP population, it was not considered significant enough, using different statistical calculations by the FDA, to be accepted for XP treatment.[16] Moreover, T4-endonuclease does not repair all kinds of UV-induced DNA lesions, particularly the very mutagenic pyrimidine 6-4 pyrimidone photoproducts, contrary to normal NER, rendering this treatment to be less safe over the years.

Cellular and Gene Therapies

Since all the XP genes have now been cloned and because skin is an easy target to be seen and modified, gene therapy could be considered as a future treatment for XP. Two strategies can be proposed: (1) introduction and expression of wild type exogenous DNA repair genes in XP using viral vectors and (2) correction directly of the germline mutation at exact site of the XP mutation by involving homologous recombination.

Viral Vectors

Two types of viral vectors have been used in gene therapy experiments: the non-integrative adenoviral vectors and the integrative vectors basically based on recombinant retrovirus constructions.

Adenoviral vectors expressing XP genes have been constructed and has been shown in vitro to be able to fully complement the UV-sensitivity of XP cells.[17-19] Indeed, XP cells, transduced by recombinant adenovirus, expressing the required wild type XP gene, are able to carry out normal NER for several months, although the vectors remains non-integrated and is only transiently present in the transduced cells. This adenoviral system has been used in vivo using Xpa[-/-] KO mice. Subcutaneous injection of adenovirus in these mice allowed the protection of skin towards UVB-irradiation for both UV-induced epidermal hyperplasia and skin tumour development.[20] Although these results are promising, but adenoviral vectors have been found to give rise to a strong immunologic response in humans and their presence in the cell is only transient, hence this technology is not yet available as a therapy for XP.

Recombinant retrovirus have also been used in several other cases of gene therapy in humans, including for the first successful assay.[21] In our laboratory, we have constructed several recombinant retroviruses able to fully complement in vitro XPA, XPC and XPD cell lines.[22-25]

In collaboration with the Research Group of L'Oréal in France, we have been able to produce skin, in vitro, using independent cultures of keratinocytes and fibroblasts of the same XP patients belonging the the the XPC group.[26,27] The cells from these patients were then successfully transduced and complemented by the wild type *XPC* gene and, in vitro, XP skins with normal DNA repair ability were produced. All the abnormalities of in vitro constructed XP skin, such as retardation of differentiation, increased proliferation of basal layer cells and invagination of epidermis inside the dermis, were fully complemented following retroviral expression.[28,29] Although in theory it is possible to graft these complemented skins on the body of corresponding XP patients and such complementation products should be active for life because the transgenic DNA is inserted in the genomic DNA which could not be diluted out due to loss of materials at each cell division. Would it work in practice is a big question. Up to now, the limitations of this technique are due to a slow inactivation of the transgene activity with time. This process, which is well known in gene therapy experiments, probably works due to the silencing of the promoter, driving the transgene and also because this is not the endogenous promoter and the gene is not located at the "right" position in the genomic DNA.[30]

The potential risk in the use of recombinant retrovirus is the insertional mutagenesis at the active site of the chromatin leading ultimately to activation of oncogene or inactivation of tumour suppressor genes as already found in human trial.[31] Indeed, recombinant retroviruses do not really integrate randomly into genomic DNA but prefer actively-transcribed DNA regions rendering this process prone to metabolic disorders. Other viruses, such as lentivirus or recombinant Adeno-Associated Vectors (AAV) have also be proposed for XP gene therapy.[32]

Homologous Recombination

Homologous recombination is a widespread tool for genome engineering, gene targeting being its most common application.[33,34] Basically an engineered DNA fragment is introduced into the target cell and its endogenous DNA maintenance systems catalyze the transfer of genetic material from this fragment to the targeted locus involving homologous recombination (Fig. 3A). However, this technology has its limitations due to its low recombination frequency which in mammals ranges between 10^{-6} and 10^{-9} per transfected cells. Therefore, various alternative methods have been developed in the last decade to try to enhance the targeting efficiency. Meganuclease-induced recombination (Fig. 3B) has emerged as the most robust and widespread technique.

Meganucleases are sequence-specific endonucleases which recognises a large (>12 bp) cleavage site.[35] In nature, meganucleases are essentially represented by homing endonucleases (HEs).[36] HEs are generally encoded by mobile genetic elements such as class I introns and inteins, which they contribute to propagate. The homing process, which allows the spreading of mobile introns into intron-free populations, relies on the cleavage of intron-less genes and homologous repair with a gene carrying an intron. This phenomenon looks very much like a gene targeting process. This technology is now employed in a growing number of applications. However, its major limitation is the requirement for a pre-existing specific cleavage site in the targeted locus. For natural,

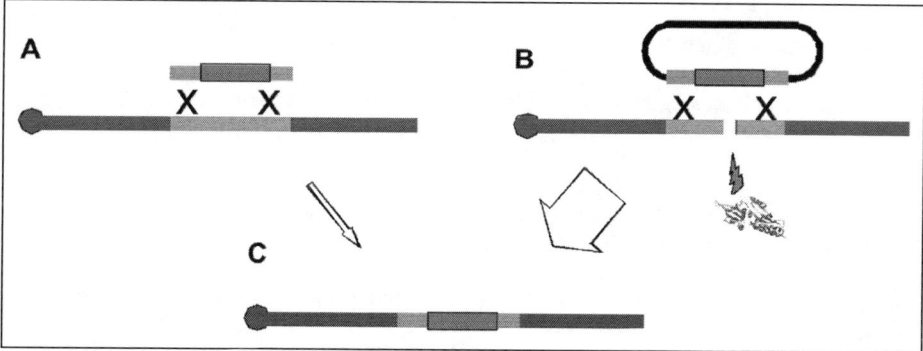

Figure 3. Different methods for homologous gene targeting. Gene knock-out is represented here as an example. Sequences of the targeted gene (blue) are replaced with a non homologous sequence, usually a marker gene (red). In the targeting construct, this marker is flanked by sequences homologous to the targeted gene. A) Classical gene targeting. B) Meganuclease-induced targeting. The meganuclease cleaves specifically and uniquely the targeted locus, which is healed by recombination with the repair matrix. C) The outcome differs only by the frequency of targeted events. Experiments in collaboration with Cellectis S.A. (France). A color version of this figure is available online at www.eurekah.com.

non-engineered loci, the probability of having a natural meganuclease cleavage sequence is close to zero. Therefore the construction of novel artificial meganucleases, with dedicated specificities, is under intense investigation.

In collaboration with the laboratory of Cellectis S. A. (Romainville, France), meganucleases, able to cleave a specific site in the *XPC* gene, have been engineered and are going to be used in our laboratory. Transfection of the genes, coding these meganucleases, should produce double-strand breaks (DSB) at unique site(s) in the *XPC* gene (Fig. 3B). Cotransfection with plasmids, containing part of the wild type *XPC* gene, will be transfected in *XPC* cells allowing the repair/recombination enzymes to repair the DSB in genomic DNA using the wild type *XPC* sequences as a template for recombination.

This process should correct the original germline mutation in the *XPC* gene via homologous recombination, leading to at least one allele of the *XPC* gene with a wild type sequence. Correction of one mutated *XPC* allele is enough to give rise to DNA repair-competent cells because parents of XP children, who are heterozygous for the XP mutation, have a normal UV-sensitivity. In theory, this repair process should produce proper wild type cells, which should remain stable for life. The corrected *XPC* gene will stay at the normal position in the genomic DNA and be regulated by the endogenous promoter, which should allow a proper regulation of this gene in the corrected cells.

Although this approach is very elegant and appealing, no experimental trial in human cells has yet been made. It is because the efficiency of recombination is fairly low. If such experiments were successful, we should be able to construct XP skin in vitro, which should be fully active for life. Such skin could be grafted on XP patients following removal of tumours from the body.

Conclusions

Due to their extreme UV-sensitivity, the best way to control XP is to completely protect the putative XP patients from UV-exposure. Alternatively it must be applied as soon as the disease has been detected, because accumulation of DNA lesions and mutations, before the onset of disease, have been known to give rise to tumours at a later stage in life. Experience has shown that early avoidance of UV exposure can delay the onset of complications and lead to a healthier and relatively happier life of the patients. The full protection and the lack of outdoor activities in sunny days are sometimes difficult to achieve both by the XP children and their family. Moreover, psychological problems are expected to develop once XP children start to grow.

In case of small skin disturbances such poikiloderma, AK or small malignant tumours, classical antitumor cream or light surgery can be employed. Once, the size and the number of tumours start to be too large, heavy surgery and autologous grafts should be envisaged with the corresponding difficult post-surgical complications.

Autologous skin grafts following gene therapy are still under investigation and expected to take several years to be operational on XP patients. Attempts to use gene therapy for other genetic skin diseases such as junctional epidermolysis bullosa, should help to design better therapies able to succeed in XP patients.[37-40]

Besides skin sensitivity, several XP patients also develop severe neurodegenerative disorders. Until now no therapy has been developed for these disorders since UV-protection after the onset of disease does not preclude the progressive neuro-degeneration. Because the knowledge about the molecular mechanisms leading to these disorders is still incomplete (see Chapter 13), it is difficult to think about any effective therapy. Perhaps, the use of antioxidant drugs, able to target brain cells, may be the next step in controlling the neurological disorders in XP patients.

Acknowledgements

This work has been partially granted by the French "Association des Enfants de la Lune" (Tercis, France) and by the "Association Française contre les Myopathies" (Paris, France). I am grateful to Dr. Vincent Laugel for reading the manuscript and Dr. Anne Stary for Figure 3.

References

1. Kraemer KH, Lee MM, Scotto J. Xeroderma pigmentosum. Cutaneous, ocular and neurologic abnormalities in 830 published cases. Arch Dermatol 1987; 123:241-50.
2. Stary A, Sarasin A. The genetics of the hereditary xeroderma pigmentosum syndrome. Biochimie 2002; 84:49-60.
3. Stary A, Sarasin A. The genetic basis of xeroderma pigmentosum and trichothiodystrophy syndromes. Cancer Surv 1996; 26:155-7.
4. Sarasin A, Blanchet-Bardon C, Renault G et al. Prenatal diagnosis in a subset of trichothiodystrophy patients defective in DNA repair. Br J Dermatol 1992; 127:485-91.
5. Hamouda B, Jamila Z, Najet R et al. Topical 5-fluorouracil to treat multiple or unresectable facial squamous cell carcinomas in xeroderma pigmentosum. J Am Acad Dermatol 2001; 44:1054.
6. Sonmez Ergun S. Resurfacing the dorsum of the hand in a patient with Xeroderma pigmentosum. Dermatol Surg 2003; 29:782-4.
7. Sakata K, Aoki Y, Kumakura Y et al. Radiation therapy for patients with xeroderma pigmentosum. Radiat Med 1996; 14:87-90.
8 . Arlett CF, Plowman PN, Rogers PB et al. Clinical and cellular ionizing radiation sensitivity in a patient with xeroderma pigmentosum. Br J Radiol 2006; 79:510-7.
9. Kraemer KH, DiGiovanna JJ, Moshell AN et al. Prevention of skin cancer in xeroderma pigmentosum with the use of oral isotretinoin. N Engl J Med 1988; 318:1633-7.
10. Kraemer KH, DiGiovanna JJ, Peck GL. Chemoprevention of skin cancer in xeroderma pigmentosum. J Dermatol 1992; 19:715-8.
11. Nagore E, Sevila A, Sanmartin O et al. Excellent response of basal cell carcinomas and pigmentary changes in xeroderma pigmentosum to imiquimod 5% cream. Br J Dermatol 2003; 149:858-61.
12. Giannotti B, Vanzi L, Difonzo EM et al. The treatment of basal cell carcinomas in a patient with xeroderma pigmentosum with a combination of imiquimod 5% cream and oral acitretin. Clin Exp Dermatol 2003; 28 Suppl 1:33-5.
13. Tanaka K, Sekiguchi M, Okada Y. Restoration of ultraviolet-induced unscheduled DNA synthesis of xeroderma pigmentosum cells by the concomitant treatment with bacteriophage T4-endonuclease V and HVJ (Sendai virus). Proc Natl Acad Sci USA 1975; 72:4071-5.
14. Lloyd RS, Hanawalt PC. Expression of the denV gene of bacteriophage T4 cloned in Escherichia coli. Proc Natl Acad Sci USA 1981; 78:2796-800.
15. Yarosh D, Klein J, O'Connor A et al. Effect of topically applied T4-endonuclease V in liposomes on skin cancer in xeroderma pigmentosum: a randomised study. Xeroderma Pigmentosum Study Group. Lancet 2001; 357:926-9.
16. Lachenbruch P, Marzella L, Schwieterman W et al. Poisson distribution to assess actinic keratoses in xeroderma pigmentosum. Lancet 2001; 358:925.

17. Muotri AR, Marchetto MC, Zerbini LF et al. Complementation of the DNA repair deficiency in human xeroderma pigmentosum group A and C cells by recombinant adenovirus-mediated gene transfer. Hum Gene Ther 2002; 13:1833-44.
18. Armelini MG, Muotri AR, Marchetto MC et al. Restoring DNA repair capacity of cells from three distinct diseases by XPD gene-recombinant adenovirus. Cancer Gene Ther 2005; 12:389-96.
19. Lima-Bessa KM, Chigancas V, Stary A et al. Adenovirus mediated transduction of the human DNA polymerase eta cDNA. DNA Repair (Amst) 2006; 5:925-34.
20. Marchetto MC, Muotri AR, Burns DK et al. Gene transduction in skin cells: preventing cancer in xeroderma pigmentosum mice. Proc Natl Acad Sci USA 2004; 101:17759-64.
21. Cavazzana-Calvo M, Hacein-Bey S, de Saint Basile G et al. Gene therapy of human severe combined immunodeficiency (SCID)-X1 disease. Science 2000; 288:669-72.
22. Zeng L, Quilliet X, Chevallier-Lagente O et al. Retrovirus-mediated gene transfer corrects DNA repair defect of xeroderma pigmentosum cells of complementation groups A, B and C. Gene Ther 1997; 4:1077-84.
23. Quilliet X, Chevallier-Lagente O, Zeng L et al. Retroviral-mediated correction of DNA repair defect in xeroderma pigmentosum cells is associated with recovery of catalase activity. Mutat Res 1997; 385:235-42.
24. Dumaz N, Drougard C, Quilliet X et al. Recovery of the normal p53 response after UV treatment in DNA repair-deficient fibroblasts by retroviral-mediated correction with the XPD gene. Carcinogenesis 1998; 19:1701-4.
25. Zeng L, Sarasin A, Mezzina M. Novel complementation assays for DNA repair-deficient cells. Transient and stable expression of DNA repair genes. Methods Mol Biol 1999; 113:87-100.
26. Bernerd F, Asselineau D, Vioux C et al. Clues to epidermal cancer proneness revealed by reconstruction of DNA repair-deficient xeroderma pigmentosum skin in vitro. Proc Natl Acad Sci USA 2001; 98:7817-22.
27. Bernerd F, Asselineau D, Frechet M et al. Reconstruction of DNA repair-deficient xeroderma pigmentosum skin in vitro: a model to study hypersensitivity to UV light. Photochem Photobiol 2005; 81:19-24.
28. Arnaudeau-Begard C, Brellier F, Chevallier-Lagente O et al. Genetic correction of DNA repair-deficient/ cancer-prone xeroderma pigmentosum group C keratinocytes. Hum Gene Ther 2003; 14:983-96.
29. Frechet M, Bergoglio V, Chevallier-Lagente O et al. Complementation assays adapted for DNA repair-deficient keratinocytes. Methods Mol Biol 2006; 314:9-23.
30. Ellis J. Silencing and variegation of gammaretrovirus and lentivirus vectors. Hum Gene Ther 2005; 16:1241-6.
31. Hacein-Bey-Abina S, von Kalle C, Schmidt M et al. A serious adverse event after successful gene therapy for X-linked severe combined immunodeficiency. N Engl J Med 2003; 348:255-6.
32. Marchetto MC, Correa RG, Menck CF et al. Functional lentiviral vectors for xeroderma pigmentosum gene therapy. J Biotechnol 2006; 126:424-30.
33. Capecchi MR. Altering the genome by homologous recombination. Science 1989; 244:1288-92.
34. Smithies O. Forty years with homologous recombination. Nat Med 2001; 7:1083-6.
35. Thierry A, Dujon B. Nested chromosomal fragmentation in yeast using the meganuclease I-Sce I: a new method for physical mapping of eukaryotic genomes. Nucleic Acids Res 1992; 20:5625-31.
36. Chevalier BS, Stoddard BL. Homing endonucleases: structural and functional insight into the catalysts of intron/intein mobility. Nucleic Acids Res 2001; 29:3757-74.
37. Magnaldo T, Sarasin A. Genetic reversion of inherited skin disorders. Mutat Res 2002; 509:211-20.
38. Magnaldo T, Sarasin A. Xeroderma pigmentosum: from symptoms and genetics to gene-based skin therapy. Cells Tissues Organs 2004; 177:189-98.
39. Hengge UR. Progress and prospects of skin gene therapy: a ten year history. Clin Dermatol 2005; 23:107-14.
40. MacNeil S. Progress and opportunities for tissue-engineered skin. Nature 2007; 445:874-880.

CHAPTER 17

Animal Models of Xeroderma Pigmentosum

Xue-Zhi Sun,* Rui Zhang, Chun Cui, Yoshi-Nobu Harada, Setsuji Hisano, Yeunhwa Gu, Yoshihiro Fukui and Hidenori Yonehara

Introduction

Xeroderma pigmentosum (XP) is a rare autosomal disorder characterized by hypersensitivity of the skin to sunlight specifically to ultraviolet (UV) which can lead to high rate of susceptibility to skin cancer and other kinds of neurodegenerative problems. Compared to normal individuals, XP patients have a more than 1000-fold increased risk of developing skin cancer on sun-exposed areas of their body. Genetic and molecular analyses have revealed that the repair of UV-induced DNA damage is impaired in XP patients owing to mutations in genes that form part of a DNA-repair pathway known as nucleotide excision repair (NER). XP is, therefore, regarded as a convincing human example of the link between DNA repair deficiency and cancer risk. However, this relationship has not been examined in detail in humans due to the limited number of XP patients and their frequent early death due to skin cancer and neurological problems. For these reasons are required the generation of equivalent animal models to determine their exact molecular mechanisms.

As described in other chapters, cell fusion studies have revealed that there are seven different genetic complementation groups in XP (denoted XPA through XPG) as well as an XP-variant form (XPV).[1] A defect in one of the seven genes (XPA to XPG), involved in the NER pathway, can cause XP and the loss of function of the eighth gene, XPV, results in clinical XP phenotype, including predispositions to skin cancer.[3,4] The recently developed technology of targeted gene replacement in mouse embryonic stem (ES) cells has provided investigators with the ability to generate mutant mouse strains defective in specific gene(s) of interest. To date, several such defective animals showing XP have been reported. These animal models exhibited characteristic features that mimic XP patients and provided a very useful experimental system for studying how DNA repair mechanisms affected tumourogenesis, development and ageing in humans.

XPA-Deficient Mouse Models

The XPA genetic group (defective in the XPA gene) comprises one of the largest groups among XP patients whose skin cancers are the most severe among all the known groups of XP.[4] XPA gene encodes a protein involved in the initial damage-recognition stages of NER and the stabilization of the multiprotein repair complex assembled at DNA damage sites.[5] Mice defective in the highly conserved Xpa gene (as well as a number of other XP mutant mice) have been generated by conventional gene targeting and have proven to be attractive models for human disease. In particular, exposure of the shaved dorsal skin of mice to UV light results in multiple skin lesions

*Corresponding Author: Dr. Xue-Zhi Sun—Regulatory Sciences Research Group, National Institute of Radiological Sciences, Anagawa 4-9-1, Inage-ku, Chiba 263-8555, Japan. Email: sun_s@nirs.go.jp

Molecular Mechanisms of Xeroderma Pigmentosum, edited by Shamim I. Ahmad and Fumio Hanaoka. ©2008 Landes Bioscience and Springer Science+Business Media.

that typically progress to tumors. A wealth of information from several Xpa-deficient strains of mice is in agreement with the epidemiological data confirming that XPA results in a marked predisposition to cancer in the sun-exposed skin.[6] The Xpa mutant mice of 8-10 wks were irradiated on the back with UV-B at a dose of 5 J/cm² three times weekly for 10 wks (total dose: 150 J/cm²). More than 30% of the irradiated mice developed tumors of melanocytic origin that metastasized to the lymph nodes. Histologically, the proliferated cells exhibited lentigo maligna melanoma or nodular melanoma. Immunohistochemistry confirmed that the tumor cells were characteristic of melanoma. Non-irradiated mice did not develop skin tumors spontaneously. This mouse model is useful for studying the photobiological aspects of human melanoma, since the mice developed melanoma from epidermal melanocytes only after UV exposure.[2]

Xpa-deficient mice are also prone to spontaneous and carcinogen-induced tumorigenesis in their internal organs besides skin cancer. For example, Xpa mice appeared to develop internal tumors with a much higher frequency and shorter latency times than normal mice upon exposure to several different carcinogens. Treatment with benzo[α]pyrene (B[α]P) by gavages resulted in an increased development of generalized T-cell lymphomas mainly occurring in the spleen, whereas exposure via the diet resulted in a higher incidence of forestomach tumors compared to their wild-type counterparts.[7-9] These experimental data are in good agreement with the epidemiological data pointing to a 10-20-fold increase in the incidence of internal neoplasms in XP patients compared to normal individuals. The fact that sensitivity of Xpa mice to carcinogens is high, it suggests that this mouse model is an excellent candidate for carcinogenicity testing.

XPC-Deficient Mouse Models

The most common XPC form of XP in North America and Europe is that associated with the genetic group C.[10] An Xpc mutant mouse was generated by replacing exon 10 of the Xpc gene with a PolII-NEO selection cassette.[11] In order to assess if deletion of the Xpc gene results in an enhanced predisposition to UV radiation-induced skin cancer, Xpc−/− animals as well as Xpc+/− and Xpc+/+ littermates were irradiated with a cumulative dose of ~140 KJ/m² of UV-B radiation over an 18-wk period. Xpc mutant mice were highly predisposed to UV-B radiation-induced skin cancer and all the Xpc−/− animals had developed skin tumors by 25 wks of irradiation while all control animals (Xpc+/− and Xpc+/+) were free of tumors.[12,13] Longer periods of observation revealed that Xpc+/− animals were at increased risk of developing skin tumors after exposure to UV radiation. The latency time for the appearance of skin cancer was reduced in Xpc+/− compared with wild type littermates, with Xpc+/− animals manifesting a 50% skin cancer incidence by 50 wks after irradiation. This is considerably earlier than Xpc+/+ mice, where the 50% skin cancer incidence reached only 90 wks after irradiation. The result that the Xpc heterozygous mutant state conferred an increased predisposition to UV-induced skin cancer has important implications for human health. The fact that no mutational inactivation of the remaining Xpc allele in tumors originating in heterozygous mutant animals has been observed, suggests that the increased predisposition associated with the heterozygous state is due to an allelic insufficiency, or haploinsufficiency. This could imply that humans carrying a heterozygous mutation in XP genes may be at increased risk to cancer associated with exposure of sunlight.

Further studies revealed that Xpc mutant mice are fertile and have no detectable developmental or neurological abnormalities.[13] Prolonged observation has also failed to find any evidence of an enhanced predisposition to spontaneous internal cancers in this Xpc mutant animals older than 1.5 years. Anatomical analyses of mice of this age and older have revealed no other gross abnormalities.

In order to examine whether Xpc mutant mice are predisposed to cancers in other tissues associated with exposure to chemical carcinogens, animals such as, Xpc−/− and littermate controls were treated with 2-acetylaminofluorene (2-AAF) or N-OH-AAF.[14] The results show that Xpc−/− animals, irrespective of the Trp53 genotype, were more predisposed to 2-AAF- or N-OH-AAF-induced liver and lung tumors. Among Xpc−/− Trp53+/+animals, 67% of the mice manifested benign or malignant tumors in the lungs or liver at 15 months posttreatment.

In contrast, 33% and 12.5%, respectively, of the Xpc+/+ and Xpc+/− littermate controls had tumors in either organ after the same time period. The observation that Xpc+/− animals exhibited a lower incidence of lung and liver tumors when compared to the wild-types was surprising in view of the opposite result obtained for UV-induced skin cancers in these mice. This suggested that haploinsufficiency of the Xpc gene did not enhance the predisposition to internal tumors after treatment with AAF. N-OH-AAF was shown to be the more potent carcinogen, since 87.5% of the Xpc−/− animals had either organ affected after N-OH-AAF administration, while 2-AAF administration affected only 43% of the tested animals.

Both Xpa- and Xpc- deficient mice are predisposed to develop carcinogen-induced skin cancer as well as internal tumors; such results have been directly associated with a defect in NER. However, the NER pathway of Xpc mice varied from that of Xpa-deficient mouse. NER is probably the most complicated and extensively studied of the DNA repair process.[15-19] It harbors a broad specificity and is able to recognize lesions that disturb the double helix conformation, such as those induced by UV light and chemicals that give rise to bulky DNA adducts and DNA crosslinks. NER consists of two distinct pathways, i.e., transcription-coupled repair (TCR), which eliminates DNA damage present in the transcribed strand of active genes that actually block transcription[20,21] and global-genome repair (GGR) that removes lesions throughout the genome and the nontranscribed strand of active genes.[22] The main difference between these two pathways is the initial assembly of proteins at the site of DNA damage. For the TCR pathway, the removal of DNA damage is initiated by stalled RNA polymerase II and includes the Cockayne syndrome (CS) caused by mutations in two genes, CSA and CSB proteins, whereas proteins such as XPC-hHR23B and the DBB complex (XPE proteins p48 and p127) are involved in GGR. The subsequent steps involved in DNA repair have been most extensively studied for GGR, but the complex which assembles in the TCR pathway is believed to include these proteins as well. All XP groups are defective in both GGR and TCR, with the exception of XPC and XPE, which are defective in GGR only.

XPG-Deficient Mouse Models

XP patients in group G show various clinical symptoms, ranging from very mild cutaneous abnormalities to severe dermatological impairments. A combination of clinical hallmarks of XP and CS has been observed in several XP-G patients. CS is also known as a repair-deficient human disease characterized by a postnatal failure of growth, a short life span and progressive neurological dysfunction.[23-25] Rare patients in three complementation groups, XP-B, XP-A and XP-G, also show characteristic features of CS, the so-called XP/CS complex.[26,27] The possible mechanisms underlying the clinical symptoms of XP or XP/CS are unclear.

Using embryonic stem cell techniques, a genetic mouse model with a disrupted XPG allele was generated by the insertion of neo-cassette sequences into exon 3 of the *XPG* gene.[28] Homozygous animals of this autosomal recessive disease displayed distinct developmental characteristics, their bodies being markedly smaller than those of their wild type littermates from postnatal day 6 and their postnatal growth failure becoming steadily more severe with time. Their life span was very short, with all the mutants dying by postnatal day 23 after exhibiting notable weakness and emaciation (Fig. 1). To confirm the reasons behind this distinctive developmental pattern, the suckling and nutrition habits of mutant littermates were examined and the mutant pups were observed to suck their mothers.[29] No obvious physical changes or abnormal suckling behaviors were noted. These mutant homozygous pups seemed to cling normally to the teats of their dams and their abdomens were full of white milk that could be seen through the pups' thinner abdominal skin. Therefore, postnatal growth retardation and premature death in mutant mice were due to the introduction of the nonfunctional *Xpg* gene. Such distinct developmental characteristics in *Xpg* mutants were similar to the typical clinical phenotypes of CS.

The *Xpg*-deficient mice also exhibited some progressive neurological signs.[29,30] As they developed, a diminished level of activity and progressive ataxia were observed, suggesting a developmental retardation of the central nervous system and progressive neuronal dysfunction in the homozygous mutants. When brain development between mutants and the normal littermates

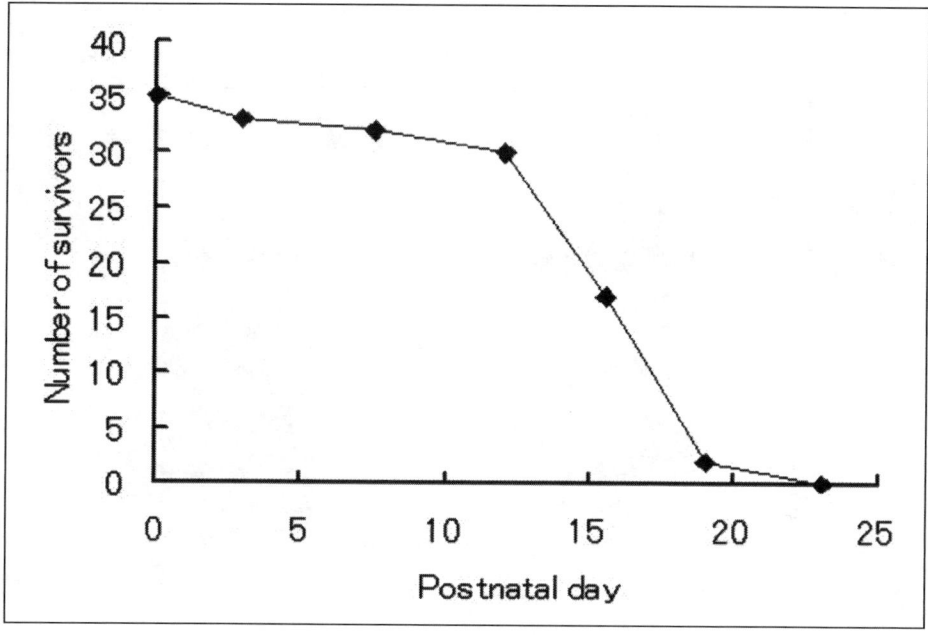

Figure 1. Survival curve of *Xpg* mutant mice after birth.

was compared at postnatal day 19, the brain sizes including those of the cerebrum and cerebellum were found to be smaller than those of wild-type mice of the same age. The average cerebral and cerebellar weights of the homozygous mice only accounted for a respective 79.5% and 66.9% of the wild type weights (Fig. 2a,b). Under the light-microscope a thinner cerebellar and cerebral cortex and smaller cerebellar foliation could be clearly observed in the mutant mice. Histological measurements further indicated that the thickness of the cerebral and cerebellar cortex differed significantly from that found in the wild type mice. Such a small brain (microcephaly), brain developmental retardation and progressive neurological signs closely resemble the symptoms observed in human XPG and/or CS patients at the early postnatal stages. Histological analysis of the cerebellum of *Xpg* mice revealed multiple pyknotic cells in the Purkinje cell layer that had atrophic cell bodies and shrunken nuclei. Further examination with immunohistochemistry for calbindin-D 28k (CaBP) showed that a large number of immunoreactive Purkinje cells were atrophic and their dendritic trees were smaller and shorter than those in wild type littermates (Fig. 3). These results indicated a marked degeneration of Purkinje cells in the *Xpg* mutant cerebellum. A study by in situ detection of DNA fragmentation in the cerebellar cortex demonstrated that, though some TUNEL-positive cells appeared in the granular layer of mutant mice, few cell deaths were confirmed in the Purkinje layer. These results suggested that Purkinje cell degeneration in the mutant cerebellum was underway, but that Purkinje cell death had not yet become apparent and that the appearance of some abnormal cerebellar symptoms in the *Xpg*-deficient mice was not due to a marked Purkinje cell degeneration, rather due to the damage of other cells. This observation may provide new insight into the roles of the XPG protein in NER and to the mechanisms underlying clinic symptoms of XPG and CS in humans.

ERCC1/XPF-Deficient Mouse Models

The human ERCC1/XPF complex is a structure-specific endonuclease with a defined polarity that participates in multiple DNA repair pathways. XPF and ERCC1 function as a complex in which the inactivation of XPF might have the same consequence as an ERCC1 knockout mice.

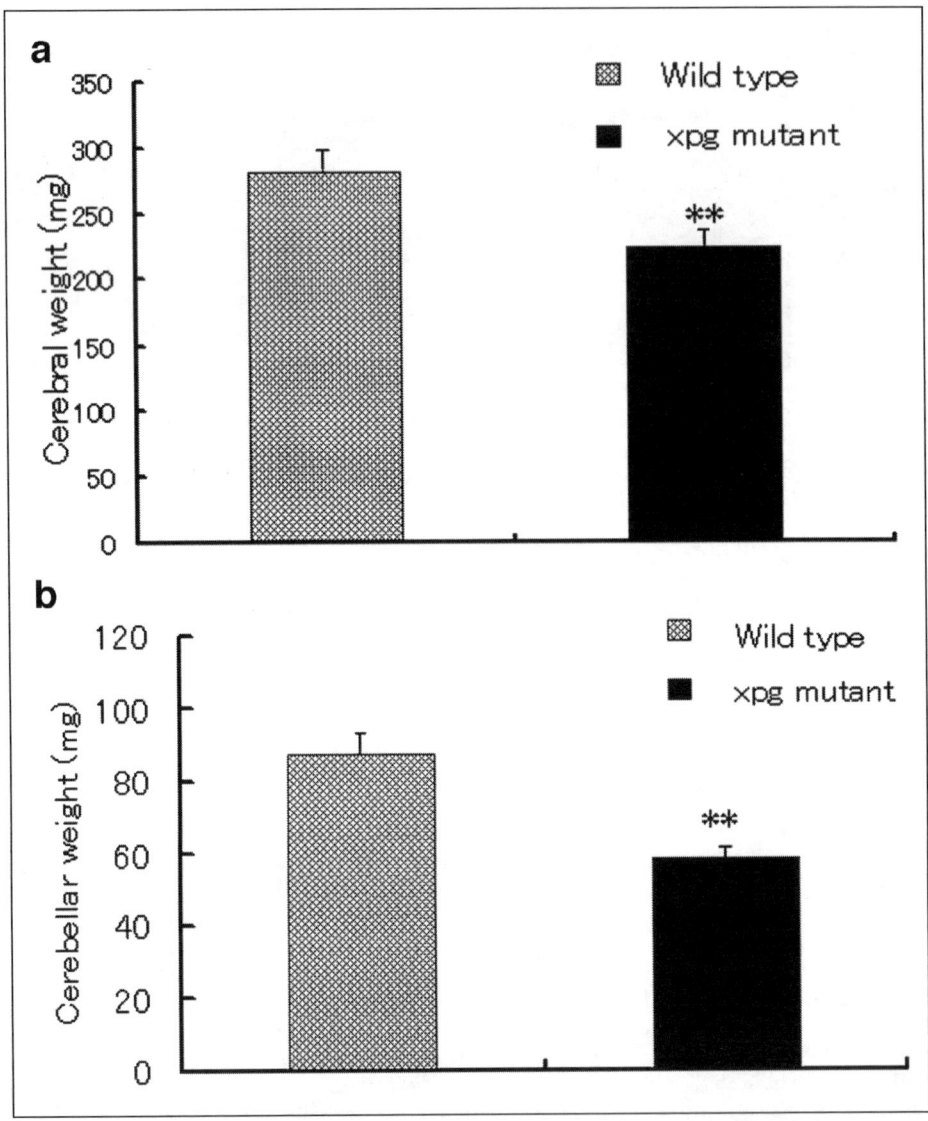

Figure 2. Brain development of wild-type and *xpg*-mutant mice at postnatal day 19. Both cerebral (a) and cerebellar (b) weight of mutants are significantly different from wild-type controls. **: P < 0.001; n = 9.

The generation of XPF-deficient mice resulted in a high incidence of embryonic lethality and mice that did survive showed a severe postnatal growth defect and died approximately 3 weeks after birth.[31] Histological examination indicated that the liver in mutant mice contained abnormal cells with enlarged nuclei. Furthermore, embryonic fibroblasts defective in XPF were hypersensitive to UV irradiation and mitomycin C treatment. These phenotypes are identical to those exhibited by the ERCC1-deficient mice and are consistent with the functional association of the two proteins XPF and ERCC1.

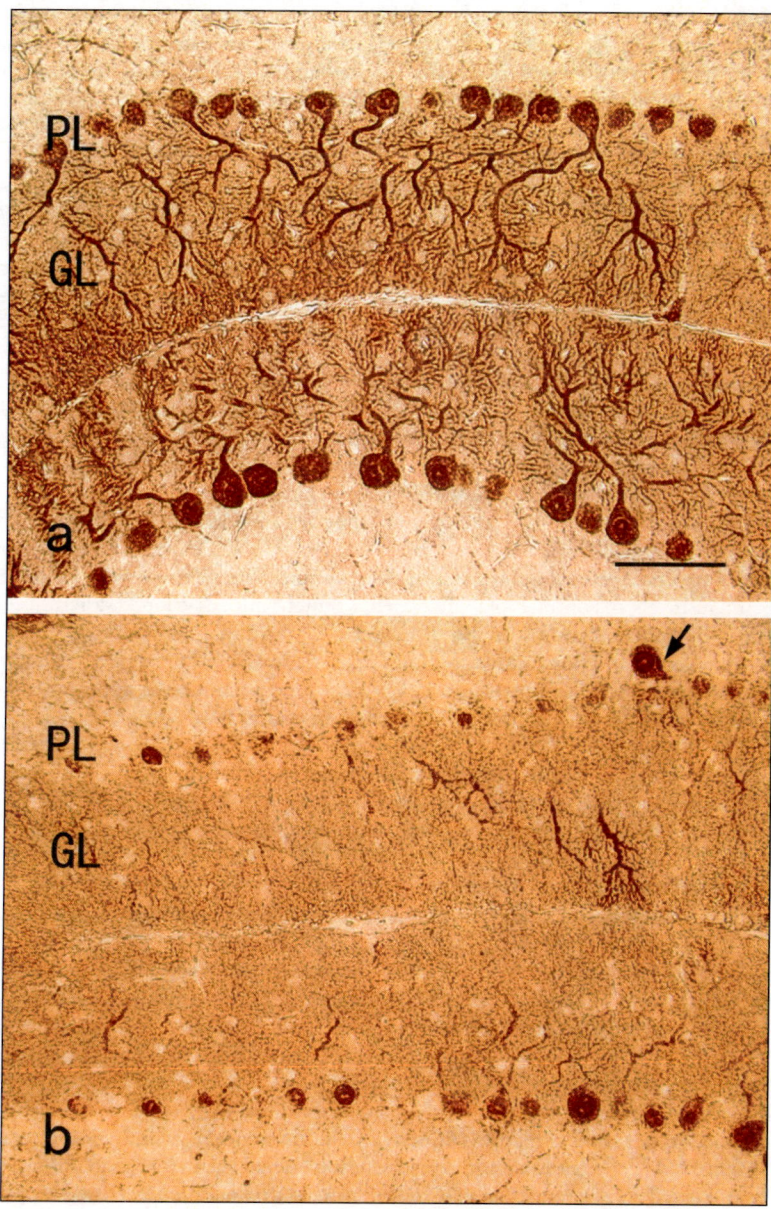

Figure 3. Sagittal sections of lobes in cerebella stained with antiCalbindin-D 28k (CaBP) at postnatal day 20. a) Homogeneously strong CaBP immunoreactivity is observed in every Purkinje cell in wild-type mouse. Characteristic morphology of normal Purkinje cell throughout is also clearly visualized in every cell. b) CaBP immunoreactive Purkinje cells are dramatically decreased both in number and in size in *xpg*-deficent littermates. Many immunoreactive Purkinje cells are atrophic and their dendritic trees are smaller and shorter than in control littermates. Ectopic Purkinje cell (arrow) can also be seen in this field. Scale bar represents 50.

In humans, XPF patients show relatively mild symptoms of photosensitivity with occasional neurological abnormalities.[32-34] The reason for this apparent discrepancy may be that the mutations found in these patients have not involved a deficiency in their ERCC1 gene and since the patients do not completely eliminate the XPF function, fibroblasts retained low levels of NER activity.

XPD Mutant Mouse Models

Mammalian NER system has been reconstituted in vitro, involving >30 different proteins.[17] Among them, the helicases ERCC2/XPD and ERCC3/XPB and two transcription factor IIH (TFIIH) subunits are responsible for opening the DNA around the site of the lesion—a crucial step in initiating the NER process. Both XPB and XPD are conserved and are essential in eukaryotes, precluding the establishment of null cell lines.[36] Hypomorphic mutations of either XPB or XPD may lead to three recessive diseases of varying severity: trichothiodystrophy (TTD), XP, or associated XP and CS.[37-39] Attempts to generate complete Xpb and Xpd in knockout mice failed because the XPB and XPD proteins were required for transcription, hence their complete absence appears to be life threatening.[40] In fact, patients with XPD-XP, XPD-XP/CS or XPD-TTD harbor point mutations in their *XPD* gene, that presumably do not lead to a complete loss of its function, explains the discrepancy between the results in human patients and the knockout mice. To circumvent embryonic lethality, a knockin mouse model was generated with a point mutation [$Arg^{722} \rightarrow Trp$ ($R^{722}W$)] equivalent to the one found in an XPD-TTD patient.[41,42] This mouse model was viable and displayed many features of TTD patients such as postnatal growth failure, impaired sexual development, skeletal abnormalities and a severely reduced life expectancy. Interestingly, in contrast to TTD patients, mutant mice developed skin cancer upon UV exposure.[42] This phenomenon was also observed in Cs mice that, like most NER-deficient mice, were viable, fertile and possessing most characteristic found in human CS patients. Both Csa and Csb mice become cancer prone upon UV exposure, although the appearance of skin cancers occurs with a longer latency time and a much higher cumulative dose of UV-B is required compared to Xpa or Xpc mice.[43,44]

Xpd mutant mice may prove a useful experimental model for further studies on the molecular basis of ageing caused by un-repaired DNA damage that compromises transcription, thus leading to a functional inactivation of critical genes and enhanced apoptosis.

XPA/CSB Mutant Mouse Models

Although the several mouse models for XP and CS have been generated through gene targeting[6,28,35,41,44,45] that reliably reflect the NER defect, they do not always represent a complete phenotype of the corresponding disorder. As mentioned above, in Xpg-deficient mice, though their susceptibility to UV- and chemical carcinogen-induced skin cancer is greatly increased, the animals fail to develop clearly detectable neurological abnormalities.[6,45] Csb-deficient mice exhibit most of the CS repair characteristics in terms of the repair defect and the manifestation of UV-sensitivity of the skin and eyes, in contrast to human CS that shows increased susceptibility to UV and chemically induced skin cancer, they showed normal development and only a mild neurological phenotype. The mechanisms underlying the difference in neurological phenotypes between Xpa-and Csb-deficient mice and corresponding human patients remain unknown.

Mice were crossed to generate *Xpa+/-Csb+/-* animals that were intercrossed to produce *Xpa+/- Csb-/-* and *Xpa-/- Csb B+/-* mice. Having mutation in both the XPA and CSB genes those mice, in contrast to single mutants, displayed severe growth retardation, ataxia and motor dysfunction during early postnatal development. Their cerebella were hypoplastic and showed impaired foliation and stunted Purkinje cell dendrites. Reduced neurogenesis and increased apoptotic cell death were seen to occur in the cerebellar external granular layer.[22] These findings suggest that XPA and CSB have supplementary roles in the murine nervous system and play a crucial role for these genes in normal brain development.

A rapidly growing number of investigations of XP and CS mutant mouse models have culminated in a breakthrough concerning our insight into the role of DNA damage and repair in the multi-step process of carcinogenesis, development and ageing. As described here, homozygous mice, deficient in one of the XP genes involved in NER, display a clear relationship between the induction of DNA damage (e.g., by UV light or chemical carcinogens) and the development of cancer. When implying such observations from mice to humans, since some results could not resolve the problem of clinical phenotypes, one should take into account that differences between human and mouse clearly exist in certain aspects of DNA repair characteristics, metabolic rates and life span. Moreover, XP and CS, like other diseases, is a complicated genetic condition and whether different DNA repair processes actually cooperate, or whether backup pathways for preventing cancer in vivo really exists, still remains to be elucidated. Further studies are required to identify whether mutant mice defective in two or more of the DNA repair genes are more susceptible to cancer. Such groundwork may help to determine which pathway is indeed critical for suppressing cancer in mice and, by extrapolation, in humans. These powerful mouse models, in the near future, will also help to reveal the complexity of the interactions between the above gene products in the maintenance of genome.

References

1. Copeland NE, Hanke CW, Michalak JA. The molecular basis of xeroderma pigmentosum. Dermatol Surg 1997; 23:447-455.
2. Hoeijmakers JH. Human nucleotide excision repair syndromes: molecular clues to unexpected intricacies. Eur J Cancer 1994; 30:1912-1921.
3. Friedberg EC, Walker GC, Siede W. DNA repair and mutagenesis. Washington, DC: American Society for Microbiology Press, 1995:633-672.
4. Volker M, Moné MJ, Karmakar P et al. Sequential assembly of the nucleotide excision repair factors in vivo. Mol Cell 2001; 8:213-224.
5. Nakane H, Takeuchi S, Yuba S et al. High incidence of ultraviolet-B-or chemical-carcinogen-induced skin tumors in mice lacking the xeroderma pigmentosum group A gene. Nature 1995; 377:165-168.
6. Yamazaki F, Okamoto H, Matsumura Y et al. Development of a new mouse model (Xeroderma Pigmentosum A-Deficient, Stem Cell Factor-Transgenic) of ultraviolet B-induced melanoma. J Invest Dermatol 2005; 125:521-525.
7. A de Vries CTM, van Oostrom PM, Dortant RB et al. Spontaneous liver tumors and benzo[α]pyrene-induced lymphomas in XPA-deficient mice. Mol Carcinog 1997; 19:46-53.
8. van Oostrom CT, Boeve M, van den Berg J et al. Effect of heterozygous loss of p53 on benzo[α]pyrene-induced mutations and tumors in DNA repair-deficient XPA mice. Environ Mol Mutagen 1999; 34:124-130.
9. Hoogervorst EM, de Vries A, Beems RB et al. Combined oral benzo[α]pyrene and inhalatory ozone exposure have no effect on lung tumor development in DNA repair-deficient Xpa mice. Carcinogenesis 2003; 24:613-619.
10. Cleaver JE, Kraemer KH. Xeroderma pigmentosum. In: Scriver CR, Beaudet AL, Sly WS, Valle D, eds. The Metabolic Basis of Inherited Disease. New York: McGraw-Hill, 1989:2949-2971.
11. Cheo DL, Ruven HJT, Meira LB et al. Characterization of defective nucleotide excision repair in Xpc mutant mice generated by targeted gene replacement. Mutat Res 1997; 374:1-9.
12. Cheo DL, Meira LB, Hammer RE et al. Synergistic interactions between Xpc and p53 mutations in double mutant mice: neural tube abnormalities and accelerated UV radiation-induced skin cancer. Curr Biol 1996; 6:1691-1694.
13. Cheo DL, Meira LB, Burns DK et al. Ultraviolet-B radiation-induced skin cancer in mice defective in the Xpc, Trp53 and Apex (HAP1) genes: genotype-specific effects on cancer predisposition and pathology of tumors. Cancer Res 2000; 60:1580-1584.
14. Cheo DL, Burns DK, Meira LB et al. Mutational inactivation of the xeroderma pigmentosum group C (Xpc) gene confers predisposition to acetyl-aminofluorene-induced liver and lung cancer and alters the spectrum of spontaneous cancer in Trp53 mice. Cancer Res 1999; 59:771-775.
15. Hoeijmakers JH. Genome maintenance mechanisms for preventing cancer. Nature 2001; 411:366-374.
16. Christmann M, Tomicic MT, Roos WP et al. Mechanisms of human DNA repair: an update. Toxicology 2003; 193:3-34.
17. Friedberg EC. How nucleotide excision repair protects against cancer, Nat Rev Cancer 2001; 1:22-32.
18. Hanawalt PC, Ford JM, Lloyd DR: Functional characterization of global genomic DNA repair and its implications for cancer. Mutat Res 2003; 544:107-114.
19. Van Hoffen A, Balajee AS, van Zeeland AA et al. Nucleotide excision repair and its interplay with transcription. Toxicology 2003; 193:79-90.

20. Mellon I, Spivak G, Hanawalt PC. Selective removal of transcription-blocking DNA damage from the transcribed strand of the mammalian DHFR gene. Cell 1987; 51:241-249.
21. Bohr VA, Smith CA, Okumoto DS et al. DNA repair in an active gene: removal of pyrimidine dimers from the DHFR gene of CHO cells is much more efficient than in the genome overall. Cell 1985; 40:359-369.
22. Murai M, Enokido Y, Inamura N et al. Early postnatal ataxia and abnormal cerebellar development in mice lacking Xeroderma pigmentosum Group A and Cockayne Syndrome Group B DNA repair genes. PNAS 2001; 98:13379-13384.
23. Khan GQ, Hassan G, Yaseen M et al. Cockayne syndrome. J Assoc Physicians India 2000; 48:1119-1121.
24. Menges-Wenzel EM, Debus O, Strater R et al. Cockayne syndrome with marked cerebral symptoms. Klin Pasiatr 2001; 213:134-138.
25. Lalle P, Nouspikel T, Constantinou A et al. The founding members of xeroderma pigmentosum group G produce XPG protein with severely impaired endonuclease activity. J Invest Dermatol 2002; 118:344-351.
26. Vermeulen W, Scott RJ, Rodgers S et al. Clinical heterogeneity within xeroderma pigmentosum associated with mutations in the DNA repair and transcription gene ERCC3. Am J Hum Genet 1994; 54:191-200.
27. Sancar A. DNA excision repair. Annu Rev Biochem 1996; 65:43-81.
28. Harada YN, Shiomi N, Koike M et al. Postnatal growth failure, short life span and early onset of cellular senescence and subsequent immortalization in mice lacking the xeroderma pigmentosum group G gene. Mol Cell Biol 1999; 19:2366-2372.
29. Sun XZ, Harada YN, Zhang R et al. A genetic mouse model carrying the nonfunctional xeroderma pigmentosum group G gene. Con Anom 2003; 43:133-139.
30. Sun XZ, Harada YN, Takahashi S et al. Purkinje cell degeneration in mice lacking the xeroderma pigmentosum group G gene. J Neurosci Res 2001; 64:348-354.
31. Ng JM, Vrieling H, Sugasawa K et al. Developmental defects and male sterility in mice lacking the ubiquitin-like DNA repair gene mHR23B. Mol Cell Biol 2002; 22:1233-1245.
32. Matsumura YC, Nishigori T, Yagi S et al. Characterization of molecular defects in xeroderma pigmentosum group F in relation to its clinically mild symptoms. Hum Mol Genet 1998; 7:969-974.
33. Sijbers AM, de Laat WL, Ariza RR et al. Xeroderma pigmentosum group F caused by a defect in a structure-specific DNA repair endonuclease. Cell 1996; 86:811-822.
34. Sijbers AM, van Voorst Vader PC, Snoek JW et al. Homozygous R788W point mutation in the XPF gene of a patient with xeroderma pigmentosum and late-onset neurologic disease. J Investig Dermatol 1998; 110:832-836.
35. Sands AT, Abuin A, Sanchez A et al. High susceptibility to ultraviolet-induced carcinogenesis in mice lacking XPC. Nature 1995; 377:162-165.
36. Egly JM. The 14th Datta Lecture. TFIIH: from transcription to clinic. FEBS Lett 2001; 498:124-128.
37. Wei Q, Cheng L, Amos CI et al. Repair of tobacco carcinogen-induced DNA adducts and lung cancer risk: a molecular epidemiologic study. J Natl Cancer Inst 2000; 92:1764-72.
38. Cleaver JE, Thompson LH, Richardson AS et al. A summary of mutations in the UV-sensitive disorders: xeroderma pigmentosum, Cockayne syndrome and trichothiodystrophy. Hum Mutat 1999; 14:9-22.
39. Lehmann AR. DNA repair-deficient diseases, xeroderma pigmentosum, Cockayne syndrome and trichothiodystrophy. Biochimie 2003; 85:1101-1111.
40. De Boer J, Donker I, de Wit J et al. Disruption of the mouse xeroderma pigmentosum group D DNA repair/basal transcription gene results in pre-implantation lethality. Cancer Res 1998; 58:89-94.
41. De Boer J, de Wit J, van Steeg H et al. A mouse model for the basal transcription/DNA repair syndrome trichothiodystrophy. Mol Cell 1998; 1:981-990.
42. De Boer J, van Steeg H, Berg RJW et al. Mouse model for the DNA repair/basal transcription disorder trichothiodystrophy reveals cancer predisposition. Cancer Res 1999; 59:3489-3494.
43. Van der Horst GTJ, Meira L, Gorgels TGMF et al. UVB radiation-induced cancer predisposition in Cockayne syndrome group A (Csa) mutant mice. DNA Repair 2002; 1:143-157.
44. Van der Horst GTJ, van Steeg H, Berg RJW et al. Defective transcription-coupled repair in Cockayne syndrome B mice is associated with skin cancer predisposition. Cell 1997; 89:425-435.
45. de Vries A, van Oostrom CT, Hofhuis FM et al. Increased susceptibility to ultraviolet-B and carcinogens of mice lacking the DNA excision repair gene XPA. Nature 1995; 377:169-173.

INDEX

Symbols

4-Hydroxy nonenal   121, 123
4-OHEN-dA   95
4-OHEN-dC   95
5' Incision   31, 33, 86, 107, 115
5'-(S)-cyclo-2'-deoxyadenosine   120, 121
(6-4) photoproduct (6-4PP)   21, 28, 31, 48, 51, 60, 61, 94, 113-115, 117, 134
8-hydrodeoxyguanine (8-OHdG)   21, 24, 121-124
8-hydroxyadenine   121
8-hydroxyguanine   121
8-methoxypsoralen   60, 72
8-oxoguanine   53, 94, 98, 121
26S proteasome   50
94RD27   88

A

Acetylaminofluorene-guanine (AAF-G)   114, 115
Actinic damage   11, 13
Alpha-spectrin II   *see* Spectrin
Amyotrophic lateral sclerosis   122
Animal model   122, 152
Apoptosis   19, 21, 22, 42, 43, 58, 60, 62, 69, 104-106, 118, 158
Apurinic and apyrimidinic (AP)   53, 70, 95, 98, 108
Arabidopsis thaliana   93, 98
Aspergillus nidulans   98
ATM   74, 104, 106
ATM and RAD3-related   104
ATPase   41, 42, 43, 116, 133
AtPOLH cDNA   98
ATR   30, 104, 105, 106
Atrophy   15, 16, 122, 132, 133
Autoradiography   3
Autosomal disorder   152

B

Basal cell carcinoma (BCC)   19-25, 139-141, 145-147
Basal ganglia   120, 123, 131, 132
Basal ganglia calcification   132
Base excision repair (BER)   4, 47, 52, 72, 87, 93, 104, 106, 108, 129, 133
Bloom syndrome   16, 42
Brain autopsy   120, 122, 123
Brugia malayi   98

C

Caenorhabditis elegans   98
CC to TT transition   21, 22, 53
Cdk-activating complex (CAK)   39, 42, 87, 105
Centrin 2   28, 47-51, 85, 103, 104, 106
Chemoprevention   146
Chemotherapy   146
Chromatin   4, 41, 60, 85, 97, 104, 118, 148
Chronic myelogenic leukemia (CML)   43
Class switch recombination   69, 71
Cockayne syndrome (CS)   3-5, 10, 15, 16, 31, 39, 40, 42-44, 83, 84, 86-90, 93, 120-123, 128-134, 140, 144, 154, 155, 158, 159
   types I, II, III   131
   connatal   131
COFS syndrome
   (cerebro-oculo-facial-skeletal syndrome)   131, 134
Complementation group   3, 10, 11, 14, 20, 31, 32, 47, 65, 68, 75, 83, 84, 87, 93, 106, 117, 121, 128, 130, 132-134, 138-140, 144, 147, 152, 154
COP9 signalosome   51, 60-62
CSB-/-   121, 158
Cyclobutane pyrimidine dimer (CPD)   3, 20, 21, 28, 31, 48, 49, 51, 60, 94, 95, 97-99, 113, 114, 117, 139, 146, 147